T0323367

Effective Coding with VHDL

Effective Coding with VHDL

Principles and Best Practice

Ricardo Jasinski

The MIT Press
Cambridge, Massachusetts
London, England

This book was set in Stone Serif and Stone Sans by Toppan Best-set Premedia Limited Printed and bound in the United States of America.

Library of Congress Cataloging-in-Publication Data

Names: Jasinski, Ricardo, author.
Title: Effective coding with VHDL : principles and best practice / Jasinski, Ricardo.
Description: Cambridge, MA : The MIT Press, [2015] | Includes bibliographical references and
    index.
Identifiers: LCCN 2015039698 | ISBN 9780262034227 (hardcover : alk. paper)
Subjects: LCSH: VHDL (Computer hardware description language) | Computer programming.
Classification: LCC TK7885.7 .J57 2015 | DDC 621.39/2—dc23 LC record available at http://
    lccn.loc.gov/2015039698

10   9   8   7   6   5   4

Contents

Preface

VHDL is notoriously hard to master. To really understand the nuts and bolts of a VHDL description, one must have a strong grasp of the language's type system, concurrency model, and signal update mechanism. On top of that, using VHDL for modeling hardware requires a substantial knowledge of digital hardware design and the synthesizable subset of the language.

Most VHDL books concentrate on teaching the language features and using the synthesis subset for modeling hardware structures. There is, however, a less explored dimension to the creation of high-quality VHDL designs. A VHDL description is also source code, and as such, it can benefit from the huge body of knowledge amassed by software developers about how to write high-quality code and how to organize and structure a solution. If we ignore this dimension, we will write code that is harder to understand and harder to work with. We will spend more time chasing bugs and less time adding interesting features to our designs.

This book addresses this unique set of skills that is also essential to the creation of high-quality VHDL designs. Rather than focus on introducing advanced language features or digital design techniques, its goal is to teach how to write high-quality VHDL code. The concepts taught here will help you write code that is easier to understand and more likely to be correct.

Who Should Read This Book?

This book is not intended as a first text on VHDL. Although most language features are reviewed and explained before diving deeper into each topic, the reader will better benefit from it after taking an introductory course or tutorial in VHDL. However, it would be inaccurate to classify this book as an intermediate or advanced text on VHDL. In fact, it is written for VHDL designers of all experience levels who want to write better code. To be accessible for readers of different levels, the book reviews all the important concepts before providing guidelines and recommendations for their use.

This book is intended for:

- design engineers or practitioners who work with VHDL,
- recently graduated students who will use VHDL professionally,
- students of electrical engineering or computer science, in digital design courses using VHDL,
- self-taught VHDL developers who want a deeper and more organized exposition of the language, and
- any developer who wants to write better VHDL code.

Book Overview and Organization

This book is organized into six parts and twenty chapters. Part I, Design Principles (chapters 1–4), starts with a brief review of VHDL aimed at designers who have not used the language in a while or who are more familiar with other hardware design languages. Then it introduces the topics of design, quality, and architecture. Next, it presents the main factors that work against a good design, such as complexity, duplicate information, dependencies, and changes. Building on these concepts, it introduces a series of well-established design principles to achieve high-quality designs, such as modularity, abstraction, hierarchy, low coupling, strong cohesion, orthogonality, the Single Responsibility Principle, and the Don't Repeat Yourself (DRY) principle.

Part II, Basic Elements of VHDL (chapters 5–8), starts with an explanation of the basic steps involved in the simulation and synthesis of a VHDL design, with special emphasis on the analysis, elaboration, and execution phases. Then it describes the basic elements of VHDL code, including design units, statements, declarations, expressions, operators, operands, and attributes. Knowing the available constructs is fundamental to writing code that clearly matches our intent and is as little verbose as possible.

Part III, Statements (chapters 9–11), covers the basic elements of behavior in any computer language. Statements specify basic actions to be performed and control the execution flow of the code. Besides covering the two kinds of statements in VHDL, concurrent and sequential, the book includes a chapter dedicated to assignment statements, a common source of confusion among newcomers to VHDL.

Part IV, Types and Objects (chapters 12–14), deals with one of the most distinctive characteristics of VHDL: its type system. It presents the categories of types in VHDL, the predefined and user-defined types, and the classes of objects created using these types.

Part V, Practical Coding Recommendations (chapters 15–18), provides useful coding advice, including guidelines for creating routines, choosing good names, and commenting and formatting the source code.

Part VI, Synthesis and Testbenches (chapters 19–20), covers these two uses of VHDL. The first chapter presents the key characteristics of code intended for synthesis and the main differences from code intended for simulation. The last chapter takes a practical approach to teaching the design and implementation of testbenches. It is structured as a series of examples that verify different kinds of models, including combinational, sequential, and FSM code.

About the Examples

VHDL requires a good amount of *boilerplate code*—any repetitive section of code that does not add to the logic of a description but is required by a language or framework. Examples in VHDL include commonly occurring groups of *library* and *use* clauses, or the code to instantiate a component in a testbench. In VHDL, even the simplest design unit may require at least a dozen lines of code.

Because this book contains hundreds of short snippets and code examples, it would not be practical (or useful) to present all of them as full-blown, ready-to-compile code. Especially for short snippets, whose main goal is to demonstrate a concept or idea, adding the required boilerplate code would be more distracting than useful.

In many cases, however, the reader may benefit from having the complete source code to experiment with. For this reason, all the examples are available for download from the book's companion website. The examples in this book can be compiled with a free version of Model Technology's ModelSim, such as ModelSim Altera 10.1e. The synthesizable examples can be compiled with the free versions of popular FPGA synthesis tools, such as Altera Quartus II or Xilinx ISE.

Part I   Design Principles

1 A VHDL Refresher

This chapter presents the key features of VHDL and a summary of the changes introduced in the 2008 version of the language. It also reviews the main constructs used in the creation of synthesizable VHDL designs.

1.1 Why This Book?

This book addresses a unique set of skills for creating VHDL models. Rather than focus on specific language features, its goal is to teach how to write high-quality VHDL code. The concepts taught here will help you write code with a higher quality, in less time, and with fewer problems. This book summarizes the best programming principles from the software field and presents them in a way that is useful to hardware designers.

To use a hardware or software language effectively, we must know more than its statements and constructs. It is possible to know everything about VHDL and still write terrible code—code that is hard to understand, hard to change, and easy to break. Writing high-quality code requires a deeper understanding of how and where the code fits in the entire design process. Before a design is complete, its source code must be read, understood, and changed countless times by human developers. Code that does not support these goals has little value in real-life projects.

Most developers, however, were never taught how to code like that. Most textbook examples are crafted to demonstrate specific language features and have different constraints from code used in real projects. There is precious little information on how to write VHDL code that is readable, comprehensible, maintainable, and testable. The existing guidelines and recommendations rarely provide enough background or rationale, and many offer subjective or conflicting advice.

To address these problems, this book puts a strong emphasis on fundamental design principles. It starts by introducing the topics of design and quality. Then it presents the main factors that work against a good design, such as complexity, duplicate information, dependencies, and changes. Building on these concepts, it introduces a series of well-established design principles, such as modularity, abstraction, hierarchy, and orthogonality. The remaining of the book reviews important VHDL concepts, clears up common misconceptions, and provides a series of guidelines and recommendations for writing higher-quality VHDL code. The recommendations are built on fundamental design principles and backed with detailed rationales, enabling the reader to adapt them to the situation at hand. Because they are based on universal principles, many of the recommendations are applicable to any hardware or software language.

The recommendations in this book are intended for use in production code and real-life projects. This is especially important for fresh programmers, who still lack a comprehensive view of the role of source code. For a novice programmer, many coding alternatives look equally appealing because they all "work" from an immediate point of view. Experienced developers, in contrast, know that few alternatives work well in practice and most create problems in the future. One of the goals of this book is to give the reader the intuition that is typical of more experienced developers. Without help, this could only be learned with years of experience, working on sizeable projects and as part of a skilled development team. The experience collected in this book will help you to learn faster and avoid part of the frustration of learning by trial and error. It will help you make decisions about your code and give you the right arguments to support your opinion. It will show you how to identify bad code and turn it into good code.

1.2   What Is VHDL?

VHDL is a hardware design language intended for use in all phases of digital system design. VHDL stands for VHSIC (Very High Speed Integrated Circuits) Hardware Description Language. Its development began in 1983 under a contract from the U.S. Department of Defense, and it became an IEEE standard in 1987. It was then updated in 1993, 2000, 2002, and 2008 to add new features, make corrections, and incorporate feedback from its large base of users. The current version of the language is specified in IEEE Std 1076–2008, the VHDL Language Reference Manual (LRM). The language version defined in this standard is called VHDL-2008.

Every language must go through periodic revisions to stay relevant over time. VHDL-2008 was the largest modification to the language since 1993, and it includes significant changes to both synthesizable and simulation code. The new features improved the language consistency, added new statements and operators, and removed unnecessary restrictions. These changes make RTL coding simpler and enable the language to be used in new simulation and verification tasks. They also reduce the verbosity that has drawn complaints since the early days of the language.

1.2.1 Key Features of VHDL

This section summarizes the key features of VHDL.

- **VHDL can be used throughout the entire design cycle.** VHDL is powerful enough to be used as a hardware design language, a simulation language, a specification language, and a functional verification language.
- **VHDL supports different levels of abstraction.** It can describe models at the behavioral, RTL, dataflow, and structural levels.
- **VHDL inherently supports parallelism.** Hardware designs are parallel by nature. VHDL has many constructs to support concurrent tasks and an execution model that provides deterministic results independently of the execution order.
- **VHDL also supports sequential programming.** Designers can write algorithmic descriptions, using constructs similar to those available in general-purpose programming languages.
- **VHDL supports different modeling styles.** There are many ways to model a design in VHDL. Designers are free to choose a style that best communicates their intent.
- **VHDL supports good design principles.** The language offers constructs for implementing hierarchy, abstraction, modularity, and information hiding, among other good design principles.
- **VHDL is strongly and statically typed.** This has always been one of the fundamental aspects of the language. It allows the compiler to perform early checks in the code and reduces coding mistakes.
- **VHDL is thoroughly planned and designed.** VHDL has its roots on Ada, one of the most carefully engineered languages in the history of computer programming. The language is also highly consistent. Once we learn its basic mechanisms, it is easy to reason about any language construct.
- **VHDL is extensible.** The basic constructs and predefined types of the language allow it to be extended with packages and user-defined types.
- **VHDL is continuously evolving.** The language is revised periodically and incorporates feedback provided by users and tool vendors.
- **VHDL is widely supported.** Designs written in VHDL are highly portable across different vendors and tools.

However, some issues make working with VHDL less enjoyable than it could be:

- **VHDL is hard to master.** A good knowledge of the language requires understanding its execution model and concepts that are not apparent in the code, such as drivers, resolution functions, scopes, and rules for hardware inference.
- **The VHDL tool ecosystem is slow to evolve.** Support for the VHDL-2008 standard is still limited six years after the standard has been published.

- **VHDL still lacks higher level programming features.** Most obviously missing are object-oriented features. Integers larger than 32 bits are another example. These and others features are expected in upcoming language revisions.
- **VHDL code is still verbose.** For those used to more expressive languages, VHDL still feels a little long-winded. Similar constructs in other languages require much less typing and boilerplate code.
- **VHDL was not designed from the ground up for logic synthesis.** One of the original VHDL requirements was to reuse as much as possible from Ada. VHDL was designed to describe the organization and operation of hardware components; however, its syntax does not convey any specific hardware elements. The fact that the language has many different uses requires a compromise, and it frustrates some users who would like to see a language more directed to synthesis.

Overall, VHDL is a great language for its purpose. It gives designers the freedom to describe hardware, from basic circuits to complete systems. It can be used at a basic level, allowing the more advanced features to be learned incrementally. Because it supports good design principles, such as hierarchy, abstraction, and information hiding, it is a good teaching tool. Concepts learned when studying VHDL can be used in other areas of hardware design and are compatible with the best practices of software development.

Although a basic subset of VHDL is enough to describe many models, most designs can benefit from a deeper knowledge of the language. Just as important, most designs can benefit from the use of good design principles and the best practices in writing high-quality code. The book proposes to address this skill set.

1.2.2   The LRM

The authoritative source of information about VHDL is the standard that defines the language, IEEE Standard 1076, the VHDL Language Reference Manual (LRM). The current version is IEEE Std 1076™-2008.

Although the standard completely describes the language, it is not recommended as a learning tool. First, it is not supposed to be read from cover to cover; any part of the text may use concepts described anywhere else. In other words, one needs to know the language to understand the standard. Second, the technical language used in standards is intended to be precise and succinct; ease of understanding for those still learning VHDL is not a top concern. Third, another problem is that the precise terminology used in the standard does not always match the terms used colloquially when the language is taught. It can be hard to find information about something when you do not know what this thing is called.

This book takes the careful approach of using the standard terminology as much as possible. It also tries to bridge the gap between the standard and the conventional use of the terms, including their use in the software field. This should prepare readers to communicate with other hardware developers and engage in meaningful discussions with software developers as well.

1.2.3 VHDL versus General-Purpose Languages

On the one hand, VHDL is not a general-purpose language and should not be approached as such. On the other hand, there is a body of knowledge about how to write and maintain source code that VHDL users would be wise to learn. It takes both hardware design knowledge and programming knowledge to write great VHDL code.

Most of the fundamental principles and programming techniques apply to any kind of source code. Any application can benefit from a clear organization into modules, source files, routines, and data structures. The principles of choosing good names are important for any code that will be read by humans, and every piece of code can benefit from techniques that make it easier to understand and maintain. Many of these qualities are indifferent to the compiler, but they make a big difference in how easy it is for developers to work with the source code.

We can break down the process of solving a problem using a computer programming language in two steps. The first is to raise the level of abstraction provided by the language until it is closer to the problem at hand. The second is to write the solution using the new abstractions. To raise the abstraction level, we create types that represent meaningful concepts from our application, such as instructions in a CPU or triangles in a graphics pipeline. Then we create meaningful operations to manipulate these concepts. If we follow this process, then the application code will be easier to write and understand; it will also read more naturally because all the low-level details will be hidden in the infrastructure code.

Although this design process is usually associated with general-purpose languages, there is no reason not to apply these and other good programming principles when we write VHDL code. As benefits, we can expect our code to be simpler, more readable, and easier to understand. Our applications will have less bugs because they will be easier to test and fix, and they will live longer because the code will be more reusable and easier to extend and maintain.

1.3 Levels of Abstraction in Digital Systems

In this book, the subject of abstraction will come up frequently. Abstraction is one of the most powerful tools in the system designer's bag of tricks. It allows us to create highly complex systems despite our human limitations in information processing.

In computer science, *abstraction* is the ability to use a construct while ignoring its implementation details. In a more general sense, using abstraction means dealing with ideas rather than concrete entities. In digital system design, however, the term has a slightly different connotation. Designers refer to a series of *levels of abstraction* that reflect different views of the system in the design process. Such perspectives differ in the level of detail, how the system is described, and the intended use for the description. Here, we take a look at the common levels of abstraction used in the design of digital systems.

A *behavioral* description is concerned only with what the system does, not with its implementation. It has no implied architecture and does not have to be directly implementable in hardware. At this level, the description is more like software, and the system is often modeled as a series of algorithms. The main advantage of working at this level is that it is faster to create a working description, and we can use all language features instead of only the synthesizable subset.

Behavioral models are used in simulation for various purposes. One is to have a preliminary representation of the system early in the design cycle. Such models allow us to run experiments, assess the correctness of algorithms, and investigate the performance and limitations of different implementation alternatives. It also allows us to simulate other components that will interface with the device being developed. Finally, later in the design process, we can compare the behavioral model against other representations and against the finished hardware design to check their correctness.

The next level down the ladder of abstraction is the *register-transfer level* (RTL). In an RTL description, the device is described as a series of storage elements (registers) interconnected by transfer functions implemented with combinational logic. An RTL description has a clear and definite architecture; it should be possible to identify all of its registers, although they may be implied in the code rather than explicitly instantiated. The combinational functions are described as meaningful computations and not necessarily as low-level bit manipulations. Only a subset of VHDL is required to write RTL models.

Another common term is a *dataflow* model or description. In this kind of model, the behavior is described by a successive transformation of values that flow from one unit of computation to the next. This term is often reserved for models using only concurrent statements, such as concurrent signal assignments, and it is often restricted to combinational logic. The term dataflow is more commonly used to describe a modeling style rather than a proper level of abstraction.

At the bottom of the ladder, the names are less clear-cut. The term *structural* is reserved for models that use only component instantiations and their interconnections. This level is equivalent to a textual description of a schematic. A structural model can be used to implement the top level in a hierarchy of subcomponents or describe a netlist that is the output of a synthesis tool. In this case, the instantiated components are basic hardware elements defined in a library, and the model is also called a *gate-level* description.

VHDL allows design entry at all of the above levels. Most synthesizable VHDL designs are created at the register-transfer level; synthesis tools are good at translating RTL descriptions to hardware automatically.

1.4   A Summary of VHDL

This section is a brief review of the main constructs used in the creation of VHDL designs. It is intended for readers who have not used VHDL for some time or those who want a quick refresher. The topics seen here will be discussed with more detail in subsequent chapters. Readers familiar with the use of VHDL can safely skip this section.

1.4.1   Entities and Architectures

A hardware module is described in VHDL with two complementary constructs: an *entity declaration* (or *entity*, for short) that describes the module's interface, and an *architecture body* (or *architecture*, for short) that describes the module's internal structure or operation. Figure 1.1a shows the schematic diagram for a half-adder module; figure 1.1b shows the corresponding entity and architecture in VHDL.

The entity declaration (upper half of the VHDL code) specifies the module's interface. The half-adder has two input ports, named a and b, and two output ports, named sum and carry. The architecture body (lower half of the VHDL code) defines the module's behavior. The value of each output is given by a boolean function of the two inputs. This is, as explained in section 1.3, a dataflow-style description.

Describing Behavior in Architectures   The architecture body is where we describe the operation or structure of a module. There are several ways to specify a module's behavior:

- In a dataflow style, using concurrent signal assignments;
- In a structural style, by instantiating other modules to do part of the work;

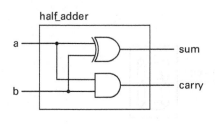

```
-- Entity declaration:
entity half_adder is
    port (
        a, b: in bit;
        sum, carry: out bit
    );
end;

-- Architecture body:
architecture dataflow of half_adder is
begin
    sum <= a xor b;
    carry <= a and b;
end;
```

(a) Schematic diagram.   (b) VHDL code.

Figure 1.1
Half-adder module.

- In an algorithmic style, as a sequence of computational steps inside a *process* statement; and
- Using a combination of the above approaches.

The statements used in the architecture of figure 1.1b are called *concurrent signal assignments*. A concurrent signal assignment is reevaluated whenever a signal used in the right side of the assignment symbol (<=) changes. Therefore, the outputs sum and carry are updated automatically whenever an input changes, and the behavior described in the code resembles a block of combinational logic.

The elements that represent the nets in a VHDL design are called *signals*. Ports declared in an entity are also signals and can be used in the architecture. The statement sum <= a xor b reads the values of signals a and b, performs the xor operation between the two values, and assigns the result to signal sum. Besides the signals defined by the entity ports, an architecture may declare internal signals for private use.

Example: Creating the Top-Level Entity of a Design   Along this chapter, we will use a worked design example to illustrate many VHDL constructs. The design is a single-digit counter that increments once every second and shows the current count value in a seven-segment display (figure 1.2). The top-level entity is called slow_counter. It has two input ports (clock and reset), and one output port (seven_seg) with a width of seven bits. The design operates with a 100 MHz clock.

Listing 1.1 shows the corresponding VHDL code. Because we are using a data type that is not built into VHDL (type std_ulogic), we need to instruct the compiler to look for the type declarations elsewhere. This is done with the *library clause* and *use clause* in lines 1 and 2. The library named sdt_logic_1164 declares the types used for the module's inputs and outputs. The clock and reset inputs are of type std_ulogic, which is intended to model logic levels more realistically than the built-in type bit.

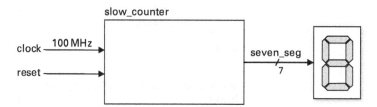

Figure 1.2
Top-level entity of the slow_counter module.

The seven_seg output is of type std_ulogic_vector, intended for modeling multibit values. The output is declared as an array of seven bits indexed from six down to zero (line 8).

We will leave the architecture body empty for now, until we review the statements needed to implement the intended behavior. The ellipsis in line 14 is not valid VHDL code; rather, it is used in this book to represent missing or omitted code.

Listing 1.1 Entity declaration and architecture body for the slow_counter module

```
1    library ieee;
2    use ieee.std_logic_1164.all;
3
4    entity slow_counter is
5        port (
6              clock: in std_ulogic;
7              reset: in std_ulogic;
8              seven_seg: out std_ulogic_vector(6 downto 0)
9        );
10   end;
11
12   architecture rtl of slow_counter is
13   begin
14       ...
15   end;
```

1.4.2 Statements in VHDL

Statements, together with declarations and expressions, are the basic building blocks of VHDL code. Statements manipulate data objects and describe the behavior of a model. Declarations create new named items in the code. Expressions are formulas that can stand for values in statements and declarations.

VHDL has two kinds of statements: concurrent and sequential. *Concurrent statements* are executed conceptually in parallel and can be used only inside an architecture. *Sequential statements* are executed in the order they are written in the code, and they can exist inside *process* statements or subprograms. Table 1.1 shows the statements available in VHDL and the sections where they are described with more detail in this book.

1.4.3 Concurrent Statements

Concurrent statements execute conceptually in parallel, have no predefined order of execution, and can be used only inside architecture bodies. Concurrent statements are useful to describe dataflow behavior and hierarchical structures.

Table 1.1 Concurrent and sequential statements and corresponding book sections

Concurrent Statements	Section	Sequential Statements	Section
process	9.2	signal assignment	11.1
concurrent signal assignment	9.6.2, 11.1.3	variable assignment	11.1
component instantiation	9.4	if	10.2
generate	9.5	case	10.3
concurrent assertion	9.6.3	loop, next, exit	10.4
concurrent procedure call	9.3	wait	10.5
block	9.6.1	procedure call	15.3.2
		return	15.3.1, 15.3.2
		assertion	10.6
		report	10.6.2, 10.6.3
		null	10.7

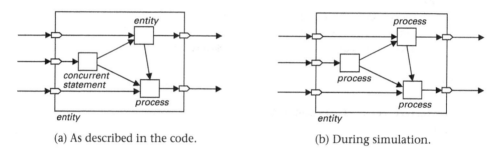

(a) As described in the code.   (b) During simulation.

Figure 1.3
Before a design can be simulated, all concurrent statements are transformed into processes.

Another way to view concurrent statements is as units of computation that are inter-connected via signals and define the behavior of a VHDL model. In fact, all concurrent statements can be reduced to one or more processes (figure 1.3); this transformation happens before a design can be simulated, in a phase called *elaboration* (which will be described in detail in section 5.5).

The most common concurrent statements are the *process* statement, the concurrent signal assignment statement, component instantiations, and the *generate* statement. We will see each one in turn.

The Process Statement The *process* statement defines a sequential region of code that is executed when a certain condition is met. Inside a process, only sequential

```
process (clock) begin
    if rising_edge(clock) then
        q <= not q;
    end if;
end process;
```

```
process begin
    wait until rising_edge(clock);
    q <= not q;
end process;
```

(a) With a sensitivity list.   (b) With a *wait* statement.

Figure 1.4
Examples of process statements modeling sequential logic.

statements can be used. However, the process itself is a concurrent statement; all processes in a design are executed in parallel.

Each process can be in one of two states: suspended or executing. The conditions
that cause a process to resume execution depend on how the process is written—with a
sensitivity list or *wait* statements. If the process has a sensitivity list, then it is executed
whenever one of the signals in this list changes. The process in figure 1.4a has a sensitivity list containing only the signal clock. If a process uses *wait* statements, then
it resumes when the condition specified in the *wait* statement happens, which may
include a signal changing value, a condition expression becoming true, or a timeout.
The condition specified in the *wait* statement of figure 1.4b is the rising edge of signal
clock.

A process executes in a loop—when the last statement is reached, it begins again
from the top. Once a process starts executing, it will run uninterrupted until a *wait*
statement is found. A process with a sensitivity list has an implicit *wait* statement as
the last statement in the process, causing it to suspend until one of the signals in the
list changes.

Processes can model combinational or sequential logic. If a process models combinational logic, then all signals read inside the process must be in the sensitivity list
because the expressions using the signal values must be reevaluated whenever one
of them changes. If a process models sequential logic, then only the clock signal and
any asynchronous control signals (such as a reset) should be in the sensitivity list. The
processes in figure 1.4 model sequential logic. The processes in figure 1.5 model combinational logic. Figure 1.5a uses the syntax of VHDL versions prior to VHDL-2008, so
the *if* statement must perform explicit comparisons with the value '1', and all signals
read in the process must be listed individually in the sensitivity list. Figure 1.5b uses
VHDL-2008 syntax. The keyword *all* means that the process should be executed every
time a signal that is read in the process changes.

The Concurrent Signal Assignment Statement   A *concurrent signal assignment* assigns
a value to the target signal whenever one of the signals on the right hand side of the

```
process (d1, d2, d3)
begin
    if d3 = '1' then
        q <= "11";
    elsif d2 = '1' then
        q <= "10";
    elsif d1 = '1' then
        q <= "01";
    else
        q <= "00";
    end if;
end process;
```

```
process (all)
begin
    if d3 then
        q <= "11";
    elsif d2 then
        q <= "10";
    elsif d1 then
        q <= "01";
    else
        q <= "00";
    end if;
end process;
```

(a) Before VHDL-2008.   (b) After VHDL-2008.

Figure 1.5
Examples of processes modeling combinational logic.

```
target <= value;
```

```
target <=
    value_1 when condition_1 else
    value_2 when condition_2 else
    default_value;
```

```
with expression select
    target <=
        value_1 when choice_1,
        value_2 when choice_2,
        default_value when others;
```

(a) Simple assignment. (b) Conditional assignment.   (c) Selected assignment.

Figure 1.6
Kinds of signal assignment.

signal assignment symbol (<=) changes. The assignments to sum and carry shown in figure 1.1b and reproduced below are concurrent signal assignments.

```
sum <= a xor b;
carry <= a and b;
```

There are three versions of the assignment statement: simple, conditional, and selected (figure 1.6). Simple assignments always evaluate the same expression and assign the resulting value to the target signal. Conditional assignments test one or more conditions until one is true and then assign the corresponding value to the target signal. Selected assignments compare the value of one expression against a number of choices and then assign the corresponding value to the target signal. In a selected assignment, the choices should cover all possible values of the expression. For brevity, the keyword others can be used to represent all the remaining choices.

Any concurrent signal assignment is just a shorthand notation for a process that contains the same assignment and has all signals read in the expressions in its sensitivity list. For an example, see section 9.2.3.

The Component Instantiation Statement The *component instantiation statement* instantiates a module inside another module, allowing for the creation of hierarchies and modular designs. For example, if we had a BCD to seven-segment decoder module called bcd_to_7seg, then we could instantiate it inside the architecture of the slow_counter module (seen in listing 1.1) with the following statement:

```
decoder: entity digit_to_7seg
    port map (digit => count_1_Hz, segments => seven_seg);
```

In the component instantiation statement above, decoder is a label given to the instance to identify it in the design hierarchy, and digit_to_7seg is the name of an existing module to be instantiated. The list in parentheses is called a *port map* or an *association list*; it defines the correspondence (or connections) between elements on each side of the arrow symbol (=>). The above example connects signals count_1_Hz and seven_seg, which exist in the instantiating architecture, to signals digit and segments, which are ports of the instantiated module.

The Generate Statement The generate statement creates zero or more copies of a group of concurrent statements. It is available in a conditional form (*if-generate*) and an iterative form (*for-generate*). The iterative form can be used to instantiate an array of components. In the example below, if signals a, b, and y are arrays of 32-bit unsigned numbers, then the generate statement implies eight 32-bit unsigned adders:

```
gen: for i in 0 to 7 generate
    y(i) <= a(i) + b(i);
end generate;
```

Example: Modeling a 7-Segment Decoder with Concurrent Statements We can use our design example to illustrate the use of some concurrent statements. For now, let us assume we have a signal named digit that holds an integer value to be shown in the display. Then we could write the 7-segment decode logic as a concurrent signal assignment. We could use any of the three forms of assignment: simple, selected, or conditional. Because the output value is chosen from a single input expression (the value of signal digit), a selected assignment is more appropriate. Figure 1.7 shows the approaches using a selected and a conditional assignment.

It is also possible to use only simple assignments. In this case, we assume the existence of a signal named bcd_digit declared with a type that allows access to individual bits, such as the type unsigned. Then we could write the boolean equations for each segment individually:

```
with digit select
    segments <=
        "0111111" when 0,
        "0000110" when 1,
        "1011011" when 2,
        "1001111" when 3,
        "1100110" when 4,
        "1101101" when 5,
        "1111101" when 6,
        "0000111" when 7,
        "1111111" when 8,
        "1101111" when 9,
        "0000000" when others;
```

```
segments <=
    "0111111" when digit = 0 else
    "0000110" when digit = 1 else
    "1011011" when digit = 2 else
    "1001111" when digit = 3 else
    "1100110" when digit = 4 else
    "1101101" when digit = 5 else
    "1111101" when digit = 6 else
    "0000111" when digit = 7 else
    "1111111" when digit = 8 else
    "1101111" when digit = 9 else
    "0000000";
```

(a) Using a selected assignment.   (b) Using a conditional assignment.

Figure 1.7
Modeling the seven-segment decoder with concurrent signal assignments.

```
alias a is bcd_digit(3);
alias b is bcd_digit(2);
alias c is bcd_digit(1);
alias d is bcd_digit(0);
...
seven_seg(0) <= a or c or (b and d) or (not b and not d);
seven_seg(1) <= not b or (c and d) or (not c and not d);
seven_seg(2) <= not (not a and not b and c and not d);
...
```

The code above uses aliases to make the expressions less verbose; however, the versions using conditional or selected assignments are still much easier to read and understand.

1.4.4 Sequential Statements

Sequential statements are executed in order, allowing us to describe computations step by step. The default execution order is from top to bottom, but it can be steered with conditional or iterative control structures. Sequential statements can be used only in two places: processes or subprograms.

The most important sequential statements are signal and variable assignments; the control structures *if*, *case*, and *loop*; the *wait* statement; and the procedure call statement. Note that a function call is not a statement because it cannot be invoked by itself in the code; it can only be called indirectly, when it is used in an expression.

Signal Assignments in Processes   The most distinctive characteristic of signal assignments inside processes is that the signal value is never updated immediately. It takes at least an infinitesimal time unit for a signal to assume its new value.

The practical consequence is that whenever we read the value of a signal inside a process, it will have the same value it had when the process started the current execution loop. All signal updates are postponed until after the process has finished executing. In the example below, although the assignment to signal copy happens after the increment to signal original, the copy will receive the value that original had when the process started execution. In consequence, copy will always be one cycle delayed with respect to original.

```
process begin
    wait until rising_edge(clock);
    original <= original + 1;
    copy <= original;
end process;
```

In fact, all signal updates are postponed until after all processes have finished executing. Only then are all signals updated at the same time. This mechanism guarantees deterministic results during a simulation, independent of the relative order of execution of the processes.

Another practical consequence for synthesizable models is that if multiple assignments are made to the same signal during a process run, then only the last executed assignment will take effect. We can think that each new assignment to the same signal overwrites any previous ones.

Variables and Variable Assignments So far our examples have used only signals to hold data values. A signal is a kind of data object meant for communication between processes, and therefore it has many special characteristics, including a history of previous values and a queue for future values. Variables are conceptually simpler. A variable is updated immediately upon assignment, and it has no queue for future values; the only information it holds is its current value.

Variables can only be declared in processes or subprograms. They cannot be used for communication between processes or entities because they are local to the process or subprogram where they are declared. Their main use is as auxiliary values in computations or to keep state in processes that imply memory.

Variables in VHDL are similar to variables in other programming languages. For example, to calculate the sum of all values in an array of integer numbers, we could write:

```
process (all)
    variable sum: integer;
begin
    sum := 0;
```

```
    for i in 0 to 7 loop
        sum := sum + integer_array(i);
    end loop;
    ...
end process;
```

Note that the variable assignment statement uses a different symbol (:=) from the signal assignment statement (<=). In the above example, variable sum is assigned and read eight times inside the loop. Because it is a variable, its value gets updated immediately after every assignment; the new value can be read immediately, in the next statement following the assignment. If it were a signal, then it would not change its value across the loop iterations; keep in mind that every time we read a signal in a process, it will present the same value it had when the process resumed execution.

Control Structures   The sequential statements *if*, *case*, and *loop* are called *control structures*. Without them, the order of execution in sequential code would always be strictly from top to bottom. Control structures allow us to specify conditional or iterative execution of groups of statements.

The *if* statement works just like in other programming languages. It executes a block of statements or not depending on the value of a condition (figure 1.8a). It can also include an *else* clause and one or more *elsif* clauses (figure 1.8b, c).

The *case* statement executes one of multiple sequences of statements depending on the value of a single expression. It evaluates the expression, compares its value with each choice in turn, and executes the corresponding statements when a match is found (figure 1.9a). Each occurrence of the *when* keyword defines an alternative, which may specify a single choice, multiple choices separated by a vertical bar (|), or a range (figure 1.9b).

Taken together, the choices in a *case* statement must cover all possible values of the expression. The last choice may be the keyword *others*, which matches all the remaining values.

```
if condition then          if condition then          if condition_1 then
   -- statements              -- statements              -- statements
end if;                    else                       elsif condition_2 then
                              -- statements              -- statements
                           end if;                    else
                                                         -- statements
                                                      end if;
```

(a) Simple *if* statement.   (b) *if* statement with an   (c) *if* statement with *else*
                                    else clause.   and *elsif* clauses.

Figure 1.8
if statement with optional parts.

```
case expression is
    when choice_1 =>
        -- statements
    when choice_2 =>
        -- statements
    when choice_3 =>
        -- statements
    when others =>
        -- statements
end case;
```

(a) One choice per alternative.

```
case expression is
    when choice_1 =>
        -- statements executed when expression
        -- evaluates to choice_1
    when choice_2 | choice_3 =>
        -- statements executed when expression
        -- evaluates to choice_2 or choice_3
    when choice_4 to choice_6 =>
        -- statements executed when expression
        -- evaluates to choice_4, choice_5, or choice_6
    when others =>
        -- statements
end case;
```

(b) Multiple choices per alternative.

Figure 1.9
case statement examples.

```
variable i: integer := 0;
...
loop
    statements;
    i := i + 1;
    exit when i = 10;
end loop;
```

(a) A simple loop.

```
variable i: integer := 0;
...
while i < 10 loop
    statements;
    i := i + 1;
end loop;
```

(b) A *while* loop.

```
for i in 1 to 10 loop
    statements;
end loop;
```

(c) A *for* loop.

Figure 1.10
loop statement examples.

VHDL provides loop constructs similar to those found in other programming languages. A simple loop (figure 1.10a) is an infinite loop; it executes indefinitely until an *exit* statement is found. A *while* loop (figure 1.10b) executes while a condition is true. A *for* loop (figure 1.10c) iterates through a range that is defined before it starts executing. Inside the *for* loop, we can use the loop parameter (a constant that assumes a different value at each iteration) in expressions.

The Wait Statement The *wait* statement suspends the execution of a process until a condition is met. The condition may include:

• A *sensitivity list,* causing the process to resume when one of the signals in the list changes;

- A *condition expression*, causing the process to resume when it evaluates to *true*;
- A *time expression*, causing the process to resume when the specified time has elapsed.

The *wait* statement has the form:

```
wait on sensitivity_list until condition_expression for time_expression;
```

In simulation code, the *wait* statement can use any combination of the conditions above. For synthesis, it is only used to describe a signal edge, as in the examples below:

```
wait until rising_edge(clock);
wait until falling_edge(clock);
wait until clock'event and clock = '1';
```

Example: Modeling the Digit Counter in the Slow_Counter Design Example
One way to develop the slow_counter module is to divide it into three steps:

1. Create a slow pulse (1 Hz) from the fast input clock (100 MHz);
2. Create a counter that increments once every second, using the slow pulse; and
3. Create the logic to convert the count value to signals driving the display.

Here we will use a process and some sequential statements to implement the second step. We assume the existence of a signal called pulse_1_Hz, which is high during a single clock cycle per second. This will be the enable of our digit counter. The current digit value will be kept in a signal named digit. We should always strive to give our signals clear and meaningful names that maximize the likelihood that they will be understood correctly when someone reads the code.

Listing 1.2 shows the code for the digit counter process. Because we want to model sequential logic, we must include the signals clock and reset in the sensitivity list (line 7). The reset is asynchronous because it is read outside the test for the clock edge (lines 9–10). All other actions are synchronized with the rising edge of the clock signal (line 11).

Two nested *if* statements control the increment of the counter. The first one (line 12) works as an enable signal, allowing the counter to be updated only once per second. The second *if* statement (line 13) checks whether the digit reached the maximum value of nine before incrementing; if so, it resets the digit counter to zero.

Listing 1.2 Digit counter process for the slow_counter example

```
1    architecture rtl of slow_counter is
2        signal digit: integer range 0 to 9;
3        signal pulse_1_Hz: std_logic;
4        ...
```

```
 5   begin
 6
 7       digit_counter: process (clock, reset)
 8       begin
 9           if reset then
10               digit <= 0;
11           elsif rising_edge(clock) then
12               if pulse_1_Hz then
13                   if digit < 9 then
14                       digit <= digit + 1;
15                   else
16                       digit <= 0;
17                   end if;
18               end if;
19           end if;
20       end process;
21
22       ...
23   end;
```

1.4.5 Subprograms

In VHDL, routines are called *subprograms*. Routines are essential to keeping the code
clean, concise, and free of duplication. Like processes, subprograms are regions of
sequential code. The most common places to declare subprograms are inside packages,
architectures, and processes.

VHDL offers two kinds of subprograms: *functions* and *procedures*. Functions always
return a value and can be used only in expressions. Procedures do not return a value
(but may modify its parameters) and are used as stand-alone statements. Both kinds of
subprogram may have zero or more parameters. Figure 1.11 shows one example of each.

Here is how the subprograms would be used in the code:

```
constant fibonacci_seq: integer_vector := (0, 1, 1, 2, 3, 5, 8, 13);
variable max, min, total: integer;
...

-- A function call; the return value 33 is assigned to variable 'total'
total := sum(fibonacci_seq);

-- A procedure call; the output parameters 'max' and 'min' are
-- overwritten with 13 and 0
find_max_min(fibonacci_seq, max, min);
```

```
function sum(vect: integer_vector)
    return integer
is
    variable total: integer;
begin
    total := 0;
    for i in vect'range loop
        total := total + vect(i);
    end loop;
    return total;
end;
```

```
procedure find_max_min(
    vect: integer_vector;
    max, min: out integer) is
begin
    max := vect(vect'left);
    min := vect(vect'left);
    for i in vect'range loop
        max := vect(i) when vect(i) > max;
        min := vect(i) when vect(i) < min;
    end loop;
end;
```

(a) A function.   (b) A procedure.

Figure 1.11
The two kinds of subprograms in VHDL.

1.4.6 Expressions

An *expression* is a formula that specifies the calculation of a value. Expressions are not self-standing constructs—they are used in statements and declarations, in places where a value is needed.

The basic elements of an expression are *operators* and *operands*. Operators are predefined symbols or keywords that can be used as functions in expressions. Examples of operators include the conventional mathematical symbols (+, -, *, /), logical operators (and, or, not), shift operators (sll, srl, ror), and relational operators (<, >, =, /=). For the full list of operators, see table 8.1. In the assignments below, the source value is given by expressions using different kinds of operators:

```
-- An expression using logical operators:
read <= en and cs and req and not busy;
```

```
-- An expression using arithmetic operators:
mac <= input * coeff + mac;
```

```
-- An expression using concatenation and shift operators:
byte <= (nibble & "0001") srl 3;
```

Operands are the values used by operators in expressions. Examples of operands include literal values, object names, aggregates, function calls, and expressions enclosed within parentheses. The example below shows an expression using two concatenation operators (the & symbol) and three different kinds of operands: a function call, an aggregate, and a bit-string literal:

```
expression :=
    to_unsigned(1234, 16) &          -- A function call.
    (sign_bit, '0', nibble, "00") &  -- An aggregate with names and literals.
    x"1234_abcd";                    -- A bit-string literal in hex base.
```

1.4.7 Packages

Packages allow us to share and reuse code across design units. Although entities can also be reused, in the sense that they may be instantiated multiple times, only packages allow reuse at the source code level. With a package, we can write constructs such as types, subprograms, constants, and other declarations and then use them in different places in a design, removing code duplication.

A package is divided into two parts: a *package declaration*, which contains the items that are publicly visible; and a *package body*, which contains the implementation details. All subprogram implementations go in the package body. The package declaration usually includes types, constants, subprogram declarations, and component declarations.

Packages are important to prevent code duplication, facilitate reuse, and extend the basic functionalities provided by the language. Many useful additions to the VHDL standard are provided in packages, such as the standard multivalue logic system `std_logic_1164` and the synthesizable fixed- and floating-point packages of VHDL-2008.

Example: Modeling the Integer to 7-Segment Decoder as a Function in a Package

By moving the logic of the seven-segment display decoder to a function in a package, we make it available for every module that needs this functionality. Listing 1.3 shows the source code for the package, including the package declaration (lines 4–8) and package body (lines 10–28). The package declaration is the public interface for users of the package; it contains only the function declaration, which is all the information that should be necessary to use the function. It is good practice to add a comment (lines 5–6) when the function name is not enough to communicate everything the routine does or when the routine imposes limitations on its input values.

The package body contains the function body. The selected signal assignment from figure 1.7a was converted to an equivalent *case* statement and multiple *return* statements.

Listing 1.3 A package and function to decode an integer digit to a seven-segment driver

```
1   library ieee;
2   use ieee.std_logic_1164.all;
3
4   package seven_segment_pkg is
5       -- Return a vector for driving a 7-segment display (active-high LEDs).
6       -- Limitation: input values outside the range 0-9 return all zeros.
```

```
 7        function seven_seg_from_integer(digit: integer) return std_logic_vector;
 8    end;
 9
10    package body seven_segment_pkg is
11        function seven_seg_from_integer(digit: integer)
12            return std_logic_vector is
13        begin
14            case digit is
15                when 0 => return "0111111";
16                when 1 => return "0000110";
17                when 2 => return "1011011";
18                when 3 => return "1001111";
19                when 4 => return "1100110";
20                when 5 => return "1101101";
21                when 6 => return "1111101";
22                when 7 => return "0000111";
23                when 8 => return "1111111";
24                when 9 => return "1101111";
25                when others => return "0000000";
26            end case;
27        end;
28    end;
```

To use this function in another module (for example, in the `slow_counter` module), we need a *use clause* to make its declaration visible there:

```
-- The use clause makes the declarations in the package available for use here:
use work.seven_segment_pkg.all;

architecture rtl of slow_counter is ...
begin
    -- Now that the function is visible, we can use it:
    seven_seg <= seven_seg_from_integer(digit);
    ...
end;
```

1.4.8   Data Types

A type is defined by a set of values and a set of operations. When we give an object a type, we are defining the set of possible values it can hold and what operations can be performed with the object. Signals, variables, and constants are examples of objects in VHDL.

VHDL is famous for its strongly typed nature. Every object must have a type, which is defined when the object is declared and checked every time it is used in the code (a language feature called *static type checking*). An object of a type cannot be used as if it had a different type, and the language performs no implicit type conversions: if we want to use an object as if it had a different type, then we need to convert it manually. Another common kind of type conversion, *type casting*, can be used only between closely related types in VHDL.

This strongly typed nature forces the developer to be aware of object types all the time. It also makes the code a little more verbose because of all the type conversions—especially if the types were poorly chosen in the first place. However, it allows the compiler to perform a series of type checks and catch bugs before they are compiled into a design.

Different from other programming languages, VHDL introduces the concept of *subtypes*. A subtype is a type together with a constraint that limits the possible values of an object. In the examples below, the subtype uint8_t has a *range constraint* and can only assume values from 0 to 255. The subtype byte has an *array constraint* and can only use indices between 7 and 0.

```vhdl
subtype uint8_t is integer range 0 to 255;      -- type + range constraint
subtype byte is std_ulogic_vector(7 downto 0);  -- type + array constraint
```

By definition, a type is a subtype of itself. Subtypes can be used just like their base types, and we can make assignments between objects that belong to subtypes of the same base type. Objects of a subtype can use the operations defined for their parent types.

VHDL offers a great set of tools for creating our own types. If we use it to our favor, then we can write code that is less verbose, is safer to use, and communicates our intent more clearly.

Categories of Types   VHDL has a strict type hierarchy with several categories and subcategories of types. The full type hierarchy is shown in figure 12.4. Here we touch briefly on the most commonly used categories: enumerations, integers, arrays, and records.

An *enumeration type* is defined by a list that contains all possible values for that type. Examples of enumeration types include the predefined type boolean and the standard multivalue logic type std_ulogic, defined as:

```vhdl
type boolean is (false, true);
type std_ulogic is ('U', 'X', '0', '1', 'Z', 'W', 'L', 'H', '-');
```

The two examples above illustrate that the values in an enumeration type can be identifiers (valid VHDL names, such as *true* and *false*) or character literals (such as 'X' and '1'). Incidentally, note that VHDL is not case-sensitive, so the identifiers false, False, and FALSE all represent the same value. However, character literals define specific character values, so they are case-sensitive: 'X' and 'x' are different character values, and 'x' is not a value of type std_ulogic.

Enumeration types are useful when we know that an object can only assume a limited number of values and we want to give each value a name. A typical example is to define the states in a finite state machine (FSM). Typically, the only operations performed on objects of enumeration types are comparisons (which are defined automatically whenever we create an enumeration type) and any new operations that we may define using custom functions and procedures.

Integer types are contiguous ranges of integer numbers. They include the predefined type integer and any other integer types that we create. Besides the comparison operators, integer types have all the arithmetic operators predefined. In VHDL, integer types and objects may have a range constraint limiting their minimum and maximum values.

An *array* is a collection of objects of the same type that can be accessed by indexing. To access an array element, we write the name of the array object followed by the index value in parentheses. To create an array *object* in VHDL, a corresponding array *type* must exist. It is not possible to create an array directly from a single-element type, as in other programming languages. An example of a common array type is std_logic_vector, already used in our examples. Here is how we could create an array of standard logic vectors (an array of arrays):

```
-- Declare an array type for a register file. The number of registers
-- and the word width can be set later, when an object of this type is created.
type register_file_type is array (natural range) of std_logic_vector;

-- Declare a register file with 16 registers of 32 bits each.
signal register_file: register_file_type(0 to 15)(31 downto 0);
...
-- Clear the first register of the register file.
register_file(0) <= x"0000_0000";
```

A *record* is a collection of objects that may have different types and are accessed by name. To access a record element, we write the name of the record object, followed by a dot and the name of that element. The example below shows how we could create and use a record type for keeping the time of day as three separate numeric fields (hours, minutes, and seconds), plus a boolean field for AM/PM (is_pm).

```
-- Define a record type representing a time of day with separate values
-- for hours, minutes, and seconds, and a boolean field for AM/PM.
type clock_time_type is record
    is_pm: boolean;
    hours: integer range 0 to 11;
    minutes: integer range 0 to 59;
    seconds: integer range 0 to 59;
end record;

-- Declare two objects of the record type.
signal clock_time: clock_time_type;
signal alarm_time: clock_time_type;
...
-- Advance clock time in one hour.
clock_time.hours <= clock_time.hours + 1;
...
-- Set alarm for 06:30:00 AM.
alarm_time <= (hours => 6, minutes => 30, seconds => 0, is_pm => false);
```

Arrays and records are collectively called *composite* types. To specify a value that can be assigned to a composite object, we use a construct called an *aggregate*. The assignment to alarm_time above uses an aggregate as the source value. Aggregates have a powerful syntax and are a big help in reducing the verbosity of VHDL code. Aggregates are covered in section 8.2.2.

Predefined and Standard Types   The previous section reviewed some of the categories of types available in VHDL: enumerations, integers, arrays, and records. These categories can be used to create types, which in turn can be used to create objects. As important as knowing the categories of types is knowing the actual types defined in the VHDL standard and available for use in our models.

To create good VHDL models, we should be familiar with at least a basic set of types, including std_ulogic/std_logic, std_ulogic_vector/std_logic_vector, signed, unsigned, boolean, integer, and (for simulation) time. It is important to have a good vocabulary of types; otherwise we may end up trying to shoehorn all kinds of data into just a few types that are not adequate for the job. The predefined and standard data types are described in section 13.1.

Example: Modeling the Integer to 7-Segment Decode Logic Using an Array   To illustrate the use of arrays and array types, we could change the implementation of function seven_seg_from_integer, shown in listing 1.3. Instead of using a *case* statement, we could simply return a value indexed into an array. The resulting code is shown in listing 1.4. The trick is to declare an array so that the input parameter can be used directly as an index. By restricting the range of the parameter digit between 0 and 9 (line 1) and declaring the array type with the same index range, the conversion becomes a simple table lookup (line 9). The array object is declared as a constant, and the values are given in the form of an array aggregate (lines 5–7).

Listing 1.4 Conversion function using a conversion array

```
 1    function seven_seg_from_integer(digit: integer range 0 to 9)
 2        return std_ulogic_vector
 3    is
 4        type conversion_array_type is array (0 to 9) of std_ulogic_vector(6 downto 0);
 5        constant CONVERSION_ARRAY: conversion_array_type := (
 6            "0111111", "0000110", "1011011", "1001111", "1100110",
 7            "1101101", "1111101", "0000111", "1111111", "1101111");
 8    begin
 9        return CONVERSION_ARRAY(digit);
10    end;
```

1.4.9 Predefined Attributes

In VHDL, anything that we declare and name is called a *named entity*. *Predefined attributes* are a mechanism to retrieve special information about certain kinds of named entities, such as types, arrays, and signals. To use an attribute, we write a name followed by the tick (') symbol and the attribute identifier. For example, if we declare a signal named byte as an std_logic_vector:

```
signal byte: std_logic_vector(7 downto 0);
```

We can find out its size with the 'length attribute:

```
byte'length   --> 8
```

Attributes vary a lot on the kind of named entity they can be applied to and on the kind of information they return. For example, the 'event attribute can be applied to a signal and returns information from the simulator. The 'range attribute can be applied to an array and returns a range. And the attribute 'high can be applied to a type and returns a value.

Attributes of types and arrays are useful because they allow us to write more generic code—for example, code that works independently of an array's length. Attributes of signals are useful in simulations and modeling clock edges. The example below illustrates the use of the attributes 'event, 'left, and 'range:

```
find_max: process (clock)
    variable max: integer;
begin
    -- clock'event is true if clock value has changed in current simulation cycle
    if clock'event and clock = '1' then
        -- vector'left is the index of the leftmost element in the array
        max := vector(vector'left);
        -- vector'range returns the range of an array type or array object
        for i in vector'range loop
            if vector(i) > max then
                max := vector(i);
            end if;
        end loop;
    end if;
    max_out <= max;
end process;
```

Table 1.2 shows some of the most commonly used attributes. The complete list is shown in tables 8.5 through 8.8.

1.4.10 Generics

A *generic* is an interface item that allows a module to be parameterized when it is instantiated. The most common use for a generic is to allow for variations in the hardware structure of a model. In the following example, the generic constant WIDTH configures the size of the port input_data:

Table 1.2 Some predefined attributes of arrays, types, and signals

Attribute	Applies to	Return type	Return value
S'**event**	Signal	boolean	True if S had an event in the current simulation cycle
S'**stable**(t)	Signal	boolean	A signal that is true if S had no events in the last t time units
S'**transaction**	Signal	bit	A signal of type bit that toggles on every transaction of S
A'**length**(d)	Array	integer	Size (# of elements) of dimension d (if omitted, d = 1)
A'**range**(d)	Array	*Range*	Range of dimension d (if omitted, d = 1)
T'**high**	Type	T	Largest value of T
T'**low**	Type	T	Smallest value of T

```
entity fifo is
    generic (
        -- Number of words that can be stored in the FIFO
        DEPTH: natural := 8;
        -- Size in bits of each word stored in the FIFO
        WIDTH: natural := 32
    );
    port (
        input_data: in std_logic_vector(WIDTH-1 downto 0);
        ...
    );
end;
```

Generics allow us to parameterize a model to work under different conditions, making the code more flexible and reusable.

Example: Final Code for the slow_counter Example with a Generic Clock Frequency
Listing 1.5 shows the final code for the slow_counter example, using a generic to allow configuring the external clock frequency when the module is instantiated (CLOCK_FREQUENCY_Hz, in line 7). The behavior is modeled in a single process (lines 19–38). Variable count works as a clock divider, incrementing once per clock cycle until it reaches the numeric value of the clock frequency (lines 27–28), thus generating a 1 Hz timebase. Every time this variable overflows, signal digit is incremented (lines 31–32). To convert from an integer value to a vector of logic levels, appropriate for driving the seven-segment display (line 40), we use the function seven_seg_from_integer from package seven_segment_pkg (shown in listing 1.3).

Listing 1.5 Final code for the slow_counter example using a generic for the clock frequency

```
1   library ieee;
2   use ieee.std_logic_1164.all;
3   use work.seven_segment_pkg.all;
4
5   entity slow_counter is
6       generic (
7           CLOCK_FREQUENCY_Hz: natural := 100e6
8       );
9       port (
10          clock: in std_ulogic;
11          reset: in std_ulogic;
12          seven_seg: out std_ulogic_vector(6 downto 0)
13      );
```

```
14    end;
15
16    architecture rtl of slow_counter is
17        signal digit: integer range 0 to 9;
18    begin
19        process (clock, reset)
20            constant COUNT_MAX: natural := CLOCK_FREQUENCY_Hz - 1;
21            variable count: natural range 0 to COUNT_MAX := 0;
22        begin
23            if reset then
24                count := 0;
25                digit <= 0;
26            elsif rising_edge(clock) then
27                if count < COUNT_MAX then
28                    count := count + 1;
29                else
30                    count := 0;
31                    if digit < 9 then
32                        digit <= digit + 1;
33                    else
34                        digit <= 0;
35                    end if;
36                end if;
37            end if;
38        end process;
39
40        seven_seg <= seven_seg_from_integer(digit);
41    end;
```

1.5 VHDL-2008

VHDL-2008 was the largest change in the language since 1993. It incorporates a lot of user feedback and contains many improvements for both synthesis and simulation code. The changes made the language more consistent, less restrictive, and less verbose. For the designer, they make the coding job more productive and less error-prone.

The following list is a sampler of the most relevant changes, focused on features that affect the designer's job. For a complete list, check the excellent book *VHDL-2008—Just the New Stuff*, by Peter Ashenden and Jim Lewis.

If you are not familiar with the previous versions of the language, then it is probably not important to know which changes were introduced in which version; it may be more productive to skip this section and proceed directly to the next chapter.

New and Improved Statements

- **Conditional and selected assignments in sequential code.** These kinds of assignments were allowed only in concurrent code regions. Now they can be used in processes and subprograms as well.
- **Conditional and selected assignments for variables.** These kinds of assignments were allowed for signals only. Now they can be used with variables as well.
- **Matching case and selected assignment.** The new `case?` and `select?` statements use the predefined matching equality operator (`?=`) for the comparisons and work with *don't care* values (`'-'`) in the choices.
- **Improved *generate* statements.** The *if-generate* statement now accepts *elsif* and *else* clauses. A *case-generate* statement was also added.

New Operators

- **Array-scalar logical operators.** Previously, logical operators did not allow for the cases where one operand is an array and the other a single element (a scalar). The new array-scalar operators can be used in such cases; they are equivalent to repeating the same operation between the single-bit operand and each element of the array.
- **Logical reduction operators.** VHDL-2008 provides new unary versions of the logical operators, which operate on a single array. These reduction operators work as a big logic gate whose inputs are the elements of an array and whose output is a single-bit value.
- **Condition operator.** The condition operator (represented by the symbol `??`) converts a value of a given type to a boolean, and it is implicitly applied to the condition expression in conditional statements such as *if*, *when*, or *while*. The upshot is that these statements now accept an expression of type `bit` or `std_ulogic`, besides `boolean`:

```
if en = '1' and cs = '1' and req = '1' and busy = '0' then … -- Before VHDL-2008

if en and cs and req and not busy then …                      -- After VHDL-2008
```

- **Matching relational operators.** The new matching operators (for example, `?=` and `?>`) propagate the original operand types in an expression rather than returning a boolean. This allows the result of a logical expression to be assigned directly to an object of a logic type.

Syntactic Improvements

- **Keyword *all* in sensitivity list**. The keyword *all* is a shorthand notation for all signals read inside a process. Using this keyword, it is not necessary to update the sensitivity list manually for combinational processes, a tedious and error-prone job.
- **Expressions in port maps**. Besides type conversions, expressions involving signal values are now allowed in port associations, eliminating the need for intermediate or auxiliary signals when instantiating a component.
- **Sized bit-string literals**. This feature makes it possible to specify the exact length for the expanded string. Previously, it should be a multiple of three for octal base and four for hexadecimal base.
- **Context declarations**. This new construct groups *library* clauses and *use* clauses under a single unit, helping to reduce verbosity at the top of source files.

Type System Improvements

- **Unconstrained element types**. Previously, if the elements of an array or record type were of a composite type, their size had to be fixed at the type declaration. For example, in a register file type, the number of registers could be unconstrained, but the size of each register had to be fixed. VHDL-2008 lifts this restriction.
- **Resolved elements**. VHDL-2008 allows arrays to be resolved one element at a time. This allowed the declaration of type `std_logic_vector` to change from an array of resolved elements to a resolved subtype of `std_ulogic_vector`. The practical consequence is that we can finally use the standard logic types as they were supposed to be: `std_ulogic_vector` when the signal has a single driver and `std_logic_vector` when it has multiple drivers. For the details, see section 13.1.

Improved Generics

- **Generic lists in packages and subprograms**. Prior to VHDL-2008, only entities could have generic lists. Now they can be used in packages and subprograms as well.
- **New kinds of generics**. Prior to VHDL-2008, only constants could be passed as generics. Now it is also possible to pass types, subprograms, and packages.
- **Reading generics declared in the same generic list**. VHDL-2008 allows using the value of a generic parameter in the definition of subsequent generics in the same list.

Other Relaxed Restrictions

- **Reading from out-mode ports**. VHDL-2008 allows ports of mode *out* to be read in a model.

Simulation and Testbench Features

- **External names.** In VHDL-2008, a testbench can access a constant, shared variable, or signal declared in another unit in the design hierarchy. For the details, see section 8.2.3.
- **Force and release assignments.** A force assignment overrides the value provided by the drivers of a signal with an arbitrary value. A release assignment returns control of the signal back to the model. This can be used to simulate an invalid state or bypass complex initialization sequences.
- **The to_string function** returns a string representation of a value. It is predefined for all scalar types (including user-defined enumeration types) and for character-based array types.

New Packages

- **Synthesizable fixed- and floating-point types.** Package fixed_generic_pkg defines fixed-point types and operations. Package float_generic_pkg defines floating-point types and operations in a format compatible with IEEE Std 754. Both can be parameterized to configure parameters such as rounding style and guard bits.
- **The std.env package** defines functions and procedures for accessing the simulation environment. The procedure stop interrupts a simulation, allowing it to be resumed later. The procedure finish terminates the simulation.

2 Design and Architecture

This book is about writing high-quality VHDL code. Why, then, should it start with a chapter on design? Isn't design something we do before writing any code? We need to talk about design because what we do at the keyboard while we are coding *is* design, and every programmer is a designer. No matter how detailed the plans for implementing a system, there are always decisions to be made at the code level. More often than not, bad code is not caused by illiteracy in a programming language; rather, it is a design failure.

Design problems are introduced in a system for different reasons, at any stage of the development cycle. Some are caused by poor planning or unforeseen interactions between the parts of a design. Others are caused by coding mistakes when the system is implemented or modified. No matter the cause, many design problems can be prevented in the same way: by following a set of good design principles when we work with the code.

But what defines a good design principle? Most of the times, we can find advice on both sides of the fence about any design issue. To remove any subjectivity, we must agree on a set of *design qualities*—desirable characteristics that we would like to see in any solution. A good design principle is a rule that consistently improves the desired system qualities.

Because design is all about tradeoffs, we cannot apply the same guidelines to all kinds of problems. No ready-made set of rules can replace a solid understanding of good design principles. In fact, this is why we need to be familiar with the theory of design: to assess any principle, practice, and recommendation in terms of their contribution to the quality of our work. This chapter introduces the topics of design and architecture and discusses the main system qualities directly related to source code.

2.1 What Is Design?

As a verb, to *design* means to create a plan for turning specifications into a working solution. As a noun, *design* can mean either a set of such plans or the way in which something has been made—in other words, the sum of all design decisions made while creating the solution.

Design is related to source code in two ways. First, we need to do some design before coding because a good plan improves the odds of achieving our goals. The simple acts of thinking out a few alternatives and choosing the most promising one result in a better solution—one that is easier to implement and works better when complete. Second, the quality of the source code deeply influences the design qualities of the finished system. Attributes such as complexity, comprehensibility, and maintainability are strongly influenced by the quality of the code we write.

The key concept in design is tradeoffs. Every design decision involves a compromise between different system qualities. Is it more important that the system be cheap or fast? Should we make it future-proof or be done with the design quickly? The designer's job is to make informed decisions and choose the alternatives that best support the overall system goals.

Take, for instance, an example from digital signal processing. To represent analog samples, a designer can choose between fixed- or floating-point numbers. Fixed-point numbers use less hardware and result in faster computations. Floating-point numbers have a larger dynamic range, are simpler to use in most algorithms, and are easier to debug. Which format should you choose? The right answer is the one that meets your design constraints (including development time) yet keeps the overall design quality as high as possible.

Even with unlimited physical resources, design would still involve tradeoffs. Design decisions cannot always be evaluated by measure of a physical parameter, and design qualities that are not immediately visible can still interfere with our ability to achieve the system goals. Complexity, comprehensibility, and maintainability are fundamental properties of a system and, in most cases, at least as important as efficiency and performance. To make a good decision, the designer needs to consider the overall context, including the constraints imposed by the internal system architecture and the environment where the system will operate.

So far everything we said about design is general enough to apply to both software and hardware. However, because many practices in this book are drawn from the software field, we need to address the question: How is hardware design different from software design? The fundamentals, as we shall see, are the same—especially when it comes to writing source code. However, some distinctions are worth exploring.

One of the main differences—with impact over the entire product development cycle—is that a software product is generally easier to change than a hardware product. Developing a new version of a hardware product involves long verification and production cycles and possibly high nonrecurring engineering (NRE) costs. Software developers, in contrast, usually have on their desks everything needed to produce a new version of the product.

As a consequence, a hardware design has a high premium for getting the product right the first time. In the hardware field, a company that does not follow a strictly controlled design process will go out of business. To improve the odds of first-time-right silicon, design houses invest heavily in precise and detailed designs and produce large amounts of documentation.

Another big difference is that hardware design requires a solid background in electronics. Besides hardware design languages, a circuit designer must be proficient in a large number of tools and techniques, including logic design and optimization, basic hardware structures, simulation, prototyping, measurement, and instrumentation. Coding is often a small part of the job. In consequence, a typical hardware designer does not have a large exposure to proper software design techniques, including good design principles and best coding practices.

Despite such differences, a hardware designer using an HDL and a software developer using a computer programming language share many common tasks, and both get the same benefits from working with high-quality code. If our code is more readable, we will communicate better with our team members, who can work faster and make fewer mistakes. If our solution has a clean design and everything is in the right place, our code will be easier to navigate, modify, and extend. And if our code is more testable, it is more likely to be thoroughly tested and safer to modify. Overall, because such goals are as important in software as in hardware, most software design principles aimed at producing high-quality code are also valid for VHDL.

2.2 Quality Measures

The quality of the code we write has a deep influence on the quality of the finished system. But as we code, how can we know whether we are making a system better or worse, especially its intangible qualities? Put another way, how can we evaluate a system and its underlying design?

Informally, we use the term *quality* as the degree of excellence of something. However, it is possible for a product to excel in one area and perform poorly in all others. A two-way radio may have the longest range in its category but at the cost of short battery life. What can we say about its overall quality?

To assess quality objectively, we break it down into a number of *quality attributes*, which can be defined more precisely and measured independently. This division allows us to compare two possible designs for a system: "System A is more *maintainable*, but system B is more *efficient*."

Another use for quality attributes is to compute an overall quality score for a product by assigning different weights to each individual attribute. The weights could be determined, for instance, from the value given to each quality dimension by management, developers, or end users. This way, even if two competing designs differ in many quality attributes, it is still possible to make an objective comparison. The remainder of this section introduces the most common quality attributes of hardware and software designs.

2.2.1 Design Qualities

The *design quality attributes* of a system (also called *design qualities*) refer to its internal and nonfunctional characteristics. A system has a good design if it scores high on its quality attributes. The main reason to study design qualities is because they help us identify good design principles. This section defines the quality attributes used to evaluate the design principles introduced in the next chapter. Although there are many others, we will focus on attributes that are closely related to source code.

Strictly speaking, some of the attributes in this section could be classified as *system* qualities, such as performance and efficiency. Nevertheless, they are mentioned here because they are deeply influenced by the system's design.

Simplicity Intuitively, the simplicity of a piece of code is a measure of how easy it is to understand. Simplicity is a desirable characteristic because it has a positive impact on many other attributes, such as understandability, modifiability, testability, reusability, reliability, and maintainability. If you can make a design simpler, then you are improving all these qualities at once.

It is hard to define precisely what makes for simple code, but some clues help us recognize it. Simple code is straightforward and flows naturally, whereas complex code requires extreme attention and a deliberate effort to understand. In simple code, every sequence of statements follows a logical order, whereas complex code is full of exceptions and special cases. Simple code solves similar problems in similar ways, whereas complex code uses hacks and ad-hoc patches in an attempt to solve a problem in the most immediate way. Simple code is also free from unnecessary mechanisms; each part contains only data and behavior that are central to its task. Finally, simple code is easy to explain to our peers. You should always be suspicious of a piece of code that you cannot explain in a few sentences.

Simplicity does not come naturally; it has to be designed. To start with, a system must be created simple. After it has been implemented, keeping it simple requires a conscious effort by every programmer who touches it. Developers must resist the temptation to do quick fixes that "get the job done"; instead, every new piece should be put in the right place, conforming to the original design. Finally, the system design and implementation should be constantly revised in search of possible simplifications.

Although simplicity takes some effort, it pays off easily, even in the short run. Complex code always requires more effort to work with—it is harder to understand, modify, and maintain. Keeping our code simple yields benefits for the entire life of the application.

Because it is so hard to characterize simplicity, quantitative measures usually deal with its opposite, *complexity*. We will have more to say about complexity in chapter 3. For now, it is enough to say that managing complexity should be the main concern of every programmer, and it should play a major role in every technical decision.

In the next few chapters, we present many rules and guidelines aimed at minimizing complexity. To assess their effectiveness, we need a way to evaluate complexity objectively. Over time, researchers have proposed different metrics for estimating code complexity. Two simple metrics are the number of statements and the number of lines of code (LOC) in a program. Despite being a little oversimplified, they correlate well with the incidence of bugs and with the effort it takes to understand a section of code.

Another simple metric is the number of modules in a system. This number has a peculiar relation with complexity. At one end of the spectrum, the system is built as a single, monolithic module. This is one of the most complex design alternatives because it is impossible to reason about the entire system at once. On the other end of the spectrum, the system has a large number of trivial modules; this alternative is also highly complex due to excessive intermodule communication. Somewhere in the middle lies the optimal number. We see some of the techniques to find the right balance when we discuss *modularity* in section 4.

Other more elaborate measures are based on the cognitive load to understand a piece of code. One of the most established metrics is McCabe's *cyclomatic complexity*, or *McCabe complexity*.[1] McCabe's complexity counts the number of paths in a section of sequential code using a simple algorithm: start with one and increment the complexity every time a decision statement is found. In VHDL, this mean incrementing the count every time we find the keywords *if, elsif, case, for, while, when,* or *select.* Figure 2.1

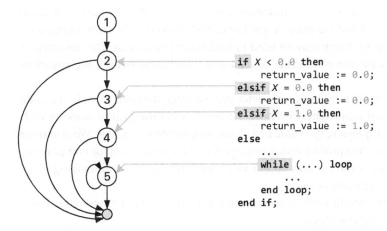

Figure 2.1
Computation of McCabe's complexity for the *ieee.math_real.sqrt* function.

illustrates the computation of McCabe's complexity for a simplified version of the *srqt* routine, from package *ieee.math_real*. This example has a cyclomatic complexity of five.

There are at least two ways to use cyclomatic complexity. The first one, during implementation, is to set an upper limit that should be observed by every module in the system. The second use is to identify problematic areas of the code after it has been implemented. According to McCabe, values between 1 and 5 are fine, and modules with a complexity above 10 should be rewritten. Modules between 6 and 10 are good candidates for *refactoring*: improving the internal structure of the code without changing its behavior.

Most of the time, the exact metric used to evaluate complexity is not important. The goal is always to minimize complexity and prevent it from deteriorating the original design.

Understandability   A system description in a hardware design language (HDL) can serve many different purposes. Among other things, it can be simulated, synthesized, programmed into an FPGA, or sent for fabrication. For all those tasks, the quality of the source code is not utterly important. As long as the source code follows the language rules, the tools will be able to understand it, and as long as the code is functionally correct, the device will work as expected.

This is, however, only part of the story. Before a design is finished, we must write its HDL description, understand it, discuss it with other programmers, and verify its correctness. After the code has been deployed, we need to maintain it. The upshot is that writing understandable code is as important as writing correct code. Unintelligible code, even when it works, is easy to break and hard to fix.

In every realistic situation, code is read much more often than it is written. In team development, programmers need to understand each other's code. In long projects, we need to make sense of what we wrote a few weeks earlier. Even if we are working solo, the clever code we write today may be hard to understand tomorrow. The net effect is that any effort to make the code easier to read and understand yields a big payoff in development time.

All this may sound obvious, but for a novice programmer, the need to make the code self-explanatory is not immediately evident. We need to feel the pain of working with messy code to fully appreciate the value of a clean, well-factored solution. We also need exposure to good and bad programming examples to have some basis for comparison. Without these preconditions, it may sound an exaggeration to insist on making every line of code utterly clear and obvious.

To test your own commitment to code readability, try to imagine how you would deal with the following situations:

- **Example #1.** A programmer writes an *if* statement, and its condition expression involves several variables and logical operations. To simplify the condition, the programmer creates two auxiliary variables, each one corresponding to one half of the original expression. The resulting code has more lines of code; however, the new condition in the *if* statement looks much simpler because it is a simple logic *or* of the two auxiliary variables. If you were the programmer, would you have made this change?
- **Example #2.** A programmer spends a few moments trying to understand a group of statements in the middle of a process. With some effort, the developer is able to figure the intent of the code: it calculates a partial value in a computation based on one variable and one signal. Although this calculation is used only once in the entire application, the programmer decides to extract the sequence of statements to a separate function, giving it a clear and meaningful name. If you were the programmer, would you make this change?

If your answer was *no* to one of the above questions, then maybe code readability and understandability are not at the top of your priorities. The changes made in the two examples minimize the time it would take for someone else (and for ourselves) to understand the code in the future. Of course real-life situations have more shades of grey, but even small changes like these make a big difference when applied to a large base of code.

Modifiability Modifiability is a measure of how easy it is to make changes in the code. This is an important attribute because in real-life projects we are constantly changing the code. We need to modify it to fix errors, add functionality, improve performance, or adapt to changes in the hardware environment, among other reasons.

A highly modifiable system has two desirable characteristics: it enables localized changes, and it prevents ripple effects. A system enables localized changes if the number of modules that need to be touched during a change is minimal. A system prevents ripple effects if no other modules need to be changed besides those directly affected by the modification. These two properties can be built into complex systems, but they require a strict design discipline. To achieve a high modifiability, each module in the system must have a single responsibility, and the modules should be minimally connected. Both design principles are explained in detail in chapter 4.

Testability Testability is a measure of how easy it is to test a system and its components. There are three main requisites to making a system more testable. First, it should be easy to make its components work in isolation. Second, it should be easy to manipulate each component and put it in a known state. Third, it should be easy to observe the system outputs and identify a failure when it happens.

The testability of physical devices, circuits, or systems is outside the scope of this discussion. Instead, we are interested in how to verify that the VHDL code we write corresponds to our expectations. To test our own code, we write programs called *test-benches*. A testbench applies a set of inputs to a design and checks whether the output values are correct. A complete VHDL design is typically accompanied by many kinds of testbenches with different purposes. *Module* testbenches test a single module in isolation. This is similar to the concept of *unit testing* in software. *Subsystem* testbenches integrate two or more components and check their interoperation. Finally, *top-level* testbenches verify the entire system from an outside point of view. In a CPU design, for instance, a top-level testbench could execute instruction code and then check the results at the CPU outputs.

The main reason to test individual modules is that, in most cases, it is not possible or practical to test the system entirely from an external point of view. In the CPU example, it is clearly not possible to test all the programs that it might run. By testing individual components exhaustively, we ensure that each small part of the system works as expected, reducing the need for exhaustive tests at the system level.

Another reason for testing individual modules is to assist in failure location. Even if a top-level testbench could test the entire system, we would still need detailed information to locate the source of a failure. For example, a top-level testbench that treats the CPU as a black box can only tell which instruction caused it to fail, while lower level tests would help pinpoint the problem to a specific module.

Finally, on a more philosophical note, there is a strong correlation between ease of testing and good design.[2] Many of the principles that make for a good design also make the system easier to test. The reverse is also true: when you find a design that is not easily testable, there is usually an underlying design problem. Keep that in mind for the next time you find yourself struggling to create a testbench for some piece of code.

Extensibility Extensibility is a measure of how easy it is to accommodate new functionality in a system or make it operate with a larger capacity. An extensible system accepts this kind of modification with minimal changes to the existing design and source code.

Extensibility is better supported at the architectural level and should be considered while the high-level design is planned. Coding style, however, can also make a system easier or harder to extend. For example, if we create modules that are generic and parameterizable, then the system will be easier to reconfigure for a larger capacity.

Reusability Reusability is a measure of how easy it is to use the code in a place other than where it was designed for. Reuse is an appealing idea because it can save development time and effort. There is also the benefit of using something that has been tested and proven.

One problem with the term *reuse* is that it can mean several different things. For example, the simple act of copying and pasting code is a crude form of reuse. A similar practice applies at the project level: instead of starting from scratch, we can build a new project using an existing design as a basis. These two practices are examples of *casual* or *opportunistic reuse*—reusing source code that was not explicitly designed with reuse in mind.

Going up one level, a common practice is to reuse the same module in different parts of a project. A system can be designed from the outset with the explicit goal of reusing some of its routines and components in several different places, possibly with different configurations. This practice can provide significant benefits, but we must be careful to avoid excessive or premature generalization. Because generic modules are harder to develop than single-use ones, we should create them only where there is a tangible benefit.

Taking it up one more level, another common practice is the creation of libraries to be reused across different projects. Such libraries are examples of *formal* or *planned* reuse because they are designed from the beginning to be used in different scenarios. Detailed documentation and extensive testbenches are a key success factor for this kind of reuse.

Finally, at the highest level of reuse, there are large macros or IP cores created specifically to be instantiated in other projects. Such macros must be portable, easy to integrate in other projects, and robust enough that their basic functionality does not have to be reverified by the client. Developing such blocks is significantly more expensive than creating a block for a single use. The *Reuse Methodology Manual* estimates that, for small blocks, the cost of developing a generic macro is ten times that of creating a module for a single use.[3]

The upside is that following good design principles is a big part of making a module more reusable. In other words, to make a module reusable, it first needs to be usable. Modules that solve a general problem, have clear interfaces, do not rely on low-level hardware structures, and are backed by good tests are natural candidates for reuse.

Portability Portability is a measure of how easy it is to migrate a system to a different environment. One of the main benefits of HDLs is the portability of the resulting designs; it is a shame to waste this advantage by writing unportable code.

A design becomes unportable when it uses features that are unique to a tool or implementation technology. A common example in VHDL is the use of compiler directives embedded in comments. Such directives are called *pragmas* or *metacomments*, and they give the compiler additional information about how to process the source code. Because this information is passed inside comments, most pragmas are not a standard part of the VHDL language; each tool vendor may define pragmas that do not exist in other tools.

As another example, each FPGA contains a set of basic hardware structures. Some of the structures are common in every FPGA, whereas others are unique and vary widely between architectures. By instantiating such structures directly, we reduce the portability of a design. In some cases, this is necessary to achieve a better performance or fine-tune the configuration of a hardware block. For portability sake, however, this should be minimized.

There are two simple yet powerful recommendations to make our code more portable. The first one is to minimize the use of nonstandard language features. Always write in standard VHDL; if you need your compiler to infer some hardware with a special characteristic, then read the tool's coding style guide. This is usually enough to make the tool infer the kind of hardware we want.

The second recommendation is to clearly separate application logic from hardware dependencies. Code that implements algorithms, performs calculations, or makes high-level decisions should never be allowed to touch hardware directly. Instead, isolate each hardware dependency inside a wrapper module whose interface exposes only the basic functionality. For example, if you need a clock multiplier, then the wrapper module could have a clock input, a clock output, and generic constants to set the base frequency and multiplication factors. In this way, if we need to switch vendors or technologies, then our application code remains intact. All we need to update is the thin wrappers encapsulating the hardware elements.

Maintainability   Maintainability is the degree to which a system is easy to maintain, including changes to fix errors, add functionality, improve performance, and adapt to a changing environment. Maintainability is the combined result of other qualities, such as modifiability, understandability, extensibility, and portability.

As a fresh programmer, it is easy to underestimate the value of maintainability, but it should play a major role in every design decision, just like complexity. Maintenance represents on average 60% of the costs in a software system. In most real-life projects, it is more important to reduce the maintenance effort than the initial implementation effort.[4]

If we break down the maintenance costs, we will find that the dominant task is "to understand the existing product." The upshot is that whatever we can do to make our code more understandable and easier to maintain will have a significant impact on the total system cost.

Performance   Performance is a measure of how well the system does its intended function. Several performance metrics are applicable to digital systems; some of the most common are response time, throughput, latency, circuit area, and energy consumption.

Performance is a real concern in complex applications, and this partly explains why developers are usually obsessed with it. However, the main problem with this fixation is that performance is best addressed at higher design levels, not by manually tweaking the code. Many programmers feel an urge to write code that is "efficient" and "optimized" and give performance a disproportionally high importance relative to other design qualities. However, trying to change the source code to improve its performance is an uncertain bet and almost guaranteed to decrease other qualities of the code. A much better approach is to keep our code clean and our designs modular and modifiable. Then, and only if our design does not meet its performance requirements, we should concentrate on improving the troublesome spots in a circuit.

This subject tends to spark heated debates, so it is important to state our terms clearly. We are referring to small-scale modifications made to the source code in an attempt to make it more efficient. This approach is often called *code tuning*.

The first argument against code tuning is that there are faster and more effective ways to address performance issues. For instance, choosing the right algorithm or pipelining a calculation are examples of decisions much more impacting on performance and also less harmful to the source code quality. In general, design changes made at the architecture level are much more significant than those made at the code level.

Another argument against code tuning is that it is hard to predict its effect on the speed and area of a synthesized circuit. In fact, most attempts to improve performance by tweaking the code will have negligible results. Others will backfire and make performance worse. The key point is that the only way to know whether a modification improves a circuit's performance is to measure it.

To illustrate this, consider an example where a designer needs to multiply an unsigned integer number by seven. The developer might think, "I need to multiply this number by seven, but multipliers are expensive, so why not do some bit manipulation instead? If I shift the input value two bits to the left, add it to the input value shifted one bit to the left, and then add it to the original input value, I will get the same result without using a multiplier." Thrilled with such cleverness, the developer sets out to write the optimized code, which might look like this:

```
-- Multiply 'a' by seven by shifting it two bits to the left,
-- adding the value of 'a' shifted one bit to the left, and then
-- adding the original value of 'a'.
y <=
    ("0" & a & "00") +
    ("00" & a & "0") +
    ("000" & a);
```

Satisfied with this "optimized" code, the developer never bothers to measure its actual performance. Had it been measured, we would find that the following straightforward version of the code

Table 2.1 Synthesis results, manually tuned code vs. straightforward implementation

Tool	Area, manually tuned code	Area, straightforward code
A	100%	100%
B	100%	100%
C	100%	91%

```
y <= a * 7;
```

gives the same or better results in several popular FPGA synthesizers (table 2.1).

The "optimized" version has other undesirable qualities: it has more places for bugs to hide, and it requires detailed comments to communicate the author's intent. This example illustrates two other points against code tuning. First, it tends to produce code that is harder to read and maintain. Second, if you change the implementation technology or compiler settings, then all your assumptions need to be revalidated because what was a gain under certain circumstances could well become a loss under different conditions. This is a strong argument against code tuning: Because you are trying to second-guess the compiler, you should provide actual proof of any assumed benefits, and this proof needs to be renewed every time one of the variables involved in the process changes.

Another point against code tuning is that it is hard to guess, during implementation, what the performance bottlenecks of the finished system will be. Especially in synchronous systems, there is no speed benefit in optimizing a circuit outside its critical path. Therefore, it is best to wait until we can actually measure the circuit's performance rather than perform speculative optimizations. Before you decide to make a change that will decrease other design qualities, including readability, you should make sure that the problem you are trying to solve really exists.

To summarize, the best way to write code that helps achieve performance goals is to have a modular design and keep the code clean and modifiable. Wait until you can actually measure the system's performance before tuning the code. Do not optimize as you go; base your actions on measurable results only and never on project mythology or folklore. Finally, keep in mind that this is not an excuse to write terribly inefficient code. This is simply an argument for deferring optimizations to the appropriate time in a project and making changes when and where they offer the best payoff.

Efficiency   Efficiency is closely related with performance; it is a ratio of the amount of useful work done by the amount of resources used. Just like performance, the efficiency of a digital system may use several metrics, such as throughput, area, and energy consumption. Thus, if two systems perform the same task but one of them uses less energy, then the latter is more *energy-efficient*. If two circuits perform the same task but one of them uses half the area, then the latter is more *area-efficient*.

Efficiency also means to make good use of scarce resources. This has two main implications. The first is that we need to know how scarce our resources are. The second implication is that, if a resource is not scarce, then it is not wise to put effort into minimizing its usage. You don't get any extra points for shaving off an extra nanosecond if your design is an alarm clock that increments only once every second—especially if it reduces other qualities of your design.

Because the scarce resources are different in synthesis and simulation, efficiency has different parameters in those two cases. In large simulations, the scarce resources are those from the machine where the simulation is run, such as CPU time and RAM. In silicon, the scarce resources include gates, area, latency, delay, memory bits, and the energy consumption of the synthesized circuit.

The same recommendations given for performance are valid for efficiency. Avoid premature optimizations and use a good design that preserves the qualities of the code, especially modularity and modifiability. Remember that development time is *always* a scarce resource.

2.2.2 Business Qualities

The quality attributes presented in the previous sections have a close relationship with system design. When the system is considered as a product, however, other qualities must be taken into account. In this section, we touch on those attributes lightly, only to clear up some misconceptions about the tradeoffs among cost, development effort, and the quality of HDL source code.

The Cost of Quality Paper accepts all ink, but a compiler accepts only legal VHDL grammar, and your QA department will accept only applications that meet their specifications and requirements. It is important to make this distinction because it helps to correct a common myth: that there is a tradeoff between source code quality and development speed.

The problem with this theory is that when we have requirements to meet, we cannot just write any code as fast as possible. The shortcuts we take today will come back to slow us down tomorrow or next month. For this reason, there is a large region in the quality versus cost curve where there is no tradeoff. In this region, increasing the quality of the source code actually *reduces* development time.

This phenomenon is observed in many human processes, whenever the cost of prevention is lower than the cost of rework. Improving the quality of a development process can save a company more money than it costs. This conclusion was made famous by Philip Crosby in his book *Quality Is Free*, where he estimated that the total cost of rework in a typical company is about 25% ("one dollar out of four").[5]

In software, Steve McConnell shows in *Code Complete* that debugging and the associated rework consume about 50% of the time in a traditional software development cycle.[6] According to McConnell, the most obvious method to save development time is to improve the source code quality, thus decreasing the amount of time spent in debugging. His conclusion is similar to Crosby's: McConnell's "General Principle of Software Quality" states that improving a software's quality reduces its development costs.

But how can we improve the quality of a design, aiming at reducing the cost of rework? If we think about the design process, there are two complementary approaches. The first is to get the design right the first time, thus minimizing the amount of rework. The second is to be constantly rethinking and reorganizing a design to make it easier to work with. This is important during maintenance as well as initial implementation. Every time we add or change something, we should look for ways to improve our design so that it incorporates the new functionality in the simplest and cleanest way.

This recommendation often surprises the less seasoned developers. Two common doubts are: "Is it really time-efficient to change the design frequently?" and "Isn't that an unnecessary risk to change the design for no obvious reason?"

To answer the time-efficiency issue, remember that one of the biggest time-consumers in software development is to understand existing code: Before making a change, we need to understand all the surrounding code. This effort can be greatly reduced by adhering to good design principles because well-designed code is much easier to work with: There are fewer surprises, fewer bugs, and less need to change unrelated parts of the code. In contrast, working with messy code slows everything down. The upshot is that the only way to maintain a good development speed is to keep our code clean and our design simple.

As for the perceived risk of changing the code frequently, there is no such risk when our design is backed by proper automated tests. A good suite of tests works as a safety net: it allows us to make changes that would be dangerous otherwise. It immediately tells us when we break anything in the existing code.

On a final note, these recommendations are not only for professional developers working in big companies. Even if you are working alone or in a class project, the conclusions still apply. The only difference is that the savings are not in the product cost but in development time. You will be able to do more and in less time if you keep your code quality high.

The Cost of Change   As the system implementation moves forward and the source code grows, the effort to make a change typically increases. This may sound obvious, but what is not so evident is that the rate at which this effort grows depends largely on the quality of our designs. In a bad design with large and excessively connected modules, localized changes are all but impossible. Everything we touch has side effects in seemingly unrelated parts of the code, and changes get harder and harder over time.

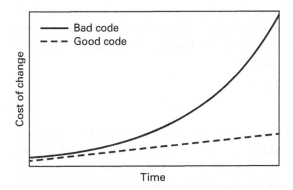

Figure 2.2
The cost of change grows over time, but the quality of the source code makes a difference.

Figure 2.2 gives a rough idea of what happens to the cost of change as the project grows. Of course, every project has its own particularities, and the distinction between "good code" and "bad code" is not so clear-cut, but the effect shown in the graph is true for virtually any project. Also, because the source code grows as the project evolves, the horizontal axis could represent either time or size.

Over the life of a project, good design and high-quality code can make a huge difference. With bad code, the effort to make a change increases disproportionately. The visible effect is that programmer productivity decreases over time.

Why is a bad design so much harder to change? A common trait in bad designs is poor modularity. Instead of small modules focused on a single task, bad designs lump unrelated functionalities into a small number of modules. As a result, the modules are larger and more interconnected than they should be. Larger modules are hard to work with because we need to read and understand a lot of code before we can make a change. They are also easier to break because there are more places for bugs to hide.

Highly interconnected designs are also bad because we cannot reason about one module in isolation. If a single responsibility is split across separate modules, then we may need to touch all of them to make a change. If a module has more than one goal, then we may have to navigate through a lot of unrelated code to make a single change.

Another common trait in bad designs is a failure to keep the design clean and well factored. In the presence of redundant information, changes are risky: We must remember to change all places where information is repeated, or we may introduce inconsistencies in the design. For more about the perils of duplicate information, see sections 3.4 and 4.8.

Finally, bad code is often poorly backed by tests. Part of this stems from the fact that bad designs are less testable than good designs. Another reason is the high correlation between bad programmers and programmers who do not write tests.

Without a good set of tests, a design tends to deteriorate. The only way around this is to reorganize the design every time we change or add something, and this is only possible when the code is backed by good automated tests. Otherwise, no one will touch the existing code for fear of breaking it.

2.3   Architecture

In the conception of a system, the term *architecture* means the highest level of design. It comprises the earliest design decisions, including the first breakdown of the system into parts. The architecture provides a framework of the overall system structure, to which the detailed components will be added later during the lower level design phases.

In simple and small systems, it may be possible to go directly from requirements to implementation. However, trying to implement a system without an upfront high-level design is an unnecessary risk. A good architecture improves the odds of meeting the system goals, and it has a profound effect on the quality of the finished system.

A planned architecture is helpful for several reasons. First, it leads to a better design. If we create the top-level division of the system following the architectural principles of simplicity, consistency, and modularity, then the resulting design will be easier to implement, test, and maintain. Second, the architecture can be used as a plan during implementation. The high-level division can be used to create a work-breakdown structure, and each block can be handed to an independent design team. It also works as a map, helping locate any functionality in the system when the developers need to add or modify anything.

Finally, the architecture description provides a simpler model of the system and a big-picture view of how it works. The architecture description is the first document that a new team member should read; collecting this information directly from the source code is nearly impossible in large systems. The recommendation is simple: do not skip the phase of architecture design by rushing into coding. Draw a diagram and explore the design alternatives on paper, where they are much easier to visualize and change.

Architecture versus Design   The line between architecture and design has always been a source of confusion. The main reason is that architecture *is* a form of design, but not all design is architecture. Therefore, the real distinction is not between architecture and design but between *architectural* design and *nonarchitectural* design (or low-level design). The following criteria help draw a line between the two:

• **Quality attributes**. An architecture defines system-level constraints and sets specific goals for each module. The low-level design of each module must meet these goals and satisfy such constraints.

- **Scope**. Architectural decisions usually have system-wide effects. Nonarchitectural decisions are local to each component.
- **Behavior**. An architecture defines the observable behavior of each major system part. However, the decision of how each component implements its behavior is nonarchitectural design.
- **Structure**. An architecture defines the major blocks of a system and how they communicate. Low-level design must adhere to any interfaces specified by the architecture.
- **Data**. Architectural design uses abstract and conceptual models of data. Nonarchitectural design uses concrete data models and structures, usually specified down to the bit level.

Examples of Architectural Decisions   The following questions are examples of architectural decisions in circuit design.

In a digital system, clocking and reset schemes involve many details and tradeoffs. Such decisions are too important to be left for the detailed design phase, so the architecture should provide the answers. Should all the system inputs and outputs be registered? How about the inputs of the internal modules? What types of storage elements are allowed? Only edge-based registers or latches as well? Should we use synchronous or asynchronous reset? How many clock domains do we need, and how do we communicate across them? If such decisions are left for the downstream developers, then the resulting system may become incredibly hard to integrate.

Another example of an architectural decision is how to distribute the processing tasks to achieve the required system performance. This also includes the allocation of resources shared by multiple modules. Can the computations be done serially, or do they need to be parallelized? Does the circuit need to be pipelined to achieve the required throughput? These architectural questions are crucial to determine the system performance and will have much more impact than decisions made at the implementation or code level.

Because one of the goals of an architecture is to divide work among independent development teams, it should define how the major components will communicate with each other. Should we use ad-hoc connections between the modules, or should we use an on-chip bus? Some complex systems may require more than one bus, with different performance requirements.

These examples highlight two key characteristics of architectural decisions. First, they are made from an overall system perspective. If the decision can be deferred to a lower level, then it is not architecturally significant.[7] Second, the architecture defines what can be changed easily and quickly and what will be hard to modify in the future. This emphasizes the importance of having a good architecture from the beginning.

Architecture Specification Document   After all the important decisions have been made and the top-level system diagram has been created, they are usually registered in an *architecture specification document* or *architecture description document*. If you are in charge of writing this document, here is a list of the typical information it should contain:

- **A big-picture view of the system.** An overview describing the system in broad terms, including its major components and their interactions. This description should be simple enough to be understood by new team members. It should also describe how the system interacts with other systems and its users.
- **High-level diagrams.** A typical architecture description includes several diagrams, showing the system in different levels of detail and from different points of view. A common example is a block diagram, including the major blocks, data paths, memories, and key signals. This diagram should identify the kind of information exchanged by the components but without much detail. Another typical drawing is a context diagram, which shows how the system interacts with its surroundings.
- **Design decisions.** The architecture document serves as a repository for early design decisions. All decisions should be backed with rationales, so that if they need to be reviewed in the future, all important factors will be considered.
- **A mapping between requirements and components.** The architecture works as a bridge between the requirements and structure of a system. The specification document should map each requirement to the corresponding system elements.
- **Constraints.** The document should include any constraints that affect the work of downstream developers. This includes resources that are allowed or disallowed, such as nonstandard design libraries. It can also limit the amount of resources available for each block.
- **Design principles.** The document should list any overarching design principles that must be respected by downstream developers. For example, an architect may have organized the system into layers. To preserve this structure, the architecture description should specify the function of each layer.

In software, documenting an architecture is a formal task. It is described in detail in the standard *ISO/IEC/IEEE 42010 Systems and software engineering—Architecture description*.[8] If you need to specify the architecture of a larger system, this standard could be a good starting point.

3   Design Challenges

Before we can talk about writing great code, we should know what makes it so hard to write. Excessive complexity, unchecked dependencies, duplicate information, and unplanned changes make any code harder to understand and work with. Furthermore, our human limitations restrict the amount of information we can absorb and manipulate at any time. A good programmer is aware of these challenges and always strives to write code that does not exceed the limits of our skulls.[1] This chapter presents the main factors that negatively affect the quality of our source code and must be kept in check to make our code more intellectually manageable.

3.1 Complexity

Complexity is the number-one enemy of high-quality code. Informally, complexity can be defined as a measure of how hard something is to understand. It is highly undesirable in any system because of its many negative effects: it increases the number of mistakes and bugs, slows down the development, makes changes harder and less predictable, and turns programming into a painful job. To make matters worse, complexity is cumulative and tends to occur naturally, unless it is kept under strict control at all times. Complexity is a major factor of technical risk in a design.

In any design, there are two kinds of complexity. First, there is an inherent amount of complexity that comes from the problem we are trying to solve. This kind of complexity is called *essential complexity*. Second, there is a significant amount of complexity that arises from the way we choose to organize our solution. This kind of complexity is called *incidental complexity*.

Incidental complexity finds its way into a design for numerous reasons. Sometimes the designers do not understand the problem completely and create a solution that is more complicated than it could be. Or there may be a failure to use good design principles to manage complexity, such as abstraction, modularity, and hierarchy. Moreover, complexity increases naturally during the lifetime or a project: we may have to add features that were not originally planned and are not easy to accommodate, thus eroding the original design. Finally, incidental complexity is also caused by general sloppiness—for instance, when a developer repeatedly settles for the first solution that comes to mind, instead of working out a solution that would preserve the original design. In any case, a lot of complexity is added inadvertently, either because the developers are unaware of it or because they lack the proper techniques to keep it under control.

Despite its pervasiveness, complexity is difficult to measure and may be hard to recognize with the naked eye; however, it can be identified by some of its common traits. For example, it increases with the number of interdependent parts in a design and with the degree of dependence between them. We can also estimate the amount of complexity from the amount of information we need to keep in our heads to work with any part of the system. For instance, at the level of statements in the code, complexity increases with the number of decisions, possible execution paths, levels of nesting, and simultaneous levels of abstraction in any section of code.

How about the number of modules in a system, is it a good approximation of the system complexity? If we imagine two systems with the same average module complexity, then a system composed of one hundred modules is more complex than one with ten modules. However, it is risky to measure complexity based on this number because one of the most effective techniques to tame complexity is to break down a large module into simpler ones. Instead, we should worry only about making each module as simple as possible and let the system have as many modules as needed. Remember that we should never have to reason about the entire system at once; most of our work should be done one module at a time.

If complexity is so harmful and pervasive, what can we do to keep our code simple? The only way is to put complexity at the center of every design decision and invest the necessary time to simplify a design, instead of settling for the first solution that possibly works. Whether we are designing, implementing, or modifying a system, we should always choose the path that minimizes complexity.

There are many heuristics to guide us in these decisions, and most of them are summarized in the form of good design principles. These principles are the subject of the next chapter, but here are a few examples. Break the application into small modules that are simple, independent, and easy to understand. Hide implementation decisions inside a module and do not let other modules use its internal parts explicitly. Restrict communication paths so that each module only has access to the information required to perform its task. Always work at the highest level of abstraction so you can think in terms of the problem you are solving rather than in low-level implementation details.

Simplicity does not come naturally in a design. When we first write a piece of code, it is never as simple as it could be. If we stop at the first solution that comes to mind, then incidental complexity will build up and start to cause harm. Plan to invest time into making a solution simpler right after you have made it work. It takes a little effort in the present but saves a lot of time in the long run.

3.2 Changes

Unlike the other design challenges in this chapter, changes are not a bad thing by and of themselves. On the contrary, changes are quite a good thing—they allow us to fix bugs, add features, and improve the code, keeping it useful for a long time. Changes are problematic only when the code is written in a way that does not afford them. Code that is hard to change is called *rigid*, whereas code that is easy to break is called *fragile*.[2] Bad code is usually rigid and fragile at the same time.

Any application may have many reasons to change. Our requirements may change, forcing us to modify or add behavior. We may have misunderstood the requirements. Sometimes we need to modify a system to optimize its performance or reduce some resource usage. At other times, we may have found a better way of doing something, and we want our code to benefit from it. If we write code that is hard to change (or code that we are scared to touch), then we will be making all those positive changes harder and less likely to happen. Even if we decide to make those changes, our work will be slower and more error-prone.

What does easy to change mean? It means that making small changes in behavior should be simple and fast—in other words, a small change in the requirements should correspond to a small change in the code.[3] Code that requires modifications in multiple places to implement a minor change in functionality is not easy to change. Ease of change also means that code which is already written should be easy to reuse in other parts of the application. Most important, it means that changes should have no unexpected side effects. An application in which distant and seemingly unrelated parts of the code break after a change is not easy to change at all.

Like simplicity, changeability has to be built into the code. However, writing modifiable code is easier than writing code that is inherently simple. With the aid of some good design practices, we can keep the code easy to change at all times. Many of the design principles presented in the next chapter have this goal. Minimizing dependencies, as described in the next section, is also essential. If we follow the right approach, writing code that is easy to change does not take longer overall. Generally, starting with a simple architecture and keeping the code clean and well factored is enough to guarantee a good changeability. In other words, there is no need to add a significant amount of code to make it future-proof. Most of the time, we do not need to guess which parts of an application will change in the future; that would be over-engineering. All we want is to preserve changeability. As the size of a project grows, keeping the code easily modifiable is the only way to continue developing at a constant speed.

Two techniques are fundamental to changeability. The first is a good set of tests, which give us safety to change the code without fear of breaking it. The basic idea is simple: as we write our VHDL models, we simultaneously write test code to verify that our models work. Each test should exercise part of the operation of a model, providing inputs and checking outputs to check whether it has the intended behavior.

These tests will be an integral part of the project and will accompany the code at all times. As we make small changes in the code, we will run the tests (or a subset related to the current change) to ensure that the new feature works as expected, and that the rest of the existing behavior did not change. In this way, we get an immediate warning if we break anything.

If you have never had this habit, then it will probably sound like more work than it really is. In reality, having our code covered by tests allows us to move faster in a project and spend less time chasing and correcting bugs.

The second technique is *refactoring*. Refactoring is the process of improving the design or internal structure of the code without changing its external behavior.[4] It is a disciplined way to clean up the code, performed in a series of small steps so that the system behavior is always preserved. Examples of such small changes include removing code duplication, renaming routines, and breaking down a module into two simpler entities. The effect of each change is small, but the cumulative effect in readability and understandability is significant.

Refactoring is typically performed immediately before or after a functional change in the code. We can refactor before making a change so that the change is easier to apply, or we can refactor after a change to remove any leftover complexity in the design. Without refactoring, the original design of an application tends to decay with every change. Refactoring gives us a chance to preserve (and even improve) the original design while keeping the code easy to change. In fact, to be effective, refactoring *requires* that the code be easy to change.

If we have good tests in place and keep the code clean and well factored, then changes are not hard or risky. In this way, we are much more inclined to make changes that improve the quality of the code. Code that is easy to understand and change is always a pleasure to work with.

3.3   Dependencies

Most systems are too complex to take in all at once, but they can be made easier to design if we divide them into simpler modules that can be developed one at a time. The flipside of this technique is that a module must cooperate with other modules to work. When module *A* needs any information or behavior from module *B* to perform its function, we say that *A depends* on *B* or *A has a dependency* on *B*.

Although some dependencies are inevitable, they can be a big problem if they prevent us from thinking about a module in isolation. If to develop module *A* we need to keep thinking about *B* all the time, then the division of work was not really effective. Dependencies can also make the code harder to change and easier to break: if *A* depends on *B*, then any changes on *B* may require that we update *A*, or they may cause *A* to stop working. Dependencies also make the code less reusable: if we want to use a module in a different place, then we may have to take all the modules it depends on as well. Finally, dependencies make a module harder to test because it cannot operate in isolation: we may have to instantiate or emulate all the modules on which it depends.

In a VHDL design, there are two kinds of dependencies. First, there is a *code-level dependency* between compilation units. If the name of unit *A* appears in the source code of unit *B*, then *B* depends on *A*. This implies that *A* must be compiled before *B*. Although the network of dependencies between design units can get quite complex in a project, they can be derived automatically from the code by VHDL tools. Second, there is a *logical dependency* between units, which is not strictly related with the code but with system organization. If we need to think about a module while we design another, then this characterizes a dependency, even if there is no direct connection between the two.

Dependencies can never be eliminated because some modules must cooperate to meet the system goals. They should, however, be minimized. In a well-structured system with a large number of modules, most modules do not need to know about each other. This can be achieved if each module focuses on a single task, knows just the minimum to do its job, and never exposes more details than necessary.

The degree of dependency between modules is called *coupling*. Creating systems with low degree of coupling, or *loosely coupled* systems, is an important goal in any design. We will see specific ways to reduce dependencies in section 4.4.

3.4 Duplicate Information

Every developer soon learns that if the same computation must be performed a dozen times, then it should be done in a loop. The advantages are self-evident: instead of repeating the same piece of code twelve times, we write it only once. Instead of a dozen places to fix when we find a bug, there is only one. Instead of reading the same statements a dozen times (to be sure that they are really the same), we read them only once.

Great programmers take this idea one step further and fight all forms of duplication in the code. They develop a zero-tolerance attitude toward duplication in statements, data, declarations, and even comments. If the same numeric value appears throughout the code, then it should be replaced with a named constant. If the same formula is repeated throughout the code, then it should be replaced with a function. If a similar sequence of statements appears in the code, even in different files, then they should be generalized and moved to a routine. Keeping the code free from duplication is the only way to ensure that future changes can be made locally—in other words, that there is no need to hunt for similar occurrences when we need to make a change. If the same or similar information appears in distinct places, then we must remember to change them all or we risk bringing the system to an inconsistent state.

Duplication may appear in the code for several reasons. Sometimes developers duplicate information without being aware of it. For example, we may fail to remember or recognize that a similar logic already exists elsewhere in the code. At other times, developers are not aware that duplicating code is so bad. For example, a designer may choose to represent a rectangle using three different points: its top-left corner, its bottom-right corner, and its center. The designer may not acknowledge the problem immediately, but the center of the rectangle is redundant information and may potentially put the object in an inconsistent state.

On other occasions, the developer is aware of the duplication and knows it is a bad thing, but he or she cannot find a way to circumvent it using the resources offered by a language. This is common among newcomers to the language. A typical example in VHDL is for a beginner to declare, say, sixteen input signals named $a0$ to $a15$ and sixteen output signals named $y0$ to $y15$. Given these objects, there is no way to avoid repeating the same statement or operation sixteen times in the code. Had the developer used an array, a loop statement could do away with the duplication. As a general rule, whenever you find yourself repeating the same statement or sequence of statements, stop and look for a way to write the code without duplication. In nearly every case, it is possible to find an intelligent solution.

Finally, in many other cases, the developer is just lazy. A lazy developer has thoughts such as, "If I need to write feature B but feature A already does something similar, why not copy the statements and make the few required changes in place?" or "If I know that the size of this array is 16, why should I create a constant to hold this value?" Numbers that appear in the code without a clear explanation are called *magic numbers*, and they are generally a sign of sloppy programming practice. They make the code unnecessarily hard to change and understand.

Duplication decreases many qualities of the code at the same time. Developing a zero-tolerance attitude against duplication is one of the simplest and fastest ways to improve the quality of your code.

3.5 Technical Debt

Technical debt is a metaphor coined by Ward Cunningham, inventor of the Wiki, and popularized by Martin Fowler, Steve McConnell, and others.[5] Fowler explains the concept of technical debt with the following example. Imagine you need to add a new feature to the code, and you identify two ways of doing it: one that is quick but decreases some quality of the code (e.g., it makes the code harder to change in the future), and another one that keeps the code clean and conforms to the original design but will take longer to implement.

Choosing the quick-and-dirty solution is like incurring financial debt: we trade the benefit of having something now for some burden in the future. Just like financial debt, we need to pay interest: the less-than-ideal code will slow us down as we develop new features, until we pay down the debt by refactoring the messy code. If we accumulate enough debt, all of our time will go into paying interest, and the project will come to a halt.

As intended by Cunningham, the term should be reserved for the cases where developers make a conscious decision to use a suboptimal design in order to make a deadline or allow an immediate solution. In this case, even though the solution is suboptimal, it should still be implemented in the best possible way. This is called "good" debt. In contrast, debt originated from sloppy coding practices is "bad" debt. Some authors would say the latter is not really a technical debt—it is just a mess. In any case, the important thing is that the metaphor works in both situations. Every time we write code that does not conform to good design principles, we incur some technical debt. In the general sense, technical debt can be defined as any pending work that needs to be done before the code is considered complete and meets quality standards.

In practice, bad or inadvertent technical debt is much more common than good debt. Typical situations in which we incur bad technical debt include:

- When we duplicate code instead of extracting the common logic to a single construct.
- When we implement a feature and do not write the corresponding testbenches.
- When we postpone documenting or commenting the code.
- When we write a feature in the easiest place instead of creating a new module or file.
- When we do not update the code after finding a simpler design alternative.
- When we delay breaking down a large file into smaller modules.
- When we keep on using an inadequate folder structure to avoid updating configuration files and scripts.

The debt metaphor is useful to discuss with nontechnical staff the need for keeping the code clean. Otherwise it may be hard to justify spending time in an activity that does not add features to the project and does not solve any immediate or obvious issue.

Excessive technical debt makes a project unpleasant to work with. It drags productivity down because the developers spend more and more time working around the issues that were not properly addressed. Fortunately, before it gets out of control, technical debt is easy to pay off: we can use incremental improvement techniques, such as refactoring, to reorganize the design and finish off the due work until the code meets proper quality standards.

3.6   Your Wetware

Wetware is the term used by novelists and cognitive scientists to refer to the wet computer that is the human brain, including its software and hardware. The human brain is limitless in certain aspects but extremely limited in others. Knowing its capabilities and limitations is essential to the work of professionals who produce information for human consumption, especially programmers.

While long-term storage capacity in our brains is virtually infinite, we are subject to severe information-processing limitations in other areas, including our attention span. We can focus our attention on only a few items at a time, and for a brief time, unless they are actively rehearsed. Such items are kept in our *short-term memory* (STM). The actual number varies among individuals and with the type of information, but two commonly cited numbers are 7 ± 2 for simple recall tasks and 4 ± 1 for more general tasks.[6] Items in STM are readily available for use in a wide variety of problem-solving and memory tasks, and STM is fundamental to solving problems with multiple steps, in which we must keep record of intermediate results. A simple usage example for STM is to keep the digits of a telephone number from the time we look it up until we dial it. A more complex use of STM is to explore the potential moves in a chess game.

The actual number of items that can be kept in short-term storage is not relevant to our discussions. The important thing is that there is a limit to the number of items we can keep track at the same time, and this number is small. If a new item comes in, then it will push out some old content. This is evidenced by a large number of studies demonstrating that when a person is forced to mentally juggle more than about five items, the number of errors increases disproportionately.

How is this related to programming? If your conclusion is that a good programmer needs to have an above average STM, think the other way round. The ability to mentally juggle a large number of items does not make a good programmer. The ability to break down a problem so that we can concentrate on only a small part of it at a time does.

Throughout this book, we will see many principles and guidelines aimed at making the source code more intellectually manageable. Here are some of the most common techniques and recommendations:

- Avoid clever programming tricks. Always prefer simple and clear code that is obvious at first sight.
- Use all techniques available for controlling and minimizing complexity, including hierarchy, abstraction, and information hiding.
- Hide lower level details so that you can concentrate on higher level tasks and goals.
- Write smaller modules so that you can think about an entire module at once.
- Keep sections of code short so that they fit entirely on a single screen, minimizing the memory load.

- Use clear and meaningful names that do not require memorization or association between names and the concepts they represent.
- Put as much information as possible in the code itself so that reading it does not require memorization. For example, instead of literal numbers and values, use named constants and enumerated types.
- Minimize the number of items you need to keep track of in any given context, including data objects such as variables or signals.
- Group related information into meaningful pieces. Use composite objects such as records to represent meaningful entities that have a direct equivalent in the problem you are trying to solve.
- Automate common tasks, such as compiling and running testbenches, to reduce the effects of task-switching and STM decay.
- Use keyboard shortcuts to keep your attention on the code and on the screen for most of the time.

One of the main problems in programming is that we make it much harder than it needs to be. If we keep in mind the goal of intellectual manageability and program within the limits of our skulls, then we will write code that is easier to understand, easier to work with, and more likely to be correct.

3.6.1   Recognition versus Recall

Recognition and recall are two different mechanisms to retrieve information from our memory. *Recognition* means remembering a fact, object, or event *with* the help of our senses. The human brain is extremely good at recognition, and it takes little mental effort. *Recall* means remembering something *without* help from the outside world. Recall is much harder than recognition, and our brains are much worse at that.

"Recognition is easier than recall" is one of the most used heuristics in user interface design.[7] It is the reason that graphical menus are generally easier to use than command line interfaces. It is the reason that multiple-choice questions are easier than essay-type questions. It is the reason that it is easy to recognize a person's face, but remembering their name may be hard. If we want to make something easier to use, then we must design it so that it relies more on recognition than on recall. In the theory of design, these two methods are often called "information in the world" and "information in our heads."[8] If we want to minimize the memory load during a task, then we must put as much information as possible in the world and leave less to be retrieved from memory.

The fact that recognition is much easier than recall has direct implications in programming. In fact, this is a good example of people making programming much harder than it needs to be. Every time we work with a program that forces us to memorize an encoding ("001" means ADD, "010" means XOR, etc.), part of our memory and attention gets devoted to this frivolous decoding task, leaving less processing power to really understand the program. As a general rule, all encodings, mappings, and associations like this should be explicit in the code, so that we never need to recall them from memory. In this way, if we need to refer to a value, then we can do it via a meaningful name rather than through an obscure literal number.

The same is valid for our choice of names. It may be easy enough to remember that w means the number of bits per word in a memory and y is the square root of the input. However, given our severe information-processing limitations, it is a terrible waste of resources to put this burden on the readers of our code. By using communicative names such as num_bits_per_word and square_root, we are saving their grey matter for where it is really necessary.

We can also put this conclusion the other way round: writing code that is less demanding on memory and mental resources is equivalent to making our readers (and ourselves) more intelligent as we read the code.

4 Design Principles

Chapter 2 introduced the attributes that allow us to evaluate the quality of a design. Chapter 3 presented some common challenges that jeopardize the quality of our designs. This chapter introduces a series of well-established design principles that help us achieve high-quality designs. We start with the most basic and consolidated design principles, which have been around since the early days of Computer Science. As we advance, the principles become more focused and may overlap with those introduced earlier. Nevertheless, all the principles are fundamental and appear frequently in discussions about the quality of source code and design.

4.1 Modularity

The principle of modularity states that a system should be designed as a number of modules that work as independently as possible. In a good modular system, each module has a clear and well-defined purpose, allowing the designer to concentrate on one part of the problem at a time.

Most hardware designers are used to decomposing a system into functional modules, and most VHDL users are aware of the language constructs for implementing modules and hierarchies. However, knowing the language constructs is not enough to create good modular designs. This section provides the principles and guidelines for structuring a design in a way that maximizes the benefits of modularity.

4.1.1 What Is a Module?

A module is a delimited part of a system, with well-defined inputs and outputs, that performs a well-defined function. This definition embraces the conventional description of a hardware module and is compatible with the term *design entity* as defined in the VHDL Language Reference Manual (LRM).[1] It is also compatible with the use of the term in software.

Another way to define a module is in terms of programming language constructs. In a computer program, a module is any group of statements that have been given a name and are bounded by explicit delimiters in the code. This definition is much more inclusive, embracing constructs such as functions, packages, or processes. When we create VHDL designs, it is useful to keep this definition in mind to remind us that structural design entities are not the only way to organize a solution. Many times, encapsulating part of the behavior inside a function or package is the best way to organize a design.

In this book, we adopt this broader definition for a module. For this reason, depending on the context, a module may represent different language constructs. When we talk about structural breakdown, a module could be a design entity. In a package, a module could be a subprogram. In an architecture, a module could be a process. In this last case, the module inputs would be the signals read inside the process, and the outputs would be the signals to which it assigns.

Another definition for a module is a "black box." A black box is a component that has accessible inputs and outputs, performs a well-defined functionality, but has unknown or irrelevant contents. The power of black boxes is that we can use them even though we know nothing about what they have inside. This is how an ideal module should be.

4.1.2   Key Benefits of Modularity

Modularity is one the most powerful principles in system design because it simultaneously improves many design qualities. This section discusses the main benefits from organizing the system into modules using the right criteria.

The System Is Easier to Understand   If the system is organized into modules with clear responsibilities, then it is easy to understand how the parts fit together to implement the system functions. If each module works independently from the others, then a developer can concentrate on a single module at a time during implementation, testing, and debugging. If the modules have clear interfaces, then we can treat them as black boxes and understand the system as a whole without looking inside any module.

Modules Can Be Developed Independently   If the modules have a single, well-defined responsibility, then each one can be implemented and tested separately from the others. This is an effective way to distribute and speed up the development work. However, the modules must be designed in a way that allows testing and validating each of them in isolation to prevent integration problems when they are put together in the system.

The System Is Easier to Change   If the system functionality is divided in a logical way, then it is easy to find the right module when something needs to change. Modifications are easier because we do not need to read and understand a lot of unrelated code, and there is less risk of breaking other parts of the system.

The System Is Easier to Maintain   When each module has a single responsibility, it is easy to find the most likely place for a bug. After the problem has been located, the developer can concentrate on the faulty module and safely ignore the rest of the code.

When the system needs to evolve (e.g, to add functionality or enhance its performance), it is easier to replace a module with an improved version. Because the rest of the system does not make assumptions about the module's implementation, it is easy to change its processing algorithms, the kind of memory it uses, or any specific data structures, all of which could improve performance.

The System Is Easier to Test   If a module has a single responsibility, then it is inherently easier to test. Modules that contain a mishmash of functionality are harder to put into a known state for testing. They are also harder to check because we need to consider which outputs are relevant and which ones can be ignored. In general, if you find yourself writing long sections of code to set up a module for testing, then it probably means the module is doing more than one thing.

It Promotes Reuse   Modules that have a single responsibility are more likely to be used in different contexts without changes. Modules with narrow interfaces (e.g., with a small number of ports) are easier to transport to a different system.

4.1.3   Effectively Decomposing a System into Modules

There are many ways to decompose a system into modules, with varying degrees of effectiveness. The best way to achieve a good decomposition is to observe the following guidelines, which describe the desired properties of individual modules and of the system as a whole.

A Module Should Have a Single, Well-Defined Responsibility   This is the most distinctive characteristic of a good module: it exists to serve a single purpose. A good module does not try to debounce an input, calculate a square root, and generate video signals at the same time. It does not mix a communication protocol with code that drives a display interface.

Of course, it is possible (and recommended) to organize a system hierarchically. Therefore, at some point, it becomes necessary to group together components with distinct responsibilities. In this case, we must observe two conditions to ensure that the higher level module still has a single responsibility. First, all of the lower level modules should support a central goal. This goal will be the single responsibility of the higher level module. Second, the parent module should not interfere with how its subordinates implement their functions. Instead of micromanaging, the superior should only orchestrate the interactions between the lower level modules.

How can we tell if a module has a single responsibility? This is best explained by examples:

- A module that receives a 4-bit BCD value and generates signals to drive a seven-segment display has a single responsibility. A module that calculates a game score and also writes it to a display has multiple responsibilities and is probably doing too much.
- Generating synchronism signals for a video display is probably a single responsibility. However, a module that generates such signals and also draws a spaceship on the screen has multiple responsibilities.
- Scanning a keyboard to determine which key was pressed is a single responsibility. Scanning the keyboard and then using the key value to change an object's position on the screen is not.

We can use the following three guidelines to help create modules with a single responsibility. First, in top-down design, we should break down each module recursively until we reach a point where a further division would create modules that cannot be treated independently. At this point, we would be creating complexity instead of reducing it. Note, however, that this point does not occur as quickly as most designers believe.[2] Most designs err on the side of putting too much functionality into a single module, rather than creating more modules than necessary.

The second guideline is to create a module for each functionality that could be developed independently. Depending on the context, a module could be an entity, a subprogram, or a process. Whatever the case, there is a big gain when we group a number of statements, give them a name, and encapsulate them inside a module. In nearly every case, it is beneficial to extract a cohesive piece of functionality from a larger module and put it in a separate module. The calling code becomes clearer and easier to understand, and the new module can focus on a smaller part of the problem.

Finally, the third guideline is to try to state the module's purpose in a single sentence. If the description uses the words *and, or, first, then, if,* or *when,* then the module is probably doing more than it should.

A Module Should Have a Clear and Narrow Interface   The only way to communicate with a module should be via its public interface. This interface should be as simple as possible to use, and it should expose only the minimum number of inputs and outputs required for the module to perform its function.

An interface consisting of few communication channels is called a *narrow interface*. Modules with narrow interfaces are easier to integrate, test, and reuse. They are also harder to misuse; it is easier to figure out how to use a module if it has few inputs and outputs—be it an entity, a subprogram, or a process.

If a module has a large number of ports or parameters, then it is possibly doing too much and might be broken into two or more modules. A large number of ports could also mean that the module does not have enough autonomy—maybe it has many inputs because it is micromanaged from the outside. In this case, the solution is to move part of the functionality from the controller module to the controlled one. Another alternative is to use hierarchy to concentrate the connections to a group of modules; in a hierarchy, only the top module is allowed to communicate with the rest of the system.

A module's name and interface should always be as self-explaining as possible. Avoid uncommunicative names such as *A*, *B*, or *Y* for the simple reason that they do not convey any meaning about the information they carry. Using more communicative names such as *divisor*, *dividend*, and *quotient* costs absolutely nothing and makes a huge improvement in readability. Ideally, a developer should be able to use a module correctly by looking only at its interface. This may not be always feasible, but it is a goal worth pursuing.

A Module Should Have a Manageable Size   There is no limit to how small a module can be, provided that it does a meaningful job independently. In other words, if you can understand what happens in a module without reasoning about what goes on inside other modules, then it is probably not too small. Large modules, in contrast, have several disadvantages. They are more complex, harder to read, and less likely to have a single responsibility. Moreover, they provide more room for bugs to hide.

It is difficult to give a hard-and-fast rule about the ideal size for a module in terms of lines of code or number of statements. There is, however, a much more useful yardstick: each module should be small enough for you to reason about its entire working at once. Does an entity have so many signals and processes that you cannot think about all their relationships? Break it down into smaller entities. Does a process have so many variables, levels of nesting, and control flow statements that you cannot understand them all at once? Break it down into smaller processes or move parts of the computation to their own functions. The resulting code will be much easier to understand.

A System Should Have as Many Modules as Necessary   Don't be stingy with the number of modules in a system. During the design phase, create a black box whenever you find a recognizable piece of functionality. Starting with the top-level block diagram, break down each block recursively until each module has a manageable size and is small enough that you can reason about its entire working at once. Let the solution have as many modules as the problem requires.

If modularity has so many benefits, then why do we see so few modules in many VHDL designs? One reason is the coding overhead to create a new module. Creating a function from a region of code involves selecting a group of statements, choosing the right parameters and a good function name, and choosing a place to put it. Creating a new entity typically involves creating another source file as well. In many situations, the developer chooses to add the new functionality in a module that already exists, only to avoid the typing inconvenience. This is bad for the system organization if the new functionality is not closely related with the module where it is implemented.

Another reason, common among less experienced developers, is the lack of perceived advantages. After putting in all the hard work to create a new module, the system behavior is still the same. So why bother? The end result is that inexperienced developers tend to write modules that are too large and systems with too few modules. Seasoned developers, in contrast, know that good modularity makes a system easier to understand, test, and maintain. For all but the most trivial projects, the typing and navigation overhead is negligible, and modularity is a big time-saver.

Use a Module to Hide Implementation Details   Ideally, each module should be treated as a black box: you can see what goes in and what comes out, but you are not allowed to look inside. This approach has two major advantages: it hides complexity and gives the module freedom to change, as long as its external behavior is preserved.

When we move a group of statements into a function and replace them with a call in the original code, we experience both benefits: the calling code becomes cleaner, and we become free to replace the implementation with one that has better performance or uses fewer resources.

Use a Module to Hide Design Decisions   Another useful rule is to hide each design decision behind a module. In this way, if you change your mind later, then the modification will be localized. As an example, suppose that there are two possible implementations for a computation: evaluation of a mathematical function or a table lookup using precomputed values. Whatever the choice, it is a good idea to encapsulate this decision inside a module; this keeps our options open and allows us to swap out an implementation easily.

Use a Module to Compartmentalize Areas That Are Likely to Change   Certain areas in a system are more likely to change than others. Typical examples of code with a higher tendency to change include vendor-dependent modules, difficult design areas, and custom data types or algorithms. It is a good idea to encapsulate the more volatile areas inside modules. This prevents the likely changes from interfering with the rest of the system.

Use a Module to Group a Data Structure and Its Manipulation Routines   Custom data types such as record types are usually accompanied by a collection of routines that work on objects of that type. Such types are called *abstract data types* (ADTs) because they raise the abstraction level of a data structure and allow the developers to handle it as a meaningful object. To keep the implementation details of an ADT separated from the client code, put the type definition and its operations inside a package. ADTs are introduced in section 13.4.

Keep in Mind the Types of Modules Available in VHDL   VHDL offers a rich set of constructs that can be characterized as modules. In structural design, a module is a design entity. In behavioral code, a module could be a procedure, function, or process. For a set of related declarations and operations, the module could be a package. Keep those options in mind and choose the right construct when you organize your design into modules. In this way, you can have the benefits of modularity with the minimum typing overhead.

4.2   Abstraction

In Computer Science, *abstraction* is the ability to use a construct while ignoring its implementation details. More generally, using abstractions means dealing with ideas rather than concrete entities. Abstractions are powerful because they free the developer from being aware of low-level implementation details all the time. Without abstractions, programmers would have to deal with an impossible amount of information.

In a computer program, there are two main kinds of abstraction. A *procedural abstraction* groups related statements into a single construct, hiding the steps involved in a computation. The typical procedural abstraction is the routine. A *data abstraction* groups together smaller pieces of data and allows them to be manipulated as a single object, without knowledge of its internal representation. A common example of a data abstraction is a file; you can think of it as a contiguous sequence of bytes or lines, ignoring how they are actually stored in the file system.

Abstraction is an important technique because it boosts programmer productivity and improves the readability of the source code. Our brain is limited in the amount of detail it can handle at a time; by manipulating complete ideas rather than loose pieces of information, our programs become more meaningful and easier to understand.

4.2.1   Levels of Abstraction

The effectiveness of an abstraction is proportional to the amount of detail it hides. Ideally, we would like to work with objects that reveal only the essential information for the task at hand. A representation that exposes little detail is said to have a *high level of abstraction*. A representation that reveals many details about the object's internal structure, or about how it performs its function, has a *low level of abstraction*.

According to this scale, a count value represented using the integer data type has a higher level of abstraction than the same value represented with the bit_vector data type. The latter exposes more detail about how the information is stored and provides many operations that are not relevant for working with integer values. Of course, this is not the only factor when choosing between the possible types for a data object. A more detailed discussion is presented in section 13.2.

We will get back to this subject in the section *Always work at the highest possible level of abstraction*. Now we need to introduce two terms to understand how abstraction can be used effectively in our designs.

Problem Domain versus Solution Domain   A system exists to solve a real-world problem. If we are designing a CPU, then the problem is to execute program instructions. If we are creating an alarm clock, then the problem is to show the time and awaken people. The *problem domain* is the set of all entities, relationships, constraints, and goals related to a specific problem.

To create a system that solves a problem, we use a set of tools that includes a programming language, the basic constructs it provides, and some abstractions we create. The set of tools used to create a solution is called the *solution domain*.

Each of the parts that compose our design may be close to or far from the problem domain. In a digital watch, statements that increment the time are close to the problem domain. Statements that perform bit manipulation for the I²C protocol between the display and the CPU are very, very far from the problem domain. A good design allows us to spend more time thinking closer to the problem domain and less time thinking about the mechanics of low-level data manipulation.

The level of abstraction increases as we work closer to the problem domain. Most designs use abstraction hierarchically or in layers, with each upper level hiding more implementation details. Figure 4.1 shows the typical abstraction layers found in a solution. Note that a solution may not need all the layers shown in the figure.

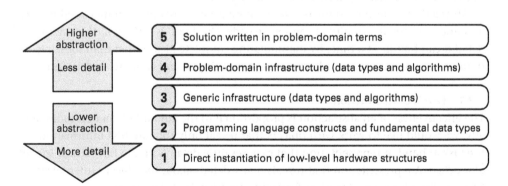

Figure 4.1
Layers of abstraction in the solution domain. Adapted from *Code Complete,* 2nd Ed.

Raising the Abstraction Level of a Programming Language   Now that we have introduced abstraction levels and the problem and solution domains, we can summarize the process of writing the code for a solution in two steps:

1. Raise the abstraction level of the language until it is close to our problem domain; then
2. Write the solution using these new abstractions.

If we fail to raise the abstraction level before creating the solution, then our code will be overloaded with details and unable to separate high-level tasks from low-level chores. It will mix up the things we do to meet system goals with things we do to satisfy the compiler. As a result, the code will be harder to follow and more likely to contain duplicate information and behavior.

However, if we try to design everything before writing the first line of code, then it will be extremely hard to foresee all the abstractions we will need. Of course, it is always good advice to think before writing code, and during the initial design phases, we will certainly identify many useful abstractions. Nevertheless, bear in mind that we will not know everything about a solution before we start implementing it. During implementation, you should always put some time aside to reevaluate and improve the original design. This practice is sometimes called *emergent design*.

4.2.2   Key Benefits of Abstraction

Now that we have introduced the key terms in abstraction, we can list its main benefits. You will notice that some advantages of using abstraction are similar to those of modularity. This is because modularity is a way of implementing abstraction; when we hide a complex operation inside a module and behind a narrow interface, we are effectively abstracting away the implementation details.

Abstraction Helps Deal with Complexity   A major source of complexity is the amount of detail required to perform low-level operations when they are translated to programming language statements. Abstraction reduces this complexity by offering a simplified view of any operation. When we write a routine, we are separating the code that uses an operation (also called the *client code*) from the code that knows how to perform it. As a result, the client code becomes cleaner, and the called code can focus on a smaller part of the task.

Abstraction Helps Programming Closer to the Problem Domain   Abstraction helps us program closer to the problem domain by providing a vocabulary of entities and actions that exist in the domain model. Instead of manipulating basic data types, we can spend most of our time performing meaningful operations with representations of real-world objects. This greatly reduces the mental burden to write and understand the code. Abstraction allows us to think in terms of decoding an instruction rather than checking a group of bits in a memory word. It allows us to operate on Cartesian coordinates rather than adding and subtracting loose numbers. Writing closer to the problem domain makes our code easier to understand and easier to relate with the steps involved in solving a problem.

Abstraction Helps Deal with Changes Because abstraction separates the internal structure of data objects from the behavior they provide, it allows us to change an object's representation without affecting the rest of the code. For this to work, the client code should use the abstraction only through its public interface. In other words, the client should never rely on information about how the abstraction is implemented. For example, if a record type represents an entity from the problem domain, then we should never manipulate its fields manually, bypassing the object's public interface. Rather, we should use the operations provided together with the data type.

Abstraction Improves Developer Productivity   When you move from a low-level programming language such as assembly to a higher level language such as C++, you experience a huge productivity boost. Tasks that used to take several lines of assembly code can be expressed with a single statement in C++. This makes the code easier to read and write, and it reduces the breeding ground for opportunistic bugs.

When you build a solution using abstractions, you get the same benefit. Instead of loose bits of information, you can manipulate meaningful objects that are closely related with the problem domain. Operations involving many intricate steps can be handled with a simple subprogram call. We can spend more time solving the real-world problem and less time doing low-level bit manipulations. Of course we still have to implement all those abstractions; however, once they are created, we can forget about their implementation and think of them as meaningful entities and operations.

4.2.3 Using Abstraction Effectively

This section provides the main guidelines to help use abstraction effectively.

Always Work at the Highest Possible Level of Abstraction The levels of abstraction apply not only to the types we create but also to the types provided by a language. VHDL offers a rich set of predefined data types. We should choose our types mindfully, rather than build every solution from basic logic types out of habit.

Start by choosing the simplest type that does the job. This maximizes the benefits of abstraction by hiding away everything you do not need. Using a type with a lower abstraction level than your application asks for results in unnecessary type conversions and in more verbose code. Novice programmers tend to think that more functionality is better, even when it is not used. Seasoned programmers know that exposing more functionality creates more opportunities to misuse an object. A narrow and simple interface makes for an object that is obvious to use and easier to change.

Keep a Uniform Abstraction Level   Reading a section of code should feel like driving on a flat road, not like a bumpy ride on a roller coaster. Modules and routines are easier to read if we keep the code at a constant level of abstraction. It is hard to concentrate on a sequence of code if it keeps alternating between high-level algorithms and low-level bit manipulations.

This is a common problem in code that has been added to an existing system without much thought about the best place to put it. On the spur of the moment, a developer may choose to patch the code in the easiest place that works, instead of creating a new module or routine. Over time, this erodes the original design, and the mixed levels of abstraction will make the code harder to read, reducing its maintainability. If you find a section of code with this problem, then extract the low-level code to its own routine and replace it with a subprogram call in the original place. For an example, see the section *Keep a uniform abstraction level in the routine* on page 382.

Use Abstract Data Types to Represent Entities from Your Problem Domain   An *abstract data type* (ADT) is a data type plus a set of operations that work on such data. In VHDL, the most common way to create an ADT is to use a package. The package contains a type declaration and the subprograms that operate on objects of that type. To use an ADT, we create an object of this new type and then use the operations defined in the package to manipulate the object. We can create signals, variables, and constants that are instances of an ADT.

When you organize a solution, do not restrict yourself to predefined data types. As a rule of thumb, consider creating an ADT for each real-world entity that your program needs to manipulate or interact with. Examples of such entities may include audio samples, CPU instructions, keyboard events, and so on. For an introduction to ADTs and a list of examples, see section 13.4.

Use ADTs to Represent Meaningful Entities from Your Solution   Occasionally you will notice that certain pieces of data are always used together. Examples include the red, green, and blue components of a color value; the x and y coordinates of a point on the screen; the exponent and fraction of a floating point number, and so on. When you identify this pattern, create an abstract data type to group together the related bits of information. This will reduce the amount of parameter passing and your operations will be more meaningful.

Avoid Leaky Abstractions   A perfect abstraction completely isolates the outside world from the inner details of a module. In other words, a perfect abstraction *encapsulates*

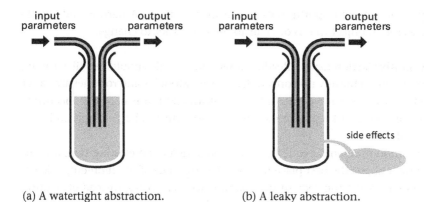

(a) A watertight abstraction.   (b) A leaky abstraction.

Figure 4.2
A leaky abstraction exposing implementation details via side effects.

the internal parts of a module. The only way to manipulate such a module is via its public interface.

A perfect abstraction is much like a bottle; we do not want any information flowing into or out of it, except through its public interface. In the case of a bottle, the public interface is its neck. In the case of a function, the public interface is its input and output parameters. Figure 4.2a illustrates the behavior of a good abstraction; the only way to communicate with the interior of the module is via its public interface.

An abstraction is leaky when it exposes part of its implementation details or when we need to be conscious of the inner details when we use the module. In any case, we lose part of the benefits of the black-box approach. In figure 4.2b, the routine's abstraction is leaky because part of its output happens via side effects rather than output parameters.

Using a floating-point number to represent a real number is an example of a leaky abstraction. We can perform most calculations without any problems, but in some cases we must be aware of the possibility of rounding errors and overflow.

Using a package to implement an ADT in VHDL is inherently leaky: to declare an object of the ADT, the client code must have access to the type declaration, and therefore it is possible to manipulate the object's data fields directly. The workaround is to use abstract data types with self-imposed discipline. Another choice, if the code is not intended for synthesis, is to use a protected type, which provides real encapsulation (see section 12.2.5).

4.3 Hierarchy

Hierarchy is one of the many possible ways to organize a modular system. In a hierarchical structure, a module can communicate directly only with the elements immediately

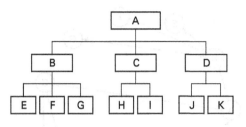

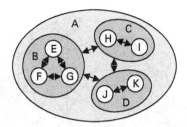

(a) A hierarchy diagram.   (b) Building a hierarchy with
                                            component instantiation.

Figure 4.3
Diagrams representing a hierarchical system.

above or below it, or in some cases with elements at the same level. An element above
a module is called a *parent*, an *ancestor*, or a *superior* of that module. An element below
a module is called a *child*, *descendent*, or *subordinate*. Figure 4.3a shows an example of a
hierarchical system.

In VHDL, design hierarchies are built with component instantiation. A parent module instantiates its immediate subordinates, which in turn may instantiate their subordinates, and so on. Figure 4.3b shows a hierarchical system created with component
instantiation according to the hierarchy from figure 4.3a. Module *A* instantiates components *B*, *C*, and *D*. Module *B* instantiates *E*, *F*, and *G*, and so on.

A hierarchy simplifies the connections in a system by restricting the communication
paths among its modules. In the example of figure 4.3, module *A* can communicate
only with its immediate subordinates *B*, *C*, and *D*. Module *E* cannot communicate with
A, *C*, or *D*, and so on. The effect of such restrictions is shown in figure 4.4. If every
module communicated with any other module, then we would have a fully interconnected system (figure 4.4a). This kind of system is terribly rigid (hard to change), and
its modules are almost impossible to reuse. This is an example of a system that is *tightly
coupled*. The precise definition of coupling is given in the next section.

If the system is organized hierarchically, then most of the connections between
modules are not allowed (figure 4.4b). Therefore, if a module needs to be modified,
then only a few components may have to change. The lower number of connections
also makes the system easier to understand. By restricting the possible connections
among modules, we exchange flexibility for clarity and maintainability.

4.3.1 Key Benefits of Hierarchy
There are several benefits to organizing a system hierarchically rather than using a
flat organization in which every module is allowed to connect to any other module.
Remember, however, that hierarchy is also a way to organize a modular system and

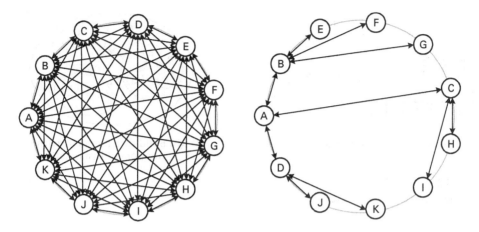

(a) Paths in a fully interconnected system.   (b) Paths in a hierarchically organized system.

Figure 4.4
Possible communication paths in different system organizations.

provide abstraction. Therefore, it offers all the benefits associated with these two principles. This section focuses on the additional benefits that result from organizing a system as a hierarchy.

Hierarchy Provides Abstraction at a Higher Level than Individual Modules   A module is an abstraction that hides implementation details and provides a defined functionality through its interface. Hierarchy does the same at a higher level: it combines the functionality of several modules and provides a narrow interface at the top of the hierarchy. This allows us to instantiate and use an entire hierarchy tree in our designs without worrying about what happens inside the lower level modules.

Hierarchy Improves Several Design Qualities   Hierarchy helps improve several design qualities. It enhances comprehensibility by providing an organized approach to understanding the system: start at the top of the hierarchy to get the big picture, and then look inside the subordinate modules to understand the details. Follow these steps recursively.

Hierarchy reduces intermodule dependence, which in turn improves modifiability and reusability. Changes are easier because their effects do not propagate to unrelated branches in the hierarchy. The code becomes easier to reuse because it has a single connection point with the client code—the top of the hierarchy.

Hierarchy also provides a way to reduce the average module size in a system. If a module grows too large, we can split it into smaller parts and still keep the same interface at the top of the subhierarchy (an example is shown in figure 4.5).

Hierarchy Makes the System Easier to Work With   In large designs, finding the right place to put each functionality can be puzzling at times. If the system does not have a clear structure, it is easier to do the wrong thing and add the functionality where it is most convenient, thus eroding the original design. A hierarchy diagram makes it easier to locate each functionality. If the right module does not yet exist, then the hierarchy also suggests the best place for a new module.

Because hierarchy makes it easy to visualize the system, it helps us reason about some of its properties. For example, if a module has many subordinates, it is probably doing too much, and we should move some of its responsibilities to a new module. In contrast, if a module has a single descendant, then it is probably not too important, and maybe the two of them could be collapsed into a single level.

Hierarchy Assists in Project Management and Development   A hierarchical structure can help in assigning responsibilities and splitting the work between developers. Modularity provides the basic packets for work assignment; hierarchy may suggest a group of modules that should be assigned to a single team. In a top-down development methodology, each branch in the hierarchy may be treated as an independent project and the results later integrated into the main design.

Hierarchical RTL design goes hand in glove with hierarchical *physical* design, a common chip design methodology. Each module is a natural design partition, and the physical hierarchy often reflects the logic hierarchy.

Hierarchy Accommodates Design Growth   A hierarchical structure can support growth without degrading the original design. As new functionality is added, a module may grow too big. Because of the many disadvantages (see section 4.1.3) of large modules, we should split them whenever they grow past an acceptable limit. Figure 4.5a shows a system where module E has grown too large and needs to be broken down. In a hierarchical system, there are two choices for arranging the new modules. One is to make them subordinate to the original module's parent. In figure 4.5b, modules W, X, Y, and Z are immediate subordinates to the original parent A.

The other choice is to keep module E in the hierarchy and to make the new modules subordinate to it (figure 4.5c). This approach has the advantage of preserving the external interface of E, and thus module A does not have to change.

Keeping the system modules within a manageable size is a highly important recommendation. This is not a perfunctory guideline or something to keep in mind only during the initial design phase. We should split large modules routinely, even if it means reorganizing the system structure. In this process, a proper set of automated tests is fundamental to avoid breaking the code while the system is being reorganized.

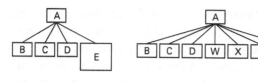 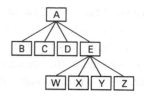

(a) Module *E* has grown larger than ideal.

(b) *E* is split into four new modules, *W–Z*, which are subordinate to *A*.

(c) Module *E* is preserved and instantiates modules *W–Z*. *A* does not need to change.

Figure 4.5
Growth in a hierarchical system.

Finally, even if our original responsibility is to design a single module for a system, this should not stop us from writing as many modules as needed to keep our code organized—we can always use hierarchy to preserve the interface expected by the rest of the system.

4.4 Loose Coupling

Coupling is a measure of how much a module knows about another module in the system and how much it relies on that module to perform its work. When a module needs information contained in another module, there is a dependency between them. The degree of coupling is proportional to the number and strength of these dependencies. If a module needs to know a lot about another module, then they are *tightly coupled* (or *strongly coupled*). If the modules have little dependence on one another, then they are *loosely coupled* (or *weakly coupled*).

We can use the same terms when referring to a system: a system is tightly coupled if its modules have a high degree of coupling. Figure 4.6a shows an example of a tightly coupled system: there are many intermodule dependencies, and some modules communicate via a common environment (global data or packages) rather than through public interfaces. Figure 4.6b shows a loosely coupled system, where direct intermodule communication is restricted and the modules rely on a common abstraction (a system bus) to communicate with one another.

A tightly coupled system has all the disadvantages of a system with excessive dependencies (see section 3.3). In contrast, loosely coupled systems are made of relatively independent parts that are easy to put together or replace.

Loose coupling and strong cohesion are typical characteristics of a well-structured system. Of course, some degree of coupling will always be necessary because the modules must cooperate to meet system-level goals. Our goal will never be to eliminate coupling, but rather to minimize it and make it more explicit.

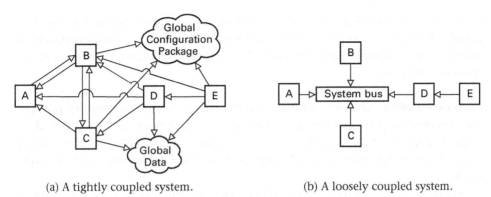

(a) A tightly coupled system.   (b) A loosely coupled system.

Figure 4.6
Two alternative designs for a system.

4.4.1 Types of Coupling
Not all kinds of coupling are equally harmful. Here we list the types commonly found in VHDL designs, starting with the most innocuous and moving to the most troublesome types of coupling.

Data Coupling This is the simplest and least harmful kind of coupling. If the only dependency between two modules is that one of them uses data values produced by the other, then they are connected via *data coupling* (also called *input-output coupling*). A typical example is a signal connecting the output port of an entity to the input port of another entity to communicate the result of a computation. Another example is a function call, if the function is pure and the input parameters are all data values.

If we can structure a system so that this is the only kind of coupling, then it will be as loosely coupled as possible. Note, however, that the values communicated between modules must *really* be data inputs and computation results. A "flag" parameter that indicates different operating modes or a synchronization signal telling a module to start a computation are not data values. They are a form of control coupling.

Control Coupling *Control coupling* happens when one module controls the internal state or the operation of another module. Typical examples include control signals between two design entities or flag parameters in a function. The problem with this kind of coupling is that the controlling module must know something about the inner workings of the controlled module. Therefore, if one of them changes, the other needs to change as well.

Common-Environment Coupling *Common-environment coupling* happens when two or more modules refer to the same data object declared elsewhere, such as a global signal in a package. The problem with common-environment coupling is that it works as an invisible interface, making it possible for one module to interfere with the operation of several other modules unintentionally. It also makes it harder to reason about the system correctness.

Content Coupling *Content coupling* happens when one module directly manipulates the internal parts of another module or uses knowledge of how that module is implemented. A typical example is when a module manually changes the fields of a data structure representing an ADT, instead of using the public interface to manipulate it.

As a more subtle example, suppose that module A knows that module B needs three clock cycles to produce a valid output. Using that knowledge, module A implements a counter to know the right time to read B's output. This implies that whenever B changes this behavior, A will have to change as well. A better solution would be to have B inform A when a new output is ready.

4.4.2 Decoupling
Coupling has many negative effects on both modules and systems, and it should always be kept to a minimum. There are two ways to reduce the level of coupling: cut down the number of interconnections or change them to a weaker form of coupling.

Data coupling can be reduced by writing modules that have a single responsibility and a narrow interface. To reduce the number of connections, we can also pass data at a higher level of abstraction.

Control coupling can be reduced by giving the controlled modules more autonomy to perform their functions. Sometimes this implies breaking down the module into smaller pieces so that each module is more focused and the parts can communicate via simple input-output connections.

Common-environment coupling can be reduced by finding a better place for the shared information and moving implicit dependencies to public interfaces. If the system is organized hierarchically, then the modules usually have a common ancestor. In this case, the ancestor is a natural candidate to host the shared information.

Another effective way to reduce coupling is to standardize the connections. Instead of having modules depend on each other via ad-hoc connections, we can make the modules depend on a shared abstraction, such as a system bus. This helps keep the dependencies unidirectional—both modules depend on the bus but not on each other. Another example is the use of buffers and queues to reduce the temporal coupling between modules. If two modules communicate through a queue, then the connection is less dependent on the input-output ratio and on the relative speed of the two modules.

4.5   Strong Cohesion

In physics, cohesion is an interactive force that attracts adjacent portions of a substance. In computing, cohesion is a measure of how strongly the code inside a module supports a central purpose.[3]

We may think of a module as a collection of processing elements. The nature of these elements depends on the kind of module: for a VHDL architecture, the elements may be process statements, instantiated components, or other concurrent statements. For a subprogram, the processing elements are the sequential statements used in the routine.

Cohesion is measured over an entire module and is only as strong as the weakest link. The module is highly cohesively only if all elements support the same purpose. If one of them is not closely related to the others, then it weakens the cohesiveness of the entire module. To increase the module's cohesion, we must move the unrelated elements to somewhere else.

For a VHDL design entity, cohesion refers to how closely its signals, processes, and statements work in the service of the same goal. If a module includes statements for debouncing inputs, running a state machine, and decoding BCD values for seven-segment displays, then it has a serious multiple personality disorder. It also has low cohesion.

For a subprogram, cohesion refers to how closely the operations in the routine are related to one another. A function that performs a single operation is more cohesive than a function that checks one argument first and then performs one of two completely disjoint operations.

Key Benefits of Strong Cohesion   Highly cohesive modules are easier to understand and reuse. They are also more likely to be fault-free. Aiming for high cohesion helps reduce the complexity in a module by forcing out functionality that does not belong with the other parts.

Strong cohesion is a key factor in effective modularity because it provides practical guidelines about what to keep inside a module and when to split a module in two. It also helps produce loosely coupled systems, as explained in the section, *Coupling and Cohesion.*

Strong cohesion also leads to better abstractions. If a module is composed of unrelated processing elements, it is hard to characterize as a meaningful entity. Incohesive modules are also harder to describe and give meaningful names.

Types of Cohesion   Cohesion is a qualitative property. When Wayne Stevens, Glenford Myers, and Larry Constantine coined the term, they identified several levels of cohesion.[4] By examining the source code (or the design) of a module, it is possible to classify it into one of several categories. Here we summarize the categories that are most relevant to VHDL designs. The levels are presented from lowest (worst) to highest (best) cohesion.

The first level is *coincidental cohesion*. This is just another way of saying that a module has no cohesion at all. The processing elements were placed in the same module for convenience or because the designer could not think of a better place to put them.

The next level is *logical cohesion*. In this case, the module still performs more than one function, but they are somewhat related. An example would be a "video" module that generates both image contents and synchronism signals. Image contents have nothing to do with video synchronization, but they were put in the same module because it is still reasonable to think of them as belonging to a same general category.

Coincidental and logical cohesion are generally unacceptable or at least undesirable; they result in code that is hard to understand, debug, and modify.

Up several steps in the ladder of cohesiveness is *sequential cohesion*. To be sequentially cohesive, a module must organize its processing elements such that the output of one element is the input of the next. Because cohesiveness is only as strong as the weakest relation between processing elements, a sequentially cohesive module can contain nothing but sequential processing steps. A sequentially cohesive module also performs less than a complete function, or it would be promoted to the next category of cohesion.

Functional cohesion is the strongest type of cohesion. To be functionally cohesive, a module must have a single purpose, and every processing element within it must be essential to this goal. Nothing could be removed or the function would be incomplete, and nothing could be added or the module would be demoted to a lower kind of cohesion. Aim for this level of cohesiveness whenever you design or implement a module.

Coupling and Cohesion   Before we can discuss the relationship between coupling and cohesion, we need to clear up any confusion in the definition of the two terms. *Coupling* is a measure that looks at a module *from the outside* and evaluates the strength of its connections with other modules in the system. *Cohesion* is a measure that looks *inside* a module and evaluates how closely its internal processing elements are related. Figure 4.7 illustrates this distinction.

The two measures are interrelated. A highly cohesive module is more likely to work on its own, therefore requiring fewer connections with other modules. In general, as the cohesion of individual modules in a system increases, the coupling between them decreases. Both measures are strong indicators of the quality of a design.

Creating Modules with Strong Cohesion   A key advantage of cohesion is that it translates easily to rules that we can follow while designing or coding. The first guideline to produce highly cohesive modules is to ask ourselves, "Does everything in this module belong together?" Whenever we add a new processing element, we should ask, "Is it

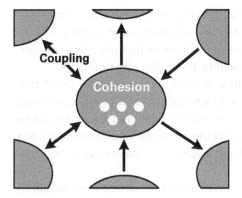

Figure 4.7
Coupling versus cohesion. Coupling refers to the connections between modules, whereas cohesion refers to the connections inside a module.

strongly related to all other elements already in the module?" Be prepared to break down a module whenever necessary. Use cohesiveness as your yardstick: when the cohesion of a module is reduced, it is time to break it down into separate modules that are individually more cohesive.

For routines, it is almost always possible to achieve functional cohesion. Do not settle for anything less. Pay attention to the relation between data and instructions: whenever a routine can be broken down into two sections that operate on unrelated sets of data, it is probably not functionally cohesive.

Finally, remember that a good module should have a single, well-defined purpose. This purpose should be easy to describe in a single sentence. If the simplest description you can imagine uses the words *and*, *or*, *first*, *then*, *if*, or *when*, then it is probably less than functionally cohesive.

4.6   The Single Responsibility Principle

The *single responsibility principle* states that each module in a system should be responsible for a single feature or functionality, or aggregation of cohesive functionality.[5] This principle is strongly related to the concept of *functional cohesion* explained in the previous section.

Another common way to phrase the single responsibility principle is to say that each module in the system should have only one reason to change. This is only possible if each module implements a single task, and this task supports exactly one of the system features. Thus, when this feature needs to change, the other modules could be left untouched. In this view, a responsibility means a possible reason for change.

The implication of this rule for typical VHDL designs is that the average module size should be much smaller than we are used to. Usually, the overhead of creating separate modules for simple tasks causes most modules to incorporate multiple pieces of functionality. This is bad for readability, maintainability, testability, and simplicity.

To write modules with single responsibilities, it is useful to remember that VHDL gives us plenty of choices of where to put new functionality. To name a few, we could use a design entity, package, subprogram, or process. The right choice will depend on other factors, such as the current module size and the size of the modification at hand. Our ultimate goal should be to separate unrelated behaviors so that we can concentrate on a single feature at a time.

In hierarchical systems, the definition of what constitutes a single responsibility increases as we move up in the hierarchy. At the lower levels, it would violate the single responsibility principle to put together raw functionality for, say, video and sound. However, as we move up, it would make sense to *instantiate* the video and sound *modules* inside a higher level module responsible for the user interface. Finally, at the top level, the system must provide for all the required features, so we could say that the top module's responsibility is to glue together the entire system functionality. As long as it does not perform other computations, a top-level module organized in this way has a single responsibility.

4.7 Orthogonality

Orthogonality is a term with its roots in geometry. In an orthogonal system, the axes meet at right angles (figure 4.8). In such a system, if you move along the x axis, then your position in the y and z axes does not change. In other words, you can change your x coordinate without affecting your y and z coordinates.

If x, y, and z are functional dimensions of a system, then this is exactly what we would like to achieve: to change one function without affecting the others. For this

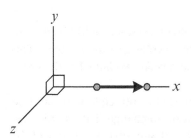

Figure 4.8
An orthogonal system of coordinates.

reason, in programming jargon, the term *orthogonal* came to mean *unrelated* or *independent*. Orthogonality arises spontaneously in a well-designed system—one that observes the principles of modularity, abstraction, information hiding, loose coupling, strong cohesion, and single responsibility.

Orthogonality is a highly desirable characteristic when we need to change or extend a system, and its biggest benefit is to reduce the risk associated with changes. Non-orthogonal systems are a maintenance nightmare: every small modification causes a cascade of unrelated changes in seemingly unrelated parts of the system. In *The Pragmatic Programmer*, Andy Hunt and Dave Thomas propose an easy test for evaluating the orthogonality of a design.[6] We should ask ourselves, "If I dramatically change the requirements behind a particular function, how many modules will be affected?" In an orthogonal system, the answer should be "one."

4.7.1 Guidelines for Creating Orthogonal Systems

If orthogonality is such a desirable property, how can we make our systems more orthogonal? Here are a few guidelines.

Minimize Dependencies If a module is self-contained, then it has no reason to change coming from the outside world. In contrast, if a module depends on many others, then each dependency is a possible reason for change. Keeping our modules loosely coupled is an effective way to create orthogonal systems. However, this should be done without violating other design principles; copying and pasting code to make a module self-contained does not make a system more orthogonal.

Eliminate Duplicate Information If we want to be able to make localized changes, then we need to keep each source of information restricted to a single place in the system. If any information is duplicated, then we need to replicate the changes in all the places where the same information appears in the system.

Eliminate Overlapping Functionality If a module has more than one responsibility, then two things happen. First, whenever we change it, we may inadvertently change the behavior of other parts in the system. Second, overlapping functionalities cause a module to change more often—if a module has two responsibilities, then it has at least two reasons for change. Observing the Single Responsibility Principle is a good way to ensure that a module is orthogonal.

Refactor Constantly Put some time aside for thinking about how to improve your design. Spend part of that time fixing the code to keep the code free of duplication and each functionality restricted to a single module.

4.8   Single Source of Truth (The DRY Principle)

The *single source of truth* principle states that every piece of data or behavior should be defined only once in a system. In other words, we should never allow any form of duplicate information in the code. If the same data or behavior is needed in different places, then we should define it in a single location, give it a name, and reference it from all other places that use it.

The proper way to enforce this principle depends on the kind of information. For data, instead of repeating numerical values or computations, we should store them in a constant, variable, or signal whenever they are used more than once. For behavior, we should identify similar sequences of statements and move them to a routine.

The idea that we should keep the code free of duplication has been restated several times in the software field. In *The Pragmatic Programmer*, Dave Thomas and Andy Hunt call it the *DRY* principle—an acronym for *Don't Repeat Yourself*.[7] In the *Extreme Programming* methodology, Kent Beck calls it the *"Once and Only Once"* rule.[8] If you duplicate information, then you are violating not only one but three design principles!

The causes for duplicate information in a system were discussed in section 3.4. Sometimes the developer is unaware of the duplication. At other times there seems to be no way around it. Most of the time the developers are not aware that duplication is a real problem, or they are just sloppy. A typical example is the practice of "copy-and-paste programming," an attempt to save some time and effort by copying similar code from elsewhere in the system.

The main problem with duplication is that it introduces potential inconsistencies in the system. Every place where the code is replicated is an independent implementation of the same behavior. If this behavior needs to change, then we must remember to update all the occurrences. It is only a matter of time until we forget to update one of those places and make the code contradict itself, resulting in bugs that may be hard to track down.

The best way to prevent duplication is to avoid it in the first place. After it has been introduced in the system, there is no easy way to find all the similar occurrences in the code. Great programmers develop a zero-tolerance attitude against any form of duplication; the DRY principle is the first rule you should master if you are serious about improving the quality of your code.

The best way to fight duplication that is already in a system is to refactor the code constantly. When we write a new feature, it is rarely clean and well factored. Make it a habit to invest some time cleaning up the code after every change. Remove any duplication that you may have introduced inadvertently or that was already in the code. Remember the Boy Scout Rule: "Always leave the campground cleaner than you found it. If you find a mess on the ground, you clean it up regardless of who might have made the mess."

Besides the improved reliability and comprehensibility, keeping your code *dry* has another benefit: it helps reveal abstractions hidden in the code. Many times, until we spot a recurring sequence of statements, we do not realize that their function could be generalized as a routine. At other times we will notice that several data objects are always used together. This is a strong hint that they should be a single object or data structure. Removing this kind of duplication helps uncover meaningful abstractions and incorporate them in the design.

Part II   Basic Elements of VHDL

5 Analysis, Elaboration, and Execution

During the design and implementation of a system, its VHDL description is processed by a number of tools. Understanding the role of each tool is important for two main reasons. First, it gives us a finer control over the process, which can be automated or tailored to our needs. Second, it helps us in locating and fixing problems: an error is easier to fix if we know in which part of the process it originated.

A common task in a VHDL project is to simulate a model to verify its behavior. Another task is to synthesize the model into a hardware circuit. For both synthesis and simulation, the code goes through the same two initial steps: *analysis* and *elaboration*. For simulation, the additional phase in which we can observe a model in operation and check its behavior is called *execution*. This chapter gives an overview of the overall VHDL design flow and then provides a more detailed view of the analysis, elaboration, and execution phases.

5.1 VHDL in the System Design Flow

VHDL is used in several phases during the design of a system. At the beginning, it can be used as a specification language or to create a behavioral description of the system that is not directly implementable. Later, this description can be refined as behavioral models are replaced with register-transfer level (RTL) descriptions, which can be automatically translated to hardware with HDL synthesis tools (for a description of the RTL design methodology, see section 19.3.1). During the entire process, the designer runs many simulations to ensure that the different descriptions are consistent and match the original design specifications.

The set of tools used to simulate and verify a design is called the *simulation toolchain*; the set of tools used to transform a VHDL description into a hardware circuit is called the *synthesis toolchain*. The term *toolchain* means that the tools are often used as a chain—the output of a tool serves as the input for the next one.

Figure 5.1 shows a simplified view of the system design flow, emphasizing the role of VHDL descriptions and their interaction with the simulation and synthesis toolchains.

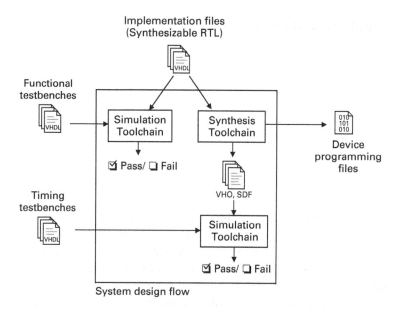

Figure 5.1

Simplified design flow, emphasizing the role of VHDL and the simulation and synthesis toolchains.

At the top are the design implementation files, written in synthesizable VHDL and usually at the RTL level. The implementation files go through the synthesis toolchain, resulting in one or more files ready for implementation in a physical device—a programming file in the case of an FPGA or mask design files for an ASIC.

In theory, it would be possible to write only synthesizable VHDL code and then generate the programming files directly, using only the synthesis toolchain. Instead of using simulations, the device would be tested in real operation. In practice, however, more than half of the VHDL code in a project is simulation code, used to verify the design and ensure that it meets its specifications.[1] Such files are called *testbenches*. A testbench applies a set of inputs to the design and then checks its outputs against expected values. If the outputs differ from the expected behavior, the testbench issues an error message. A good set of testbenches tests all the behavior described in the design specifications, emulating the behavior of other entities that interact with the device if needed, and then produces a pass/fail result.

Simulations can be used in several places in the design flow. We can simulate the RTL files directly without passing them through the synthesis toolchain. This is called an *RTL simulation*. If we pass the design files through a synthesizer and then simulate the resulting description, we will perform a *gate-level* or *post-synthesis* simulation. We can also classify a simulation according to the presence of timing information provided by a synthesis tool. If we simulate the design without any timing information, we have a *functional simulation*. If we use the detailed models provided by the synthesizer and annotated with delay information, we will be performing a *timing simulation*.

The design flow shown in figure 5.1 uses two kinds of simulation. The first one (on the top left corner) is performed directly on the RTL files, without passing through a synthesis toolchain. This is, therefore, a *functional RTL simulation*: it only verifies the design functionality, independently of any implementation device or technology. When we are confident that the design meets its functional specifications, we can use a synthesizer to implement the design in the chosen hardware technology. Besides the programming files, the synthesis tool produces a set of output files that can be used to simulate the design in a more realistic way, including timing information of the physical device. This second simulation is, therefore, a *post-synthesis timing simulation*.

If a timing simulation is always more detailed and realistic, why should we perform functional simulations using the RTL code? In practice, most of the simulations are performed at the RTL level for many reasons. First, they allow us to check that we correctly translated the specifications to RTL before we add more variables to the process. Second, in an RTL simulation, we are dealing directly with the code we wrote before it gets transformed by a synthesis tool, so it is easier to relate any problems with our original design. Third, RTL simulations offer better debugging capabilities: we can single-step through sequential code, inspect the values of variables and signals, and observe the effects of each executed statement. Finally, they are much faster than gate-level simulations because the tool can work with simpler data objects and operations. For example, in an RTL simulation, a 32-bit addition could be simulated as a single operation in the host machine, whereas a gate-level simulator would emulate the behavior of each gate that composes a 32-bit adder.

Besides RTL and functional simulations, the verification of a design typically includes post-synthesis and timing simulations. These simulations verify that the synthesizer correctly recognized the hardware structures described in the code and that any optimizations performed did not change the design functionality. They also verify that the design meets its timing specifications and respects the timing characteristics of the target device, such as setup and hold times.

In all but the smallest and simplest designs, we can save a significant amount of time by running the VHDL code in a simulator before testing it in a physical device. We will get back to this subject when we discuss testbenches in chapter 20.

5.2 The Design Processing Flow

We now take a closer look at the set of tools used to process VHDL files, starting with a parallel between the tools used to compile a software program and the tools used to process a VHDL description. In this comparison, we need to be careful with terminology because the terms may have different meanings for practitioners of different fields. Even among hardware designers, certain terms such as *compilation* and *synthesis* mean different things to different people, so our first task is to define them properly.

A *compiler* is a program or set of programs that translate human-readable source code to a lower level representation that a machine can read and execute. Unfortunately, the term *compiler* is highly overloaded. In the software field, the compiler is generally understood as a tool that converts human-readable source code to an equivalent program in assembly language. In VHDL, the term *compilation* is used with two different connotations: some take it to mean the same as *analysis*, whereas others regard it as the entire process of translating the source code to one or more device programming files. The same happens with the term *synthesis*. Some take it to mean a specific phase of design processing, in which an elaborated design gets converted to basic hardware elements—the first step in the physical implementation of a model. Others take it to mean the entire process of generating hardware from a VHDL model, up to the creation of programming files.

The terms *compiler* and *compilation* are not frequently used in the LRM—the Language Reference Manual that defines the VHDL language. Whenever the LRM refers to specific phases in the processing of a design, it uses more precise terms such as *analysis* and *elaboration*. As for synthesis, the LRM does not offer a definition; however, it defines a *synthesis tool* as "any tool that interprets VHDL code as a description of an electronic circuit and derives an alternate description of this circuit."

In this book, we use the terms *compiler* and *compilation* in a general sense, meaning the translation of VHDL code to a format that a machine can execute or implement in a physical device. As for the term *synthesis*, we use it as the general process of converting VHDL code to a circuit; for the specific step of translating HDL code to basic hardware elements, we use the term *HDL synthesis*.

Figure 5.2 shows the sequence of steps involved in the translation of human-readable source code to a lower level representation. Figure 5.2a shows the translation of a software program to executable code that a processor can run. Figure 5.2b shows the translation of HDL code to a binary file that can program a device (e.g., an FPGA).

In the software flow, the code first passes through a preprocessor, which performs basic textual transformations in the source text. The modified code is then passed to a compiler, which typically operates in two stages. The first phase, called *analysis*, breaks up the source code into smaller pieces conforming to the rules of a programming language, and it produces an intermediate representation of the code in a format that is more convenient for the tools (usually, a *syntax tree*). The second phase, called *synthesis*, translates this representation to another program in a lower level language. In the figure, the compiler outputs assembly code, which is then translated by the assembler

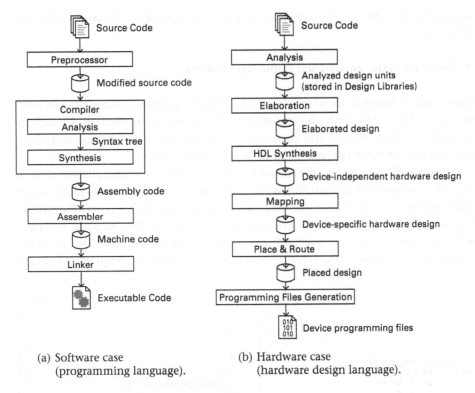

(a) Software case
(programming language).

(b) Hardware case
(hardware design language).

Figure 5.2
Translation of human-readable source code into machine-readable code.

into machine code. Finally, a linker puts together the individual pieces of machine code, producing an executable program that can be run in the target machine.

In the hardware flow, the actual sequence depends on the task at hand, synthesis or simulation. Figure 5.2b shows the steps for an FPGA synthesis toolchain. As in the software case, the process starts with an analysis phase, which translates the source code to an intermediate representation. Next comes a step specific to VHDL, called *elaboration*. Elaboration effectively creates the hierarchy of a design, instantiating subcomponents, checking the interconnections for correctness, and creating declared objects such as signals and variables.

Once the design has been elaborated, translation to a hardware circuit can start. In the first phase, *HDL synthesis*, language constructs are converted to basic hardware elements, such as multiplexers, adders, and storage elements. In the next phase, *mapping*, those basic elements are translated to the set of hardware resources existing in the target device. Next, the *place and route* phase (also called *fitting*) fits the design into the resources available in the target device, giving each hardware element a specific location. Finally, one or more programming files are generated for configuring the chosen device.

5.3 Overview of the Analysis and Elaboration Process

Analysis and elaboration are the two steps that must be performed before we can work with any VHDL design, for both simulation and synthesis. In this section, we provide an overview of how analysis and elaboration fit in the overall design processing flow. In the next sections, we take a closer look at each of these steps.

A typical VHDL design is split across a number of files. The compilation process starts with the *analysis* of each file. A file contains one or more *design units*, which are the smallest part of VHDL code that can be individually analyzed. In other words, we cannot give the compiler isolated lines of code, even if they describe complete functions or processes: to be analyzed, every piece of code must be inside one of the design units existent in VHDL. Figure 5.3 shows the main kinds of design units available

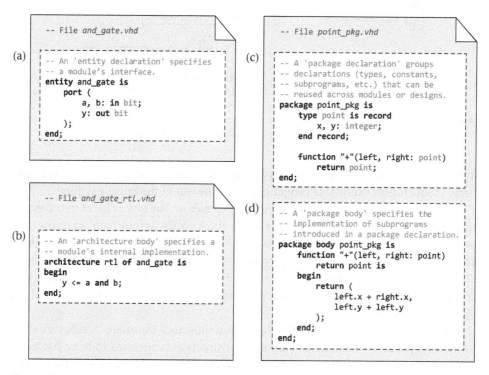

Figure 5.3

Examples of design units (dashed boxes) and design files (shaded boxes).(a) Entity declaration. (b) Architecture body. (c) Package declaration. (d) Package body.

in VHDL: (a) an *entity declaration*, (b) an *architecture body*, (c) a *package declaration*, and (d) a *package body*. In the example, the four design units are split across three design files, demonstrating that it is possible to have more than one design unit per file. The opposite, however, is not true: it is not possible to split a design unit across multiple files.

During the development of a VHDL design, its source files will be compiled numerous times as we add new features, fix bugs, and run testbenches to verify our implementation. To save time and prevent errors, this compilation process should not be done manually. If you use command line tools, you will probably want to create scripts to automate the process. If use an IDE, the tool will take care of compiling your design; you just need to tell it which files are part of your project.

In any case, the first step in processing a VHDL design is to analyze all the design files. VHDL specifies that each analyzed design unit must be placed in a special container called a *design library*. A design library is an implementation-dependent storage facility for analyzed design units; typically, it is either a directory or database in the host operating system. Figure 5.4 shows an example in which three design files containing six design units are analyzed into two design libraries.

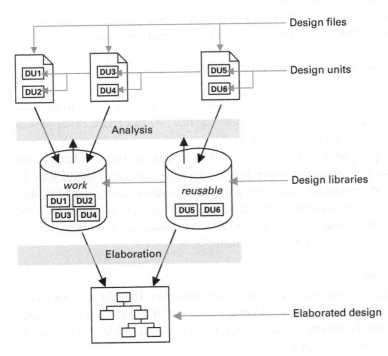

Figure 5.4
Overview of the analysis and elaboration process.

Design libraries are a way to organize our designs and projects. We can create as many libraries as we want; when we analyze a source file, we must specify a target library where the analyzed units will be stored. When a design unit is put into a library in analyzed form, it becomes a *library unit*; it is then available to be used by other designs units.

Once all source files have been analyzed, the design is ready to be elaborated. Elaboration effectively creates the hierarchy of a design, instantiating entities recursively, checking the interconnections for correctness, and creating the objects defined in declarations, such as signals and variables.

One difference about the analysis and elaboration processes is the main input passed to the tools. The analysis tool is invoked several times; each time, a design file is specified as the main input. Elaboration, in contrast, is performed only once; the main input for this step is a single library unit, which will be the root of the elaborated design hierarchy.

Another difference is that after a design unit has been analyzed, it stays in the library and only needs to be reanalyzed if the corresponding design file is changed. Elaboration, in contrast, allocates memory for signals, variables, and processes, and therefore must be run every time we start a simulation.

After a description has been elaborated, it is ready for simulation or to proceed with the synthesis process where it will be transformed into a hardware circuit.

5.4 Detailed View of the Analysis Phase

Now that we have a better understanding of how analysis and elaboration work together and where they fit in the design processing flow, we can take a closer look at the tasks performed during the analysis phase.

In a typical compiler, analysis is divided into three stages. In the first phase, called *lexical analysis*, characters from the source text are grouped into units called *tokens*. In the second phase, *syntax analysis*, tokens are grouped in syntactic units, such as expressions, statements, and declarations. In the third and final phase, *semantic analysis*, the compiler uses the entire program structure and all the language rules to check the program's consistency. We review each phase in turn.

5.4.1 Lexical Analysis
At the lowest level, any source code is composed of characters. The first analysis phase, *lexical analysis*, is conceptually simple: it groups characters from the source text into meaningful units called *tokens* or *lexical elements*. For this reason, the lexical analyzer is also called *tokenizer*, *scanner*, or *lexer*.

VHDL has only four basic lexical elements: *identifiers*, *literals*, *comments*, and *delimiters*. This means that each group of characters in the source text must belong to one of

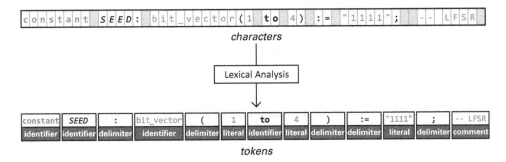

Figure 5.5
Lexical analysis of a line of VHDL code.

these categories. To separate the tokens, a language defines a set of special blank characters called *spacers* or *separators*. A typical lexer strips all the spacers from the code and passes to the next phase a sequence of tokens, tagged with their respective categories. Figure 5.5 shows the lexical analysis of a line of VHDL code. At the top, the line is shown as a sequence of characters; the shaded boxes are spacers and will be removed by the lexer. The output is a sequence of tokens; each token is a sequence of characters tagged with the kind of lexical element it represents.

A lexer can only detect basic errors because it does not know the language syntax rules. For instance, it does not make a distinction between object names and reserved words. Error messages from the lexer usually concern missing delimiters or invalid literals, which are generally easy to spot and fix.

5.4.2 Lexical Elements
Lexical elements are the categories identified during lexical analysis; they are the most basic building blocks of VHDL code. We review each kind of lexical element in turn.

Identifiers An *identifier* is a sequence of characters and underscores used with one of two purposes: to name an element declared in the code (such as entities, signals, or subprograms) or as reserved words of the language (such as *begin*, *case*, or *variable*). The two categories are treated indistinctly by the lexer. VHDL is not case-sensitive, so the identifiers begin, Begin, and BEGIN all have the same meaning in the code.

Literals A *literal* is a value written directly in the code. VHDL allows several kinds of literals. *Numeric literals* are integer or real numbers, and they can be written in base 10 (default) or any other base from 2 to 16.

```
-- Examples of integer literals:
33, 1e6, 2#1111_0000#, 16#dead_beef#
```

```
-- Examples of real literals:
33.0, 1.0e6, 2#1111_0000.0000#, 16#dead.beef#
```

A *character literal* is a notation for a value of the *character* type. It consists of a character within single quotes (apostrophes).

```
-- Examples of character literals:
'a', 'b', ' ', 'X', '0', '1', '2', '½', '@', '''
```

A *string literal* represents the value of an array of characters, or a *string*. In the source code, it consists of one or more characters enclosed within double quotes (quotation marks). In VHDL, string literals may specify the values of text strings or logic vector types (such as *bit_vector* or *std_logic_vector*). In the following examples, although some of the strings have a single character, they are still string literals because of the double quotes used as enclosing symbols. Therefore, it would not be possible to assign such values to objects of type *character*.

```
-- Examples of string literals:
"abc", "a", "123", "1", "2", "11110000"
```

Finally, a *bit-string literal* is just another way to write a string literal, furnished with conveniences for specifying values of logic vector types. It is possible, for instance, to use a base other than binary to specify a value. It is also possible to use characters other than the bit values '0' and '1', such as underscores, 'Z', '-', or 'X'. Bit-string literals also provide automatic sign bit extension for signed and unsigned values.

```
-- Examples of bit-string literals:
b"10100101", b"1010_0101", x"f0", 8d"165"
```

Note that each kind of literal has a distinctive shape, so the lexer has no problem distinguishing among them. Literals play an important role in writing code that is clear and easy to understand, so we review them in more detail in section 8.2.1.

Delimiters A *delimiter* is simply a character or sequence of characters belonging to one of the groups in table 5.1. Despite their name, delimiters serve for more than marking the beginning and ending of language constructs. Delimiters are also used as operators, as punctuation symbols, and to define the general shape of VHDL constructs.

Table 5.1 Symbols used as delimiters in VHDL

Single-character delimiters	&	'	(	)	*	+	,	-	.	/	:	;	<	=	>	`	|	[	]	?	@
Compound delimiters	=>	**	:=	/=	>=	<=	<>	??	?=	?/=	?<	?<=	?>	?>=	<<	>>						

Comments   Versions of VHDL prior to VHDL-2008 allowed only one kind of comment in the code, called a *single-line comment*:

```
-- This is a single-line comment spanning an entire line
entity top_level is   -- This is a single-line comment spanning part of a line
```

VHDL-2008 introduced *delimited comments*, similar to those existing in Java and C++:

```
generic map (clk_freq => 50_000 /* in kHz */, pulse_duration  => 10 /* in ms */)
```

Typically, comments are stripped out of the code by the lexer and are not passed to the subsequent analysis phases.

5.4.3   Syntax Analysis

The second phase of the analysis process is called *syntax analysis* or *parsing*. A parser uses a notation called a *grammar* to describe the language syntax. The grammar rules for parsing VHDL code are specified in the Language Reference Manual (LRM). A parser groups lexical tokens into higher level syntactic units, such as statements, expressions, and declarations. The parser output is an intermediate representation of the code, in a format that is more convenient for the subsequent compilation phases.

Figure 5.6 illustrates the syntax analysis of a variable assignment statement (a). The corresponding language rule for this kind of statement is shown in (b). The first step is to find the parts in the statement corresponding to *target* and *expression*, as indicated by the rule. The rule specifies that the two parts are separated by the variable assignment symbol (:=). The next step is to expand the rules for *target* and *expression* recursively until all expansions result in terminal nodes (e.g., a literal or an identifier). These rules are not shown here, but they can be seen in Annex C of the LRM. Eventually, the compiler produces an intermediate representation of the code, such as the tree shown in (c).

A compiler is free to use any data structure for the intermediate representation, but most use trees because they map well to the source code structure. Figure 5.7 shows various examples of syntactic units identified during the analysis of an architecture body. The units are organized hierarchically, with the architecture body at the top.

(a) `lfsr_input := lfsr_state(3) xor lfsr_state(4);`

(b) ```
     variable_assignment ::=
         target := expression;
     ```

(c)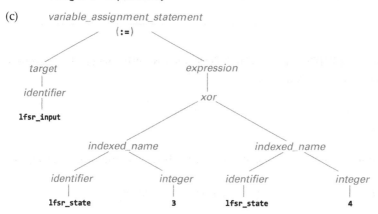

**Figure 5.6**
Syntax analysis of a variable assignment statement. (a) Statement. (b) Grammar rule. (c) Syntax tree.

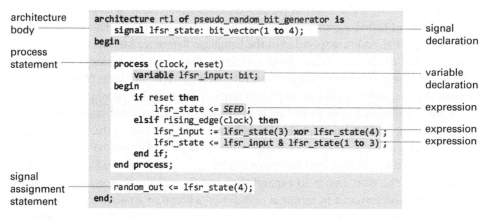

**Figure 5.7**
Examples of syntactic units recognized during syntax analysis.
     ```

Syntax analysis is able to distinguish between language keywords and object names, so it can provide more detailed information if an error is detected. Examples of syntax errors include unmatched parentheses, misspelled keywords, or the use of a sequential statement where a concurrent statement was expected. However, a parser does not check object types, and it does not use information from previously analyzed syntactic units (e.g., from a previous statement). These tasks are typically performed at the next phase, semantic analysis.

5.4.4 Semantic Analysis

The final step of the analysis process is *semantic analysis*. Before we delve into the details of semantic analysis, we should discuss the distinction between *syntax* and *semantics* because the two terms often appear together and in contrast to each other.

Many terms used in linguistics also apply to programming languages. Syntax and semantics are two examples. *Syntax* is the set of rules defining the valid arrangement of elements in a language, whereas *semantics* define the meaning of the language constructs. In the context of compilers, *syntax analysis* checks individual units of the source text according to the rules of the language grammar, whereas *semantic analysis* checks the code as a whole, taking into account the entire context of the analyzed program.

The main reason for syntax analysis and semantic analysis being two separate phases is that certain checks cannot be done by looking at individual elements of the source code. For instance, using only syntax analysis, it is not possible to check whether a name is used before it is declared, whether the same name is declared twice in a single scope, or whether a function call uses the right parameter types. These kinds of checks require semantic analysis. From a practical point of view, this also draws the line between what is considered syntax or semantics: what can be described by the language grammar alone is syntax, whereas what uses contextual information is semantics. In practice, any check that requires the compiler to consider more than one syntactic unit at a time is a semantic check.

The semantic analyzer uses as input the syntax tree produced during syntax analysis. It traverses the tree, gathering information about *symbols* when they are declared. Examples of symbols in VHDL include data objects, subprograms, and design units. When a symbol appears for the first time, an entry is created in a data structure called the *symbol table*. Then, whenever a symbol is referenced in the code, the analyzer checks whether it was used correctly. For a data object, it checks whether its usage is consistent with its type. For a subprogram, it checks whether the number and type of the parameters are correct. Because of the strongly typed nature of VHDL, type-checking is actually easy for the compiler: each object has an explicit type, there are strict rules to determine the type of an expression, and the compiler can abort the process in ambiguous situations. Additionally, there are no automatic or implicit type conversions.

Besides type checking, certain statements require additional semantic checks. For example, a *next* statement should only be used inside a loop; the target of a signal assignment should be a signal; a subprogram declared in a package declaration should be defined in the package body, and so on. All those remaining checks are performed during semantic analysis.

The outputs of a semantic analyzer are usually an augmented syntax tree and a symbol table. The augmented syntax tree is a modified version of the tree generated during syntax analysis, annotated with information such as types and links to related nodes, to make the tree more convenient for the subsequent compiler phases.

5.5 Detailed View of the Elaboration Phase

After all the design units have been analyzed, they still need to be put together before the design can be synthesized or simulated. Elaboration effectively creates the hierarchy of a design, instantiating entities recursively, checking the interconnections for correctness, and creating the objects defined in declarations, such as signals and variables.

The inputs to the elaboration phase are the library units used in the design, plus the selection of a top-level entity that will be the root of the design hierarchy. The output is a design reduced to a collection of processes connected by nets. After this transformation, the design is ready to be executed in a simulator or converted to a hardware circuit by a synthesis tool.

The elaboration process has complex rules and numerous details for each kind of construct to be elaborated. The best way to understand what happens in the elaboration of a design is with a concrete example. In this section, we will follow the elaboration of a design hierarchy step by step. The model is an edge detector circuit that operates with synchronous inputs and produces a single-clock pulse on its output whenever the input signal changes. The design is composed of three design units, whose source code is given in figure 5.8. Figure 5.8a shows the top-level entity `edge_detector`, which instantiates the two other entities, called `dff` (figure 5.8b) and `xor_gate` (figure 5.8c).

To start elaboration, we invoke the appropriate tool in our suite, giving as an argument the name of the top-level entity, `edge_detector`. It is not necessary to provide any additional information because the entity will be elaborated recursively; all lower level entities will be fetched from the libraries and incorporated into the design as needed.

The first step in elaborating an entity declaration is to create its interface, which usually consists of ports and generics. Our `edge_detector` entity has only ports, so these are created when the entity declaration gets elaborated (figure 5.9a). The next step in the elaboration of a design entity is to select an architecture body for it. If the code does not specify an architecture explicitly, the most recently analyzed architecture is

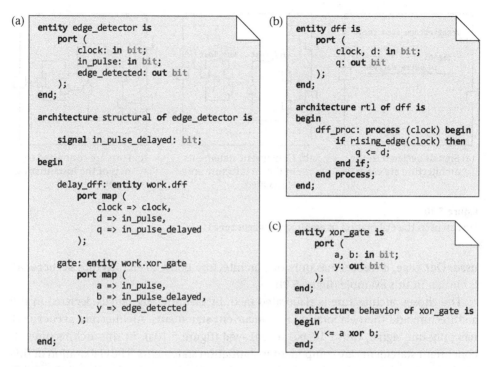

(a)
```
entity edge_detector is
    port (
        clock: in bit;
        in_pulse: in bit;
        edge_detected: out bit
    );
end;

architecture structural of edge_detector is

    signal in_pulse_delayed: bit;

begin

    delay_dff: entity work.dff
        port map (
            clock => clock,
            d => in_pulse,
            q => in_pulse_delayed
        );

    gate: entity work.xor_gate
        port map (
            a => in_pulse,
            b => in_pulse_delayed,
            y => edge_detected
        );

end;
```

(b)
```
entity dff is
    port (
        clock, d: in bit;
        q: out bit
    );
end;

architecture rtl of dff is
begin
    dff_proc: process (clock) begin
        if rising_edge(clock) then
            q <= d;
        end if;
    end process;
end;
```

(c)
```
entity xor_gate is
    port (
        a, b: in bit;
        y: out bit
    );
end;

architecture behavior of xor_gate is
begin
    y <= a xor b;
end;
```

Figure 5.8
Source code for the edge detector design. (a) Top-level entity. (b) dff entity. (c) xor_gate entity.

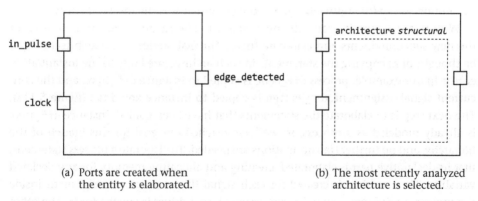

(a) Ports are created when
the entity is elaborated.

(b) The most recently analyzed
architecture is selected.

Figure 5.9
Elaboration of design entity edge_detector.

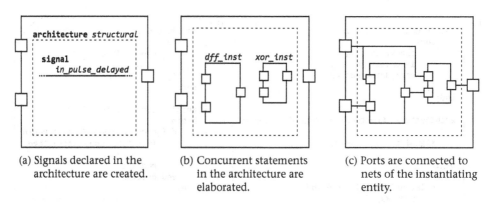

(a) Signals declared in the architecture are created.

(b) Concurrent statements in the architecture are elaborated.

(c) Ports are connected to nets of the instantiating entity.

Figure 5.10

First steps in the elaboration of architecture structural.

used. Our edge_detector has only one architecture body, so architecture structural is chosen in this example (figure 5.9b).

The chosen architecture is elaborated next, by creating any signals declared in the architecture and then elaborating its concurrent statements. Architecture structural has only one signal, called in_pulse_delayed (figure 5.10a). In this architecture, all concurrent statements are component instantiation statements. The elaboration of this kind of statement is a recursion of what was done for the top-level entity: first create a black box for each instantiated entity, and then create its ports (figure 5.10b). The next step is to connect the ports with nets from the parent entity (figure 5.10c). This step was not necessary for entity edge_detector because it was the top-level entity.

With this structure in place, the next step is to choose an architecture for each of the new subcomponents and elaborate them. The elaboration of an architecture can be thought of as copying the statements in each architecture body to the instantiating entity. In our example, process dff_proc is copied to instance dff_inst, and the concurrent signal assignment xor_assign is copied to instance xor_inst (figure 5.11a). The next step is to elaborate the statements that have been copied. Instance dff_inst is already modeled as a process, so we have reached our goal for this branch of the hierarchy, and no further transformations are needed. To elaborate a process statement, first its declarative part is elaborated, creating and allocating memory for any declared variables. Then a driver is created for each signal that receives an assignment inside the process. In dff_proc, signal q gets assigned, so a driver is created for it. The other subcomponent, instance xor_gate_inst, is modeled as a concurrent signal assignment. Every concurrent statement can be rewritten as one or more processes; therefore, elaborating a concurrent statement consists of replacing it with the equivalent process or processes. In the case of a concurrent signal assignment, the equivalent process

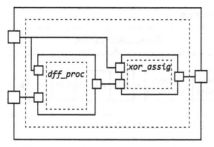

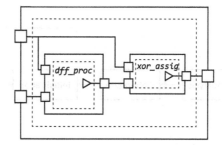

(a) Statements in subcomponents'
architectures are copied to the
instantiating entity.

(b) Drivers for the signals are created
as each process is elaborated.

Figure 5.11
Final steps in the elaboration of architecture structural.

contains a single statement, which is a copy of the original assignment. This equivalent
process is then elaborated, creating a driver for signal y (figure 5.11b).

Here we have reached the point where the design has been reduced to a collection
of processes and nets, thus concluding its elaboration. The resulting model is ready to
be executed using a simulator or converted to a hardware circuit using a synthesis tool.

5.6 The Execution Phase

Starting from the VHDL source code, a design goes through three distinct phases when
it is simulated: analysis, elaboration, and execution. The analysis phase needs to be
performed only once unless there are changes to the design. The elaboration phase is
typically repeated every time we start a simulation because this is when all variables
inside processes are initialized. Once a design has been elaborated, the execution phase
can start.

Execution is the most visible part of a simulation; during this phase, we can see a
model in operation. Execution consists of an initialization phase, followed by repeated
iterations of a simulation cycle (figure 5.12).

Before we can discuss the execution phase in more detail, we need to introduce two
basic concepts: the mechanism of discrete-event simulation and the concept of drivers
in VHDL. These are the subjects of the next two topics.

5.6.1 Discrete-Event Simulation

The behavior of a VHDL simulator is described precisely in the Language Reference
Manual (LRM). VHDL simulators are *discrete-event simulators* (also called *event-driven*
simulators). In a discrete-event simulation, only the moments when the state of a

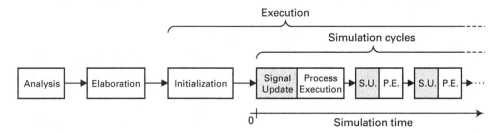

Figure 5.12
Steps required for a simulation, with a detailed view of the execution phase.

system can change are simulated. Because actions can happen only in response to events, the simulator can skip the time interval between one event and the next. No computational resources are wasted while the state of the model does not change. (Note that in this discussion, the term *event* is used in the general sense, and not with the VHDL meaning of a change in the value of a signal.)

Another important concept is *simulation time*, which means the time seen by a model during a simulation. It is essentially a counter, set to zero when the simulation starts and incremented whenever the simulator advances toward the next event. Simulation time is equivalent to the timing that a physical device would face in the real world; it has no relation with how long it takes to run a simulation.

In VHDL, the agent that manages the occurrence of events, controls the execution of user-defined processes, and propagates signal changes is called the *kernel process* or *simulation kernel*.

5.6.2 Drivers

To understand how signal values are generated and updated, we need to introduce the concept of drivers. In a VHDL model, a *driver* is a conceptual device that sources a value into a signal. Drivers are inferred from signal assignments in the code; a driver is created during elaboration and exists until the end of a simulation. During this time, the driver is always sourcing a value into the signal. The only exception is *guarded signals* (seen in section 9.6.1), which are rarely used.

In the simulated model, each driver has a *projected waveform* and a *current value* (figure 5.13). The projected waveform contains all future transactions for the driver. Each transaction has two parts: a value component, which determines the value to be sourced to the signal when the transaction matures, and a time component, which determines when the transaction will become the current value of the driver. Transactions are ordered according to their time component.

The initial value of a driver is taken from the signal's default value, which is given at the signal declaration. The current value of a driver is updated by the simulation

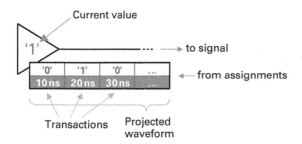

Figure 5.13
Simulation model of a driver.

kernel. As simulation time advances, the value component of each transaction in the projected waveform becomes the current value of the driver.

When the current value of a driver is overwritten with the value component of a transaction, the driver is said to be *active* in the current simulation cycle. It does not matter whether the current value is overwritten with the same value or a different value. In the cases where the new value is different from the old value, the signal also has an *event* in the current simulation cycle.

A driver is implied in the code when a process or concurrent statement makes an assignment to a signal. If the target is a scalar, then a single driver is created for the signal; if the target is an array or record, then each scalar subelement of the target will have its own driver. In any case, each process implies *at most one driver* for a scalar signal or subelement. All assignments made to the same signal within a process share the same driver.

In the following example, process p1 implies one driver for signal a and one driver for signal b. The fact that signal a has more than one assignment in p1 is irrelevant. Process p2 implies a single driver for signal b. Because b has multiple drivers (one in p1 and one in p2), it must be a resolved signal. In this example, b is resolved because it was declared with type std_logic, which is a resolved type (rather than std_ulogic, which is unresolved). However, if we tried to make an assignment to a from a process other than p1, it would be an error because a is unresolved.

```
architecture examples of drivers is
    signal a, w, x, y, z: std_ulogic;
    signal b: std_logic;
    signal select_1, select_2, select_3: boolean;
begin
    p1: process (all) begin
        -- Assignment to 'a' implies one driver:
        a <= '0';
```

```
        -- Second assignment to 'a' reuses previous driver:
        a <= w when select_1;
        -- Assignment to 'b' implies one driver:
        b <= x when select_1 else 'Z';
    end process;

    p2: process (all) begin
        -- Assignment to 'b' in a different process implies another driver
        -- for the same signal; signal must be resolved.
        b <= y when select_2 else 'Z';
    end process;

    -- Concurrent assignment implies a 3rd driver for 'b':
    b <= z when select_3 else 'Z';
end;
```

When a resolved signal has multiple sources, the values of all sources are grouped in an array and passed to a *resolution function*, which is invoked by the simulator and determines the resulting value based on a resolution algorithm. Resolved signals and resolution functions are covered in section 12.4.

An important thing to note is that a driver is not associated with a signal declaration but rather with one or more assignments. Every signal assignment is associated with a driver, and it affects this driver only. The effect of an assignment is to insert values into the driver's projected waveform. The following example would add a series of transactions to the driver of signal a, yielding a projected waveform similar to the one shown in figure 5.13. When the assignment is executed, the time value of each waveform element is added to the current simulation time, and then the transactions are placed in the queue in increasing time order.

```
a <= '1', '0' after 10 ns, '1' after 20 ns, '0' after 30 ns;
```

If a time component is not specified for a transaction, then an implicit after 0 ns is assumed:

```
a <= '1'; -- same as: a <= 1 after 0 ns;
```

However, even if an assignment is specified for the current simulation time, the value of a signal is never updated immediately. The earliest time in the future for which a transaction can be scheduled is the next simulation cycle. Hence, this assignment would be scheduled for an infinitesimal time unit in the future, called a *delta time unit*. We will get back to this subject when we discuss the simulation cycle later in this section.

One final point to discuss about drivers is their use in subprograms. Assignments in subprograms do not imply drivers; they borrow drivers from a process. According to the scope and visibility rules of VHDL, a subprogram can access objects that are visible at the place where it is declared, even if the objects are not passed as parameters. Hence, if a subprogram is declared local to a process, it can make assignments to signals using the process drivers. In contrast, if we write the subprogram in the architecture declarative part, it will be able to read signals values but not assign to them. In this case, the solution is to pass the signal to the subprogram as a signal parameter, and then the subprogram will use the driver from the calling process.

5.6.3 Initialization Phase

After a description has been elaborated, it is ready to be executed (see figure 5.12). However, before the simulation enters the iterative phase of executing simulation cycles, the model and the simulation kernel must go through an initialization phase. During initialization, the simulation kernel:

1. Sets the simulation time to zero,
2. Initializes all signals in the model,
3. Executes each process once, until it suspends, and
4. Calculates the time of the first simulation cycle.

The initial value of a signal is calculated considering the initial value of its driver (which corresponds to the default value given at the signal declaration), any resolution function if the signal is resolved, and any type conversions and conversion functions if the signal is connected to a port. For a more detailed description of the signal initialization algorithm, see section 14.7.

Calculating the Time of the Next Simulation Cycle   The time of the first (or next) simulation cycle is calculated as the earliest of:

- The maximum value allowed for type time (i.e., time'high),
- The next time at which a driver or signal becomes active, or
- The next time at which a process resumes.

If the calculated time is the same as the current simulation time, then the next cycle is called a *delta cycle*. Otherwise it is a *time-advancing cycle*.

5.6.4 The Simulation Cycle

Following the initialization phase, a model is ready to be simulated. Simulation consists of the repetitive execution of a simulation cycle, during which the following tasks are performed:

1. Simulation time advances to the time of the next activity, calculated at the end of the previous cycle (as described in the previous section).

2. All drivers with transactions scheduled for the current simulation time are updated. Such drivers are said to be *active* during the current simulation cycle.

3. All signals connected to nets with active drivers are updated. This may generate events (a change of value) on the signals.

4. All processes scheduled for resumption at the current simulation time are resumed. All processes sensitive to a signal that has had an event in this cycle are also resumed.

5. All resumed processes are executed, in no predefined order, until they suspend.

6. The time of the next simulation cycle is calculated.

The process repeats until there are no more pending transactions (a situation called *event starvation*) or simulation time reaches `time'high`.

In summary, in each simulation cycle, signals are updated in batch, and then processes are executed in batch. These two steps are often called the two halves of the simulation cycle: *signal update* and *process execution*. During process execution, new assignments can be made for the next delta cycle or for some arbitrary time in the future. If new assignments are scheduled for the following delta cycle, the process repeats without advancing simulation time. When all remaining assignments are for some measurable time in the future, simulation time advances, and the process restarts.

Figure 5.14 shows the succession of simulation cycles in an example simulation. It also shows how the advance of simulation time can be represented in a two-dimensional graph. Some cycles happen after an increase in the simulation time and are called *time-advancing cycles*. Other cycles only advance a *delta time unit*, without changing the simulation time. Such cycles are called *delta cycles*. In the figure, delta cycles move the simulation time downward along the vertical axis, whereas time-advancing cycles increase the simulation time along the horizontal axis. The first cycle (at 0 ns and delta = 0) is the initial cycle. The other cycles with delta = 0 are time-advancing cycles. The remaining are delta cycles.

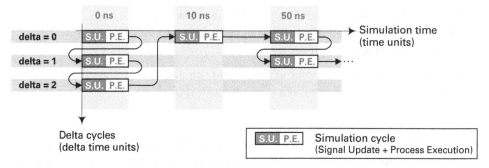

Figure 5.14
Succession of simulation cycles in a simulation.

6 VHDL Design Units

In VHDL, statements and declarations cannot exist in a vacuum: each line of code must be part of a *design unit*. VHDL provides several kinds of design units. If the unit represents a hardware module with well-defined inputs and outputs, then it is called a *design entity*. If the unit contains algorithms and declarations to be reused across different entities, then it is called a *package*.

Different design units offer different ways to organize our designs. If we choose the right kind of unit for each situation, our code will be easier to read and write. This chapter introduces the main design units in VHDL and provides guidelines for their use.

6.1 Design Units

Structurally, a VHDL design is composed of a series of modules organized hierarchically. In terms of code, a design may also include constructs that allow for source code reuse and organization. VHDL offers different constructs for the creation of modules, hierarchies, and reusable units of code. Each of these constructs is called a *design unit*. A design unit is also the smallest part of the code that can be individually analyzed and stored in a design library for future use by other units.

There are four main kinds of design units in VHDL:

- An **entity declaration** describes a system, subsystem, or component from an external point of view. It specifies a module's interface, including its name, inputs, and outputs.
- An **architecture body** describes the internal operation of a module. It specifies the module's behavior or internal structure.
- A **package declaration** groups a collection of declarations that can be reused in other design units at the source code level. The package declaration only defines the interface to a package and specifies no actual behavior.
- A **package body** contains the implementation of a package, providing the actual behavior and operations described in the package declaration.

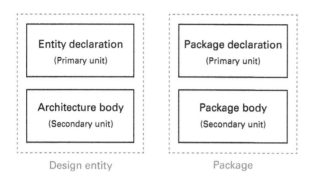

Figure 6.1
The four main design units in VHDL.

Figure 6.1 shows the four main design units available in VHDL. Each pair of design units can be grouped into a larger conceptual unit. A *design entity* is defined by an entity declaration together with an architecture body. A *package* is defined by a package declaration and an optional package body, required if the package contains subprograms. For a source code example, see figure 5.3.

The other kinds of design units available in VHDL are much less used in practice. The four main design units shown in figure 6.1, plus configuration declarations, have existed since the first version of the language. VHDL-2008 introduced three new design units: package instantiation declarations, context declarations, and Property Specification Language (PSL) verification units. Our discussions will focus only on the four units shown in the figure.

Figure 6.1 shows that design units can be classified as primary (entity declaration and package declaration) or secondary (architecture body and package body). In a nutshell, a primary unit defines the interface to a module, whereas a secondary unit defines the module's implementation. As we will see in the next section, this creates a dependency from the secondary to the primary unit that determines the order in which they must be analyzed.

A source code file containing one or more design units is called a *design file*. Typically, a design file includes a single design unit or a single conceptual unit (an entity-architecture pair or a package). Although it is possible to include more design units in a single file, this complicates the design management and should be avoided.

6.1.1 Entity Declaration
An *entity declaration* describes a module from an external viewpoint. It specifies the interface that allows the module to be used in a system, including its name, input and output channels, and compile-time configuration. In VHDL, the communication

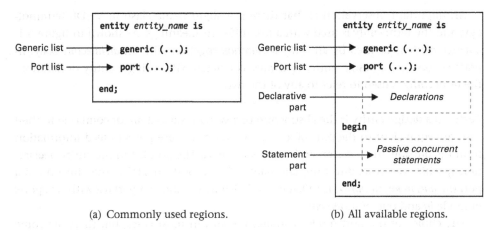

(a) Commonly used regions.   (b) All available regions.

Figure 6.2
Structure and regions of an entity declaration.

channels whose value can change while an entity is in operation are called *ports*, whereas static configuration values that must be specified at compile time are called *generics*.

The kind of module described by an entity is similar to a hardware component. An entity may represent an entire system, a subsystem, a component, or a single logic gate. Entities can be nested to create a hierarchical description of a system.

Figure 6.2 shows the general structure of an entity declaration in the source code. The declaration in figure 6.2a shows the most commonly used elements: a *generic list* and a *port list*. Both lists are optional, but any entity that is intended for synthesis needs ports. In contrast, entities used for test and verification are often self-contained and do not require ports or generics. This kind of entity is described in chapter 20.

The entity declaration of figure 6.2b has all the optional parts. It contains a declarative part, which can include most declarations allowed in an architecture—the only exceptions are component declarations and configuration specifications. A declaration made in this region will be visible from any architecture associated with the entity. In practice, the declarative part of an entity is not used very often. One of the few reasons to put declarations in an entity rather than in an architecture would be to avoid code duplication when there are multiple architectures for the same entity.

The entity declaration also includes a region for concurrent statements. The only statements allowed here are passive statements—statements that do not modify any signals. This statement part is also little used in practice. Possible uses include monitoring the model behavior during a simulation, reporting informative messages during elaboration, or checking that the objects declared in the port list and generic list are consistent and valid.

An important observation is that there is a slight name clash in VHDL terminology, and the term *entity* is used with a few different meanings. As shown in figure 6.1, a *design entity* is defined by an *entity declaration* together with an *architecture body*. In practice, because a design entity is always associated with a single entity declaration, the term *entity* is used to refer to any of the two.

Ports   If a design entity is the abstraction of a system, subsystem, or component, then *ports* are equivalent to the pins of a chip or circuit. We use ports to feed information from the outside into a design entity and vice versa. The port list in the entity declaration specifies the type, direction, and name of each port. In VHDL, the direction of a port (*in, out, inout,* or *buffer*) is called its *mode*. Figure 6.3 shows a port list with two ports of mode *in* and one of mode *out*.

The choice of the port mode is straightforward in most cases, but there are some caveats. Here is a brief description of when each mode should be used:

- in: Used to feed data into the entity. Statements inside the architecture can read the port value in expressions.
- out: Used to output data computed inside the entity. Statements inside the architecture can assign values to the port. In versions of VHDL prior to 2008, it was not possible to read the value of an *out* mode port inside the architecture. From VHDL-2008 onward, the port can be read as well as assigned.
- buffer: An output port that can be read inside the entity. After VHDL-2008, the same behavior can be achieved with an *out* mode port.
- inout: Used to model bidirectional ports. In such ports, part of the time the entity provides a value to the outside, whereas at other times it reads a value provided by another entity.

There has been some confusion about recommending the use of *buffer* ports. Until VHDL-93, *buffer* ports had severe restrictions that prevented their use in many cases. For example, outside of an entity, a *buffer* port could only be connected to a signal or another *buffer* mode port. Also, a *buffer* port could not have more than one source. This

```
entity counter is
    port (
        clock : in  std_ulogic;
        reset : in  std_ulogic;
        count : out std_ulogic_vector(7 downto 0)
    );
end;
        name  mode  type
```

Figure 6.3
A port list with two input ports and one output port.

caused problems because at some point in the design hierarchy, we would like to connect a *buffer* port from an inner component to an output port in the top-level entity. This has been fixed in VHDL-2002 by allowing buffer ports to connect to *out, inout*, or *buffer* ports. Therefore, if you use VHDL-2002 or higher, then *buffer* ports can be used unrestrictedly. If you use VHDL-2008, then you can also read directly from *out* mode ports.

One problem with the analogy between VHDL ports and hardware pins is that pins in a digital circuit are limited to one bit of information. VHDL, in contrast, has a rich type system that allows us to work at higher levels of abstraction. At the top level of a design we may be required to use basic logic types (such as `std_ulogic` and `std_ulogic_vector`), but in lower level entities we are free to use the types that make the most sense for our design, including constrained integers, arrays, and record types. For a detailed discussion about the use of logic types, see the discussion about `std_ulogic` and `std_logic` on page 316 and about the types `std_ulogic_vector` and `std_logic_vector` on page 318.

Generics   A *generic* is an item that provides static information to an entity when it gets instantiated. Because this information is set when the entity is instantiated, it can be used to allow variations in the hardware structure of a model.

In an entity declaration, the generic list comes before the port list, allowing generics to be used to parameterize the entity interface. In the following example, the generic constant `WIDTH` is used to configure the size of port `input_data`:

```
entity fifo is
    generic (
        -- Number of words that can be stored in the FIFO
        DEPTH: natural := 8;
        -- Size in bits of each word stored in the FIFO
        WIDTH: natural := 32
    );
    port (
        input_data: in std_logic_vector(WIDTH-1 downto 0);
        ...
    );
end;
```

Generics enable parameterizing a model to work under different conditions, making the code more flexible and reusable. VHDL-2008 enhanced generics to allow generic types, subprograms, and packages, besides constants.

6.1.2 Architecture Body

An *architecture body* specifies the operation of a module, describing its internal behavior or structure. An architecture body is always associated with an entity declaration, and it has access to the entity ports, generics, and other declared items. An architecture uses processes, component instantiations, and other concurrent statements to define the behavior of a module. The term *architecture body* is often abbreviated as *architecture*.

Figure 6.4 shows the general structure of an architecture body. The construct has two major parts: a *declarative part* and a *statement part*. The inner boxes show examples of declarations and statements that can be used in an architecture body.

The statement part in the figure shows all the concurrent statements available in VHDL. Communication between concurrent statements is usually done via signals declared in the architecture. Concurrent statements are covered in chapter 9.

An architecture body is always associated with a single entity declaration. The reverse, however, is not true; it is possible for an entity to be associated with multiple architectures. These architectures may describe different versions of a circuit, such as one intended for synthesis and another for simulation. In that case, VHDL offers several mechanisms to select which architecture should be used with the entity declaration. For details, see section 9.4.

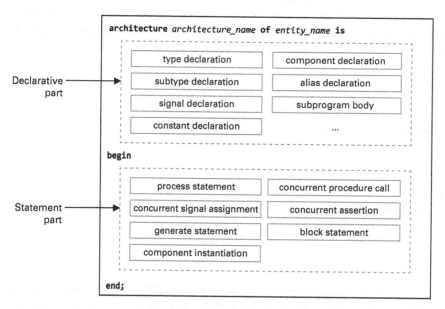

Figure 6.4
Parts of an architecture body, including examples of declarations and concurrent statements.

6.1.3 Package Declaration and Package Body

A package declaration and the corresponding package body define a *package*. Packages are a mechanism for sharing and reusing code across design units. Although entities can be reused in the sense that they may be instantiated multiple times, only packages allow reuse at the source code level. With a package, we can write constructs such as types, subprograms, constants, and other declarations and then reuse them in different places in a design.

The package declaration contains the part of a package that is publicly visible. The most common items in a package declaration are types, constants, subprogram declarations, and component declarations. Anything that should not be publicly visible should be in the package body instead. This includes the implementation of subprograms declared in the package declaration and any other declarations that are not relevant for clients of the package.

Listing 6.1 shows a package declaration and the corresponding package body for a custom data type and associated operations. The data type is a synthesizable representation for a complex number (lines 5–9). The package declaration also includes constant declarations (lines 11–13) and function declarations for working with objects of the custom type (lines 15–16). The package body (lines 20–28) contains the bodies of the subprograms specified in the package declaration.

Listing 6.1 Package declaration and package body for a custom data type and its operations

```
 1 library ieee;
 2 use ieee.float_pkg.all;
 3
 4 package synth_complex_type_pkg is
 5     -- Complex data type based on synthesizable floating-point numbers
 6     type synth_complex_type is record
 7         re: float32;   -- Real part
 8         img: float32;  -- Imaginary part
 9     end record;
10
11     constant BASE_1: synth_complex_type := (re => to_float(1), img => to_float(0));
12     constant BASE_J: synth_complex_type := (re => to_float(0), img => to_float(1));
13     constant ZERO: synth_complex_type := (re => to_float(0), img => to_float(0));
14
15     function "+"(left_op, right_op: synth_complex_type) return synth_complex_type;
16     function "="(left_op, right_op: synth_complex_type) return boolean;
17     ...
18 end;
19
20 package body synth_complex_type_pkg is
```

```
21    function "+"(left_op, right_op: synth_complex_type) return synth_complex_type is
22    begin
23        return (left_op.re + right_op.re, left_op.re + right_op.re);
24    end;
25
26    function "="(left_op, right_op: synth_complex_type) return boolean is
27        ...
28 end;
```

Because all the information that can be used directly in the client code is in the package declaration, when we compile the client code, only the package declaration must have been previously analyzed. However, to elaborate and simulate the client code, the subprogram implementations are also necessary, so by then the package body must also have been analyzed.

In practice, every design uses packages. The predefined VHDL types such as bit, boolean, and integer are declared in a package called standard. This package is included by default in every design unit we create. Other standard packages provide useful types and operations, such as std_logic_1164 for multivalued logic types and numeric_std for signed and unsigned arithmetic using vector-based types. The standard packages provide types that are commonly needed in a large number of designs, as well as the operations to work with these types.

Packages are useful to organize the code in a design. We should not think that only large pieces of code deserve a package. If any piece of information is needed in two different design units, then this is enough reason to create a package. For example, declaring a port of a user-defined type requires a package. Declaring a type that needs to be shared by the RTL code and its testbenches also requires a package. No package is too small if it serves a useful purpose.

Although we can put any set of declarations in a package, it is good programming practice to group only declarations that are logically related. Remember that a package is also a module; as such, it should have a single and well-defined purpose, and internally it should be strongly cohesive.

Global Objects   An object declared in an entity can only be used by statements inside that entity. In contrast, objects declared in a package can be accessed by any unit that uses the package. In this case, all units will see and interact with the same instance of the object. The two kinds of global objects in VHDL are *global signals* and *global shared variables*. Although constants declared in a package use the same mechanism, the term *global constant* is not commonly used.

Global signals are a way to bypass an entity's public interface (its port list) and to read or assign values to a signal that can be shared by many entities. At first thought, global signals seem like a good way to make the interfaces cleaner. In reality, they tend to make a design more obscure because it is hard to guess where the signal is coming from. This practice is also dangerous because any entity in the design could interfere with a global signal.

In practice, global signals and global shared variables are recommended only in simulation code to provide an abstract communication path between the different entities that compose a testbench. For example, if one entity generates test stimuli and another entity verifies it, then the two could exchange information via a global signal. This leaves the top-level entity of the testbench uncluttered and makes it easy to add new signals to the abstract data path that connects the testbench entities.

6.2   Design Libraries and Library Units

A *design library* is a collection of compiled design units. When the analyzer processes a source file, the results are stored in a design library for future use. VHDL does not define a specific format or location for the analyzed design units; each tool is free to choose its own storage format. In practice, most tools use a database or a folder in the host file system.

Whenever we analyze a VHDL file, a design library must be specified as the target of the compilation. This may be done automatically if we use an IDE or via a configuration file or command line option if we use the command line. If the compilation succeeds, then each analyzed design unit is transformed into a *library unit*, stored in the specified design library, and made available for use by other units in the design.

After a unit has been put in a library, we may refer to its contents by using its full name in the code. To use a declaration from a library, we start with the library name, followed by the name of the library unit and then the declaration we want to access. For example, if a package named `constants` has been compiled into a library named `work`, then we can access its declarations using the notation:

```
work.constants.pi   -- library 'work', package 'constants', object named 'pi'
```

Contrary to popular belief, a *use* clause is not strictly necessary to use items from a library. We will get back to the subject of *use* clauses in section 7.2.5 after we have introduced the concepts of *scope* and *visibility*.

Because all the references between design units are made via design libraries, we will never see the name of source files in VHDL code. The only way to access the contents of another file is via a design library after the file has been analyzed. Moreover, the choice of where to put the analyzed units is not made in the code; it is a tool configuration.

The VHDL environment comes preconfigured with two libraries. Library std is where the fundamental language items are located. It contains only the packages stan-dard, textio, and env. Library ieee is where the mathematical, multivalued logic, and synthesis packages are located. A user or tool should not add new library units to those libraries.

Libraries can be used to organize a design. The chosen approach depends on house rules, design size, and personal preference. In large designs, common practices include the use of distinct libraries for reusable components, different ven-dors, or different target technologies. In small designs, it is possible to avoid the overhead or managing libraries at all by compiling all the design units into a single library.

Library *work* To use a design library in VHDL, we need to know its *logical name*. A logical name is a simple name used in the code to refer to a design library that exists in the host system (the machine running the VHDL tools). A tool must provide a way to map between logical names (used in the code) and physical names (existing in the host). This mapping is done outside of the VHDL source code and can be configured by the user.

One of the logical names, however, is special: the name *work* is not permanently associated with any library in the host system. In VHDL code, the name *work* acts as a pointer or an alias. During the analysis of each design unit, it gets temporarily associ-ated with the library where the results of the current analysis will be placed. This target library is called the *working library*.

The fact that the name *work* is a pointer to a library that can change at each compila-tion may cause some confusion. Suppose you created a physical library and gave it the logical name *work*; how would other libraries refer to it? They can't because when they use the name *work* in the code, it will point to the current working library, not to the one you have created and named *work*.

Let us clarify this with a more concrete example:

1. Create a package named *p1*; compile it into a library named *components*.
2. Create a package named *p2*; compile it into a library named *work*.

How can you use a declaration from package *p2* in package *p1*? You can't! When you write *work.p2* inside package *p1*, the analyzer will understand that you are actu-ally saying *components.p2* because during that analysis *work* is an alias for *components*. Anything you compile into a library named *work* becomes inaccessible to libraries with a different name.

There are two recommendations to avoid making this situation any more confusing than necessary. First, if we organize our design using a single library, then it does not matter what we name it. If we choose to name it *design_lib*, then we can refer to any declaration in it by *design_lib.declaration* or *work.declaration*. We can even name this library *work*, as some tools actually do. However, it is probably a good idea to use a different name to prevent any more confusion.

Second, if we organize our design into multiple libraries, then we should use the logical name *work* only with the meaning of "the same library where the current unit is being analyzed." In this case, avoid giving the logical name *work* to any library you create.

Compilation Order   VHDL allows a design unit to use information from other design units. For example, an architecture may use a function declared in a package. In this case, to process the architecture, the analyzer will need to check whether the function was used correctly, including its parameters and return types. Because all interaction between design units happens via a library, it follows that the package declaration must have been analyzed before the architecture. In other words, the architecture *depends* on the package.

Any kind of unit (primary or secondary) may depend on a primary unit. Moreover, a secondary unit always depends on the corresponding primary unit. However, no unit can depend on a secondary unit. This makes sense because secondary units contain private implementation details that should not be relevant to other parts of the design.

Dependencies have two consequences in the compilation of a design. First, the order in which design units are analyzed cannot be random. Analysis must happen one file at a time, in an order that respects the chain of dependencies between the design units. Second, if a library unit is changed (by recompiling the corresponding design unit), then it will be necessary to reanalyze all the dependent units. Finding out which units depend on which other units may be a complex task in large designs, so most tools provide a way to do it automatically.

In any case, splitting a design into files containing at most a single design unit or one conceptual unit (an entity/architecture pair or a package declaration/package body pair) is a good way to organize a design. It also reduces the number of unnecessary recompilations.

6.3   Guidelines and Recommendations for VHDL Design Units

Each design unit is a kind of module: an entity is a module similar to a hardware component, and a package is a module similar to a software library or abstract data type. Therefore, all the recommendations about creating good modules also apply to entities and packages. If necessary, refer to section 4.1.3 for guidelines about how to divide a system into modules and to the discussion about the principles of loose coupling (4.4), strong cohesion (4.5), and orthogonality (4.7).

Draw a Block Diagram   A block diagram is a great tool to explore a design before writing any code. The diagram does not need to be detailed to the bit level; at this point, we are more interested in evaluating the amount of communication and the interaction between design entities. The most important thing is to show the nature and direction of the exchanged information.

If the diagram shows a lot of connections between the modules, then they are probably too tightly coupled. If you see a large number of lines going in and out of an entity, then it is probably doing too much and should be broken down into simpler modules. If you cannot see a logical flow of information from one edge of the diagram to another, then maybe the design is too complex; look for another design approach that streamlines the information flow. Finally, if you find it hard to draw a diagram of the intended design, then perhaps you do not know enough about the problem or solution yet. Do not start coding until you have sorted out the design structure.

Keep in Mind the Guidelines for Creating Good Modular Systems   Section 4.1.3 presented detailed guidelines on how to break down a system into modules following good design principles. Here we summarize the main recommendations.

The most important consideration when creating a module is to give it a single, well-defined responsibility. For an entity, it means that all inputs, outputs, and behavior should support a central function. If you can identify smaller, independent tasks, consider putting them in a separate module. For a package, having a single responsibility means that all declarations and operations should support a central purpose. Again, if part of a package uses only part of the declarations or the parts have clearly distinct purposes, then look for a way to break the package into smaller, more focused units.

When designing a module, pay special attention to its interface. An ideal interface is narrow, is simple to use, and exposes the minimum number of ports (in the case of an entity) or operations (in the case of a package). It should be clear and unambiguous to use.

A good module should have a manageable size. If the source code of a module spans several pages and forces you to scroll up and down all the time, then it is probably too long. A module should be small enough for you to reason about its entire operation at once. Can you think about all the signals in an architecture at the same time? Can you keep the role of each process in your mind at the same time? If not, try to break the module into more intellectually manageable chunks.

Besides the above recommendations, which are given from the viewpoint of a module, other guidelines refer to the system as a whole. Try to distribute functionality across modules in a way that minimizes the number of interconnections; systems that are minimally connected are easier to integrate and test. However, the system should have as many modules as necessary (this recommendation is elaborated in the next topic).

Finally, a module is a great way to achieve information hiding. Create a module when you want to hide implementation details or design decisions. A common example is to create a package containing a type and all the operations performed on that type. This hides the implementation details from the code that uses the package, making it easier to see the big picture in the client code. Another example is to create an entity to wrap a vendor-specific module so it is easier to change if needed.

Divide a System in as Many Entities and Packages as Needed   A system should have as many modules as necessary. A good rule of thumb is to create a module for each recognizable piece of functionality that can be developed and tested separately. Starting with the top-level block diagram, break down each block recursively until each module performs a well-defined task and is small enough to allow reasoning about its entire working at once. The final system should have as many modules as the problem asks for.

Novice programmers are often reluctant to create new modules and design files; some go through great lengths to keep the entire design within a single file, no matter how long. The common pretexts are the overhead of creating and managing files, the increased amount of typing, the effort of navigating across files while editing the code, and the difficulties of setting up the compilation process. These are valid concerns; however, experienced programmers think about a project in the long run and give each factor different weights. In projects with small, well-factored files, each file is smaller, easier to understand, easier to test, and easier to change. In team projects, small files minimize the number of conflicts that happen when developers try to edit the same file. Finally, larger modules are more likely to have bugs, and changes are more likely to have inadvertent side effects.

Modern IDEs and source code editors offer great support for navigating between files and automating the compilation process. Good editors allow us to navigate the code as if we were using a web browser: a click on a name takes to its declaration, even if it is located in a different file. Back and forward buttons can switch between recently used files, and so on. Do not waste your time using an editor that is not appropriate for editing code. Remember, do not hesitate to create a file or module if your design asks for it. In all but the most trivial designs, the advantages far outweigh the costs.

Consider Other Forms of Modularizing the Code besides Structurally   As explained in the previous recommendation, putting the entire system behavior in a single unit is a bad idea for all but the most trivial designs. This would leave the designer with a big, monolithic block of code that is hard to work with. One approach to make the code more manageable is to split the functionality across multiple design entities. In some cases, however, the design could benefit from moving part of the functionality into packages as well.

Packages offer several interesting benefits. First, package operations are often implemented as functions, which can be used directly in expressions by the client code. If the same behavior were provided in an entity, then it would require a component instantiation and signal declarations. Second, if all the behavior is implemented in architectures, then the application code becomes mixed with low-level details, making it hard to see the big picture of a design. If we move the lower lever tasks to a package, then the application code becomes cleaner and allows us to ignore the implementation details. Third, if we restrict ourselves to using entities, then we may have to spread logically related operations across separate design units. If we use a package, then we can put all the related behavior in the same place. Packages and functions are also easier to test because they do not require instantiation and do not include the time dimension, keeping the testbenches simpler and cleaner. Finally, packages make it easier to share declarations across multiple entities, promoting code reuse and helping reduce duplication.

A good way to extract a logically related part of the code into a dedicated unit is to create an abstract data type (ADT). An ADT is a type together with a set of operations for working with objects of that type, usually implemented in a package (see listing 6.1 for an example). With an ADT, the client code becomes cleaner because it can use the operations while ignoring the low-level details. The ADT code is also simpler because it can focus on solving specific tasks and safely ignore the rest of the application. Abstract data types are covered in section 13.4.

Create an Adaptation Layer between the Top-Level Entity and the Hardware Environment Many designs must be incorporated in a hardware environment over which we do not have much control. Inputs and outputs may use inadequate types and logic levels. They may also have cryptic names that would make our code less readable if used inside our design.

Generally, there is no need to carry bad names and types to the inner levels of a design. In such cases, consider encapsulating the design in a second-to-top-level entity, which is free to use the types, logic levels, and names that make more sense in your code. Then do the necessary conversions in the topmost level of the design. The idea is to create a clear boundary between what is an intentional part of our design and what exists only because of the surrounding hardware environment. This will prevent unnecessary type conversions and make the code more streamlined, less verbose, and easier to understand. The top level (or levels) of a hierarchy could also instantiate any device-specific hardware resources. In this way, if we ever need to move the design to a different environment, then there is no need to touch any of the application code.

To make the conversions in the top level less verbose, consider using conversion functions directly in the instantiation port map, as shown in the example of figure 13.4.

Restrict Generics to a Range That the Model Is Prepared to Accept If a model works correctly only for a limited range of a generic parameter, then make it explicit in the code by declaring the generic with a range constraint:

```
generic (NUM_DELAY_CYCLES: natural range 2 to 4);
```

If a parameter can assume only a limited number of choices, then consider creating an enumeration type to restrict the possible values that can be configured:

```
type key_length_type is (key_length_128, key_length_256, key_length_512);
```

This guarantees that a module will never be passed a value that it is not prepared to handle.

Prefer Generics to Constants in a Package for Parameterizing a Design Entity If a constant value is needed in an entity, then there are at least three different places where this value could be defined: as a constant in the architecture declarative part, as a generic in the entity interface, or as a constant in a package. If the value is used to parameterize the design entity, then make it a generic in the entity interface.

There are three main reasons for this choice. First, if we use generics, then we will be able to configure two or more instances of the same entity differently; with a constant, they would all have the same value. Second, generics are a visible part of the entity interface; they make it clear that the unit is configurable. In contrast, a constant declared in a package would be communicated to the architecture surreptitiously. Third, with generics, the entity can be reused in different designs. Using constants in a package, we would need to move the package to the new design as well or at least the necessary declarations. Source code "reuse" via cut and paste is never as good as code reuse using the proper language mechanisms.

7 Statements, Declarations, and Expressions

This chapter introduces the basic constructs used in VHDL code. We start with a discussion about statements, the basic elements of behavior in any program. Contrary to popular belief, not every construct in the source code is a statement. The two other major constructs are declarations and expressions. In a nutshell, declarations create and name new items in the code, statements perform basic operations and control the program flow, and expressions are formulas that stand for values. This chapter introduces these three basic constructs and presents guidelines and recommendations for their use.

7.1 Statements

Statements are the basic units of behavior in a program; they specify basic actions to be performed when the code is executed and control its execution flow. In VHDL, statements and declarations cannot be mixed in the same region of code. Figure 7.1 shows where the two constructs commonly appear in the code; statements are used only in the shaded areas. Expressions do not appear in the figure because they are not self-standing constructs; they can only be used as parts of statements or declarations.

VHDL has two kinds of statements: sequential and concurrent. *Sequential statements* are executed one after another in the order they appear in the source code. The default order of execution is from top to bottom, but it can be changed using control flow statements, which allow conditional or iterative execution. Sequential statements can only exist in sequential code regions, namely, inside processes or subprograms. The execution of a sequential code region is triggered by a subprogram call or process activation. Sequential statements in VHDL are similar to statements in other general-purpose programming languages. All sequential statements are covered in chapter 10.

Concurrent statements have three important characteristics. First, they execute side by side, conceptually in parallel. Second, they execute continuously. We can think of concurrent statements as if they executed in loops or separate threads. In each loop, the statement waits for a certain trigger condition, performs its task, and then suspends

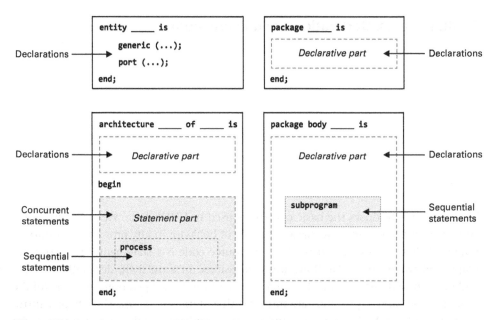

Figure 7.1
Usual places for statements and declarations in VHDL code (statements are allowed only in the shaded areas).

again. Third, concurrent statements have no predefined order of execution. Their execution is controlled by the simulator, which monitors all trigger conditions and executes the concurrent statements that are due in each simulation cycle. The simulator may execute the statements in any order or even in parallel. This means that the order in which they appear in the source code is irrelevant. Concurrent statements are only used in the statement part of an architecture.

Although the order of execution of concurrent statements may vary, the results of a simulation should not. We want our models to be predictable, and simulations should produce the same results every time. To make this possible, VHDL provides an elaborate communication mechanism between concurrent parts of the code, using communication channels called *signals*. While any concurrent statement is being executed, the simulator freezes all signal values. This means that all concurrent statements will see the same value for a signal during a simulation cycle; only after all statements have finished executing are signal values updated. This allows processes to be executed in any order and still produce deterministic results. All concurrent statements are covered in chapter 9.

In any programming language, it is also possible to classify statements into simple or compound. A compound statement can include other statements in it, whereas a simple statement cannot. An example of a compound statement is the *if* statement, which includes a body of statements that will be executed only if a condition is true. Likewise, a *process* is a compound statement that includes a declarative part and a statement part. In VHDL, every statement ends with a semicolon; this can help with sorting out where a statement ends and the next one begins.

Finally, another common item that is often mistaken for a statement is a *clause*. A clause is an incomplete fragment of a statement, declaration, or other higher level construct. For example, an *if* statement often has two clauses: an *if* clause executed when the condition is true and an *else* clause executed when the condition is false.

"Executing" Statements   When we describe the effects of a statement, it is common to refer to actions that are performed when the statement is *executed*. For example, when discussing conditional statements, we may say that a condition expression is evaluated and then a body of statements is selected for execution. In an iterative statement, we may say that a block of statements is executed a number of times.

This may accurately describe what happens in a simulator. However, it is important to keep in mind that in a synthesized circuit, individual statements do not exist, so it is not possible to control when a statement gets executed or when an expression gets evaluated.

Consider the case of an *if* statement, for instance. In the simulator, it is possible to evaluate and test the condition first and then choose between the *if* or *else* clauses for execution. In hardware, the two branches and the condition expression would be computed concurrently, and the conditional value would select one of the branches, possibly using a multiplexer. The role of a synthesis tool is to infer the behavior of a description and create a circuit that produces the same visible effects as if the statements were executed in a simulator. There is not a necessary correspondence between each and every statement in the code and a hardware element in the circuit. Moreover, if we want the synthesized hardware to perform a series of actions in a temporal sequence, then we must design a circuit that exhibits this behavior using sequential logic, such as an FSM.

The in-order execution of sequential statements and the side-by-side execution of concurrent statements are what we call a *programming model* for the developer. Even if this model does not correspond literally to what is implemented in a device, it may be easier to think about a piece of code algorithmically, with each statement corresponding to a well-defined operation that changes the abstract state of the system. In any case, when we write synthesizable code, we should have a clear picture of the hardware that we intend to model. After writing the code, we should be able to draw a block diagram of what the code physically represents, and it should match our original intent. We should also check our assumptions by comparing our expectations with the actual synthesis results.

7.2   Declarations

When we write a VHDL description, each element in the code represents a different concept in our solution, including pieces of data, behavior, and structure. A *declaration* is a construct that creates a new item in the code, defines its meaning, and gives it a name. An item created with a declaration is generically called a *named entity*. Here the term *entity* is used in a general sense and should not be confused with more specific uses in VHDL, such as the terms *design entity* and *entity declaration* seen in the previous chapter.

A declaration can only appear in a declarative code region. Many VHDL constructs are divided into a declarative part and a statement part. The declarative part comes first and introduces named entities that can be used in the statement part. In such constructs, the two parts are separated by the keyword `begin`. Objects created in a declarative part are called *explicitly declared objects*; such declarations always end with a semicolon.

Another place where objects can be declared is an *interface list:* a list that specifies the parameters in a subprogram, or ports and generics in an entity declaration. When we write such a list, the declarations are not *terminated* with a semicolon. However, two adjacent items in a list must be *separated* with a semicolon. This explains why in a port or generic list the last element does not have a trailing semicolon—the semicolon is a list separator and not a terminator associated with each declaration.

7.2.1   Scope

In the conventional Computer Science meaning, the *scope* of a declaration is the region of the source code where it is possible to use the declared item. In VHDL, the terminology is a little different: the scope of a declaration is a region of the code determined by the place where the declaration occurs. Whether a name can be used at a given place in the code also depends on *visibility* and *overloading* rules.

In VHDL, the scope of a declaration generally extends from the beginning of the declaration to the keyword end of the region where the declaration appears. For example, the scope of a declaration created in the declarative part of an architecture extends until the end of the architecture. Similarly, the scope of a declaration made in a process extends until the end of that process. There are, however, a few exceptions. For example, declarations made in an entity extend beyond the corresponding end keyword and can be used in all architectures associated with that entity.

As an example, figure 7.2 shows two declarations. The scope of declaration #1, made in the architecture declarative part, extends until the end of the architecture (from line 4 to line 17); the scope of declaration #2, made in a process declarative part, extends until the end of the process (lines 10–15). When there are nested code regions, the scope of the outer declaration also encompasses the inner regions.

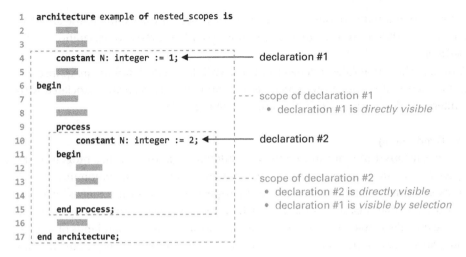

Figure 7.2
Scope and visibility of declarations.

7.2.2 Visibility

Visibility is the property of a declaration to be reachable from a given location in the code. Visibility is related with the question: from all possible meanings of a name, which one should be used? For each occurrence of a name, the compiler must determine the declaration to which it refers. This choice takes into account the scope of declarations as well as the language visibility and overloading rules.

Because names can be reused and scopes may overlap, the same name can have different meanings in different parts of the code. In figure 7.2, the two declarations use the same name N, and their scopes overlap from line 10 to line 15. In this case, we say that the innermost declaration *hides* the outermost declaration.

To allow access to objects in this and other situations, VHDL defines two kinds of visibility: *direct* and *by selection*. A declaration is *directly visible* within its entire scope, except where it is hidden by a nested declaration that uses the same identifier. In that case, where the scopes overlap, the outer declaration is *visible by selection*. Visibility by selection is also used to access declarations made inside other named entities, such as those located inside packages.

In figure 7.2, declaration #1 is *directly visible* in lines 5–10 and 16–17. In lines 11–15, declaration #1 is *visible by selection* and declaration #2 is *directly visible*. Note that per the language rules, a declaration only becomes visible after the semicolon.

If a declaration is directly visible, then we can refer to it by its simple name. If it is visible by selection, then we need to use a *selected name*. A selected name is written with the notation prefix.suffix, where the prefix is the name of a visible construct, and the suffix is a name declared within that construct. In the example, the architecture name is visible from the entire code, so we could also refer to declaration #1 by example.N.

Although the actual rules that determine the visibility of declarations are quite complex, in practice, it is enough to remember that an identifier is always associated with the innermost declaration within the structure of the code.

7.2.3   Overloading

Overloading means using the same name with different meanings. From a programmer's perspective, we can use overloading to create subprograms or enumeration literals that reuse existing names. From the compiler's perspective, overloading means selecting one of the possible meanings of a name when multiple declarations using the same name are directly visible. In VHDL, overloading is available for certain kinds of names, subprograms, and enumeration literals. The most commonly overloaded items are subprograms, operators, and enumeration literals.

Enumeration literals are overloaded when we declare enumeration types that have one or more values in common. In the following example, the identifier locked is used as a value in two enumeration types:

```
type pll_state_type is (free_running, capture, locked);
type door_state_type is (opened, closed, locked);
```

Overloading of enumeration literals is highly common in practice because the character literals '0' and '1' are members of types character, bit, and std_ulogic at the same time. Every time we write such values, the compiler needs to figure out their actual type. The actual overloading rules are quite complex, but luckily there is no need to memorize them. In most cases, the compiler is able to disambiguate between the names automatically, and in the remaining cases, we can use type qualification to explicitly state the type of a value. An example of an overloading rule is the requirement that an assignment source and the corresponding target must have the same type. Using this rule, the compiler is able to disambiguate the assignments below:

```
signal door_state: door_state_type;
signal pll_state: pll_state_type;
...
pll_state <= locked;           -- Ok, 'locked' is pll_state_type locked
door_state <= locked;          -- Ok, 'locked' is door_state_type locked
```

Overloaded versions of a subprogram use the same name but have at least one distinguishing feature that allows the compiler to disambiguate between them. Examples of such features include the number, type, and names of parameters and the return type in case of a function. These rules allow two subprograms to have the same name and still be directly visible without causing a conflict. In the following example, the two subprograms differ in the type of parameters they accept:

```
function compare(left_op, right_op: integer) return integer is ...
function compare(left_op, right_op: character) return integer is ...
```

When the compiler finds a subprogram call, it will try to disambiguate it using the overloading rules. The following subprogram calls are unambiguous because there is only one possible interpretation for each. The first call uses the version of the subprogram that accepts integers, and the second call uses the version that takes characters:

```
if compare(-1, 1) = 0 then      -- Ok, use the first version (integers)
if compare('a', 'b') = 0 then   -- Ok, use the second version (characters)
```

However, if we declare a third version of the subprogram using a type that may be interpreted ambiguously, then we will incur an error:

```
function compare(left_bit, right_bit: bit) return integer is ...
...
if compare('0', '1') = 0 then   -- Error, subprogram "compare" is ambiguous
```

The subprogram call is ambiguous because the parameter values are overloaded: '0' and '1' could be interpreted as bits or characters. In this case, a possible solution is to use a qualified expression to explicitly state the type of a parameter:

```
if compare(bit'('0'), '1') = 0 then   -- Ok, use the third version (bits)
```

Another possibility is to use parameter names to disambiguate. Note that the two function declarations use differently named parameters (left_op, right_op vs. left_bit, right_bit). Hence, we could use named association to disambiguate the function call:

```
if compare('0', right_bit => '1') = 0 then -- Ok, use the 3rd version (bits)
if compare('0', right_op => '1') = 0 then  -- Ok, use the 2nd version (characters)
```

A common use of subprogram overloading is to define overloaded versions of the predefined operators to work with custom types. This is called *operator overloading*. In the following example, operator "+" is overloaded to work with the user-defined type synth_complex_type:

```
type synth_complex_type is record
    re, img: float32;
end record;

function "+"(left_op, right_op: synth_complex_type) return synth_complex_type is
begin
    return (left_op.re + right_op.re, left_op.re + right_op.re);
end;
```

Operator overloading is discussed in section 8.1.

7.2.4   Scope, Visibility, and Overloading

Because the concepts of scope, visibility, and overloading can be a bit confusing, we wrap up this subject with a comparison among the three of them. The three mechanisms come into play when deciding the meaning of a name:

- *Scope* determines the regions of the source code where it is possible to use each declaration.
- *Visibility* uses scope to determine which versions of a name are *directly visible* or *visible by selection*. If only one version of a name is directly visible, then there is no need to use the overloading rules.
- If two or more interpretations of a name are directly visible, then *overloading* uses a set of rules to choose a single declaration from the context in which the name is used. If only one interpretation is possible, then the compiler uses this version. Otherwise the name is ambiguous and the code is in error.

Another tip to distinguish between scope and visibility is to remember that scope is determined from the viewpoint of a declaration, looking at the code. Visibility is determined from the viewpoint of the code, looking at all possible declarations.

7.2.5   *Use* Clauses

Now that we are familiar with the concepts of scope and visibility, we can discuss a mechanism to control the visibility of declarations that exist in different scopes. A *use clause* turns declarations that were visible only by selection into directly visible declarations. So far, we have been describing use clauses informally as a means of "importing" declarations from a package into the current design unit:

```
-- Declare the name of a library so that we can use it in the code.
library ieee;
```

```
-- "Import" declarations from package std_logic_1164 into the current design unit.
use ieee.std_logic_1164.all;
```

```
-- "Import" declarations from package textio into the current design unit.
use std.textio.all;
```

However, the formal explanation is that the declarations inside the packages were visible by selection only, and the use clause made them directly visible. By doing this, we can refer to the declarations without using selected names. Suppose we have the following package declaration:

```
package complex_pkg is
    type complex_type is record
        re, img: real;
    end record;

    function add(op_a, op_b: complex_type) return complex_type;
end;
```

Simply by compiling the package into the working library, we can already refer to its types and operations in our code. There is no need for a library clause because library work is always visible, and there is no need for a use clause because we can use selected names to refer to the declarations. Therefore, an entity using this package could be written as:

```
entity complex_adder is
    port (
        a, b: in work.complex_pkg.complex_type;
        y: out work.complex_pkg.complex_type
    );
end;

architecture rtl of complex_adder is begin
    y <= work.complex_pkg.add(a, b);
end;
```

Note that no library or use clauses were needed. However, we had to use selected names because only the "root" name (the library name work) is directly visible. Hence, the selected names start with the library name, followed by the package name and a name declared inside the package.

We can make this code less verbose with a use clause. A first solution is to make only the package name visible but not the declarations inside it:

```
use work.complex_pkg;   -- Make only the package name directly visible
...
a, b: in complex_pkg.complex_type;
y: out complex_pkg.complex_type
...
y <= complex_pkg.add(a, b);
```

This helps make the code a little less verbose: because the package name is now directly visible, selected names need not start from the library name. However, because the declarations inside the package were not made directly visible, we still need to use selected names.

To clean up the code even further, we could specify one by one the declarations that we want to make directly visible in the current design unit:

```
use work.complex_pkg.complex_type;
use work.complex_pkg.add;
...
a, b: in complex_type;
y: out complex_type
...
y <= add(a, b);
```

However, when we use many declarations from the same package, it is more convenient to make all declarations directly visible by using the keyword all as a suffix in the use clause:

```
use work.complex_pkg.all;
...
a, b: in complex_type;
y: out complex_type
...
y <= add(a, b);
```

In practice, this is the most used solution. It makes all declarations in a package directly visible in the current design unit and then lets the visibility and overloading rules resolve the names if there is any ambiguity. If the declarations are not overloadable (as is the case of objects, types, and subtypes), or if there is no way to distinguish between overloaded declarations, then we can always use a selected name to explicitly state which one should be used.

On a final note, because the declarative region of an entity declaration extends to any associated architectures, there is no need to repeat the library or use clauses before the architecture body. This is also true for package headers and package bodies.

Furthermore, although the most common place for use clauses is immediately before an entity declaration or architecture body, they can be used in any declarative part, including inside a process, architecture, or subprogram. We can use this to minimize naming conflicts by making declarations visible only where they are needed.

7.2.6   Live Time and Span

Although scope is a property of a declaration, an object's *live time* and *span* are determined by its usage in the code. The *live time* of an object is the number of lines between the first and last references to the object. Live time is sometimes called *total span* or *extent* of an object. The *average span* of an object is the average number of lines or statements between two consecutive references to the object.

There are two possible interpretations for calculating the live time of an object: we may start counting from the line where the object is declared or from the line where it is first used. In our discussions, we will count from the declaration only when it provides an initial value that is relevant to the algorithm. Otherwise, we start counting from where the object is first assigned.

In figure 7.3, variable element is initialized in line 7 and is last used in line 11, giving a live time of 5. Because the variable is used every other line in this interval, it has an average span of 1. The other variable, product, is first assigned a value in line 11 and is last used in line 14, giving a live time of 4. Because the variable is not used between these lines, its average span is 2.

The metrics of live time and span have an important role in the intellectual manageability of the code. If live times are small, then the programmer can concentrate on a smaller region of code while working with a variable. It also means that the declaration or initialization is closer to where the object is used. Because all the occurrences are

```
 1  function "*"(A: matrix_type; B: matrix_type) return matrix_type is
 2      variable element: integer;
 3      variable product: matrix_type(A'range(1), B'range(2));
 4  begin
 5      for i in A'range(1) loop
 6          for j in B'range(2) loop                    Variable element:
 7              element := 0;
 8              for k in B'range(1) loop                  ) span = 1
 9                  element := element + A(i,k) * B(k,j);          live time = 5
10              end loop;                                 ) span = 1
11              product(i,j) := element;
12          end loop;                                   average span = 1
13      end loop;
14      return product;
15  end function;
```

Figure 7.3
Calculation of a variable's *live time* and *span*.

more likely to fit within a single screen, short live times make less of a demand on the reader's memory. Besides minimizing the live time of each object, we should also minimize the number of live objects at any point in the code because those are the items we need to keep in our minds while reading the program.

If object spans are short, then we do not need to jump around in the code as we read it. Also, we do not need to push many intermediate values to our mental stack while we debug long sequences of statements. Short spans also reduce the chance of inadvertently changing a variable that is used later in the code. Overall, minimizing object live times and average span is an effective way to make the code easier to read and understand.

7.3   Expressions

An *expression* is a formula that specifies the calculation of a value.[1] Contrary to declarations and statements, an expression is not a self-standing construct. Instead, it is used in statements and declarations wherever a value is needed. For example, an expression can be used in a declaration to determine the value of a constant or in a signal assignment statement to determine the new value of the signal.

The basic elements of an expression are *operators* and *operands*. Examples of operators include the conventional mathematical symbols (+, -, *, /), logical operators (and, or, not), and relational operators (<, >, =, /=). In VHDL, each operator takes one or two operands and returns one value. Examples of operands include literal values, object names, aggregates, function calls, and expressions enclosed within parentheses. Operators and operands are discussed in detail in chapter 8.

To produce a value, an expression must be *evaluated*. In this process, object names are replaced by their current values, function calls are executed and replaced by their return values, and operators are applied to these results. The order in which this happens is specified by the language syntax. If needed, it is possible to control which operator gets associated with which operands by using parentheses, just like in a mathematical formula.

Every expression has a *type*, which is the type of the value produced when the expression is evaluated. The type of an expression is fixed, and it is determined only from the types of its operands and the return type of the used operators. Remember that operators can be overloaded, so their return type also depends on the type of the operands; in other words, the operator symbol alone is not enough to determine its return type.

Static versus Dynamic Expressions   In VHDL, expressions are classified based on when they are evaluated when a design is analyzed, elaborated, and executed. An expression whose value is calculated before execution is called *static*. Expressions whose values can change during execution or that are calculated only after elaboration are called *dynamic*.

There are two types of static expressions. Expressions of the first type, called *locally static*, are evaluated early in the compilation process when the design unit in which they appear is analyzed. This kind of expression is called locally static because its value can be determined locally, with no need to consider other design units. An example of a locally static expression is the value of the two constants in the following example. The value of the first expression is locally static because it is a literal, and the second expression is locally static because it only uses literals, predefined operators, and constant values that are also locally static:

```
architecture rtl of memory is
    constant ADDRESS_WIDTH: natural := 12;
    constant MEMORY_DEPTH: natural := 2 ** ADDRESS_WIDTH;
    ...
```

The second kind of static expression is called *globally static*. Expressions of this type are evaluated later in the design processing, when the instance in which they appear is elaborated. This category includes, for example, expressions that use generic values. This kind of expression is called globally static because its value can only be defined globally; in other words, its value cannot be determined from the viewpoint of an isolated entity. In the following example, the expressions are globally static because their formulas depend on the value of the generic constant NUM_ADDRESS_LINES:

```
generic (NUM_ADDRESS_LINES: natural := 12);
...
architecture rtl of memory is
    constant ADDRESS_WIDTH: natural := NUM_ADDRESS_LINES;
    constant MEMORY_DEPTH: natural := 2 ** ADDRESS_WIDTH;
    ...
```

Finally, if an expression is not locally or globally static, then it is called a *dynamic expression*. The value of this kind of expression can change after the design has been elaborated (e.g., during simulation). An expression is dynamic if it includes any element whose value cannot be determined during analysis or elaboration, such as the name of a signal or variable.

It is important to know this classification because certain constructs require a specific kind of expression. For example, the range limits in a type definition must be locally static, the choices in a *case* statement must be locally static, and the condition in an *if-generate* statement must be globally static. The following examples illustrate these restrictions:

```
generic (GEN: natural := 12);
...
-- The following type declaration is an error because the range in a
-- *type* definition must be locally static:
type int_type is range 0 to GEN;

-- The following subtype declaration is Ok because the range in a
-- *subtype* definition must only be globally static:
subtype int_subtype is integer range 0 to GEN;

signal S: boolean;
...
if_gen: if S generate -- Error: condition in if-generate must be globally static.
    ...
end generate;

case value is
    when GEN =>    -- Error: choice in case alternative must be locally static.
        ...
end case;
```

7.4 Guidelines and Recommendations for Statements, Declarations, and Expressions

This section presents guidelines and recommendations about the basic elements of VHDL code seen in this chapter: statements, declarations, and expressions. Other recommendations about the elements used in expressions—operators and operands—will be seen in the next chapter.

Put Each Declaration in the Tightest Scope Possible   Constants, variables, and signals should be declared in the place that makes their scope as small as possible. If a constant is only used in an architecture, then do not declare it in a package, do it in the architecture declarative part. If a signal is only used inside a *generate* statement, then do not declare it in the architecture, do it in the statement declarative part. If a value is only used inside a process, then do not make it a signal in the architecture, use a local variable instead.

When we move a declaration from a broad to a narrower scope, the original declarative region becomes cleaner, making it easier for a reader to understand the organization of the code. It also puts the declaration closer to where the object is used, reducing the need to jump around in the code as we read it. The code is also safer because the object cannot be changed except in the small region where it is declared and is intended to be used. If something does not work as expected, then the region of code that needs to be debugged is smaller.

Anyway, remember to keep the constant declarations where they make sense. In many cases, a constant is used only several levels down the hierarchy, but it represents a higher level configuration value. In this case, it should be declared in one of the top-level entities, where it makes sense, and then passed to the lower level modules as a generic.

Minimize Object Live Times Even when objects are already in the tightest scope, it may be possible to organize the code so that their live times are minimized. Short live times make the code easier to understand because we need to keep less information in our head when we read a section of code. To minimize live time, we need to arrange the statements so that references to an object are near to each other. This is valid for both sequential and concurrent code. If a group of signals is related, then it makes sense to place the statements that generate and use those signals close to one another.

Initialize Each Variable Close to Where it Is First Used This is a corollary of the previous recommendation. The place of a declaration does not necessarily determine the live time of an object. Some languages, VHDL included, require that all declarations be made in a declarative region, separate from statements. However, the live time of an object effectively starts when it gets the first value that is relevant to the algorithm. Hence, it is a good practice to assign the initial value of a variable close to where it is first used. The benefits are the ones explained in section 7.2.6 about minimizing object live times.

Another interpretation of this recommendation is that if the code is short, then it is acceptable to initialize a variable at its declaration. Despite increasing the live time a little, this helps make the code more compact, which is also a good thing. However, as the code grows, it may be better to initialize the variable with an assignment in the statement part.

Put Each Declaration on a Separate Line Most declarative constructs allow us to declare multiple items at a time. However, as a general rule, it is better to make a single declaration per line. Writing each declaration on a separate line makes it easier to scan the code for a name because we can read the declarations vertically rather than from top to bottom *and* from left to right. It also makes the code easier to modify because we can add, delete, comment-out, or move a declaration without affecting the others. Functional aspects of a declaration, such as its type or initial value, are also easier to change. It is also easier to add a comment about a specific item when needed. Finally, because compilers usually provide a line number with any error message, errors are easier to locate in the code.

Put Each Statement on a Separate Line The same recommendation about giving each declaration its own line is also valid for statements. Most of the advantages are the same: the code is easier to scan because we can follow its left margin and read one state-ment at a time. It is easier to add, modify, or reorder statements when needed. It is also easier to delete or comment-out an individual statement while debugging.

Besides these common advantages, there are two other reasons to put each state-ment on a separate line. First, it makes it easier to perform single-step debugging. Second, it does not make a program look less complex or artificially shorter than it really is.

Break Down Complicated Expressions into Simpler Chunks   If an expression is long, has many factors, or uses many operators, then separate it into simpler expressions and assign them to variables with meaningful names. Such variables are called *explaining variables*. This makes the logic of an expression easier to follow and simplifies the test when the expression is used in a condition.

For example, if we find an expression such as:

```
if (sync_counter = 63 and (data_in_sync(7 downto 0) = "01000111" or
    data_in_sync(7 downto 0) = "00100111") and (current_state =
    acquiring_lock  or current_state = recovering_lock)
) then
    locked <= '1';
    ...
end if;
```

we could identify the meaning of each part of the expression and break it down into simpler chunks:

```
last_sync_bit := (sync_counter = 63);
waiting_lock := (current_state = acquiring_lock) or (current_state = recovering_lock);
current_byte := data_in_sync(7 downto 0);
preamble_detected := (current_byte = "01000111") or (current_byte = "00100111");
```

Besides making the computation more self-documenting, it also simplifies the condition:

```
if waiting_lock and last_sync_bit and preamble_detected then ...
```

For other ways to cleanup condition expressions, see the recommendations for writing *if* statements in section 10.2.1.

Use Extra Parentheses to Clarify Expressions   VHDL has a strict evaluation order for expressions: each operator is assigned a precedence level, and for operators that can be chained, the evaluation is performed from left to right. However, a reader should not have to remember all the rules to understand an expression. Who knows in which order the operations in the following expression are evaluated?

```
not a xor b srl 1 + 3 * 2 ** n;
```

Using extra parentheses, we make the expression clearer and accessible to everyone:

```
(not a) xor (b srl (1 + 3 * (2 ** n)));
```

This is especially important when developers need to switch between languages that have different rules for evaluating expressions.

8 Operators, Operands, and Attributes

In the previous chapter, we have seen the basic elements that can exist as standalone constructs in the code: statements and declarations. We have also introduced expressions, defined as formulas that can stand for values in statements and declarations. Here we take a closer look into the parts that compose an expression: *operators* and *operands*. Then we describe *attributes*, special constructs used to retrieve information about named entities in the code. We close the chapter with guidelines and recommendations about the use of the three constructs.

8.1 Operators

Operators are predefined symbols or keywords that can be used as functions in expressions but allow a more convenient notation similar to mathematical formulas. In VHDL, all operators take one or two operands and return a single value. Moreover, all operators are pure functions, meaning that they cannot modify their operands, must always return the same value for the same operands, and cannot have any side effects.

The main advantage of operators is that they can be used more naturally than function calls in expressions for two main reasons. First, they can use symbols that would not be allowed as ordinary function names, such as conventional mathematical symbols. Second, operators can use prefix or infix notation and do not require the parentheses used in function calls. Prefix notation is used when the operator has a single operand; in these cases, we write the operator name followed by a value. Infix notation is used when the operator has two arguments; then the operator comes in the middle of the two operands. This allows us to write mathematical formulas and boolean expressions that look cleaner and read more naturally. Compare:

```
mul(add(1,2),3)
```

with:

```
(1 + 2) * 3
```

Table 8.1 Categories of operators (listed in order of decreasing precedence)

Operator Category	Operators
Miscellaneous	** abs not
	Unary version of and or nand nor xor xnor [1]
Multiplying operators	* / mod rem
Sign operators	+ -
Adding operators	+ - &
Shift operators	sll srl sla sra rol ror
Relational operators	= /= < <= > >=
	?= ?/= ?< ?<= ?> ?>=
Logical operators	Binary version of and or nand nor xor xnor
Condition operator	??

[1] In VHDL-2008, the new unary logic operators (the single-operand versions of and, or, nand, nor, xor, and xnor) have the highest precedence level).

This is only possible because the set of names and symbols reserved for operators is predefined in the language. Consequently, we cannot create new operator symbols or names. We can, however, overload the existing ones to work with our custom types (operator overloading is discussed in section 8.1.6).

Without a well-defined set of rules, an expression could admit multiple interpretations. For this reason, the language specifies a precedence among operators. During the evaluation of an expression, operators with higher precedence are associated with their operands first. VHDL does not give each operator an exclusive precedence level; rather, the language defines eight categories of operators. Operators within the same category have the same precedence. Table 8.1 lists the operators available in VHDL, in order of decreasing precedence (highest precedence at the top).

The language specifies that when a sequence of operators with the same precedence appears in an expression, they are evaluated from left to right. However, following one operator with another of the same category without any parentheses is not legal for many categories. In fact, chaining is only allowed for adding operators, multiplying operators, and the associative logical operators (and, or, xor, and xnor). Here are some examples of what is allowed and what is not:

```
a and b and c and d;      -- Ok, and is associative.
a or b or c or d;         -- Ok, or is associative.
a xor b xor c xor d;      -- Ok, xor is associative.
a xnor b xnor c xnor d;   -- Ok, xnor is associative.
a nand b nand c nand d;   -- Error, nand is non-associative; chaining is disallowed.
a nor b nor c nor d;      -- Error, nor is non-associative; chaining is disallowed.
```

```
a and b or c;          -- Error, mixing of same-precedence logical operators is disallowed.
not a and b;           -- Ok, not has higher precedence, evaluates as (not a) and b.

i + j + k + m; -- Ok, addition is associative.
i - j - k - m; -- Ok, subtraction is non-associative but chaining allowed, calc. from left to right.
i * j * k * m; -- Ok, multiplication is associative.
i / j / k / m; -- Ok, division is non-associative but chaining allowed, calculate from left to right.
i + j - k * l / m; -- Ok, uses precedence and left-to-right. Operations order is: *, /, +, -
```

The previous examples demonstrate that although it is illegal to chain nonassociative logical operators (nand, nor), it is legal to chain and mix arithmetic operators, even those that are nonassociative. Finally, the operator not is classified as a miscellaneous operator only for precedence purposes. For all other purposes, it is effectively a logical operator.

8.1.1 Condition Operator (??)

The *condition operator* (represented by the symbol ??) converts a value of a given type to boolean. The operator is predefined for type bit, for which it converts '1' to true and '0' to false. It is also defined in package std_logic_1164 for type std_ulogic, for which it converts a '1' or an 'H' to true and all other values to false.

The condition operator can be called in an expression to convert a bit or std_ulogic value to a boolean, as in the following example. It can also be used as a type conversion function (e.g., when a subprogram expects its arguments to be of type boolean and the objects used in the model have logic types).

```
signal req, ready: std_ulogic;
signal start_condition: boolean;
...
start_condition <= ??(req and ready);
```

However, the main use for the condition operator is not to be called explicitly in expressions. Introduced in VHDL-2008, it helps reduce verbosity in conditional statements, such as *if, when,* or *while*. The language requires the expression in such statements to produce a value of type boolean. The change in VHDL-2008 is to apply the conditional operator implicitly to condition expressions, in certain situations. If the compiler cannot interpret the expression as producing a boolean, then it looks for condition operators that can convert the possible expression types to a boolean. If only one is found, then this operator is applied implicitly.

A typical use case is in expressions where objects of logic level types (such as bit or std_ulogic) had to be compared against '1' or '0' explicitly:

```
if en = '1' and cs = '1' and req = '1' and busy = '0' then ...
```

With the implicit application of the condition operator, the expression can be rewritten as:

```
if en and cs and req and not busy then ...
```

The difference between the two expressions is that the first one uses the boolean version of the logical operators, whereas the second one uses the std_ulogic version. Consequently, the first expression produces a boolean value (which can be used immediately as the condition of the *if* statement), whereas the second one produces an std_ulogic (which requires a conversion to boolean). In VHDL-2008, the condition operator is applied implicitly in the second case. Keep an eye out for opportunities to unclutter the code using this new feature.

8.1.2 Logical Operators

As for the number of operands, operators in VHDL can be classified into unary or binary. Operators that take a single operand are called *unary operators*. Operators that take two operands are called *binary operators*. This distinction is important to understand the two kinds of logical operators. In this discussion, we will refer to the operands of logical operators as "bits" for sake of brevity, but the operators are available for types bit, boolean, and standard logic.

Prior to VHDL-2008, binary logical operators could only be used between two single-bit values or between two 1D arrays of bits. VHDL-2008 extends the use of logical operators to the cases where an operand is a single bit and the other is a 1D array. It also allows the use of logical operators with a single operand that is a 1D array. In this case, the logic operation is applied in cascade across all elements of the array, returning a single bit value, and the operator is called a *reduction operator*. Table 8.2 summarizes the logical operators available in VHDL.

For the case where both sides are a single bit (first row in the table), the result follows the conventional truth table for the corresponding boolean operation. For the cases where one side is a single bit and the other is a 1D array (rows 2–3 in the table), the result is an array with the same index range of the array operand. Each bit of the resulting array is calculated by repeatedly applying the operator between the single-bit value and the corresponding element of the array operand:

```
'1' xor '0'   --> '1'      Use bit × bit version (called scalar-scalar operator)
'1' xor "010" --> "101"    Use bit × array version (called array-scalar operator)
```

For the case where both sides are 1D arrays (fourth row in the table), the operation is applied pairwise between elements in the corresponding positions of the two arrays.

Table 8.2 Logical operators

Operator						Left Operand	Right Operand	Result	VHDL Version
and	nand	or	nor	xor	xnor	Single bit	Single bit	Single bit	Any version
and	nand	or	nor	xor	xnor	Single bit	1D array	1D array	VHDL-2008
and	nand	or	nor	xor	xnor	1D array	Single bit	1D array	VHDL-2008
and	nand	or	nor	xor	xnor	1D array	1D array	1D array	Any version
and	nand	or	nor	xor	xnor	none (unary)	1D array	Single bit	VHDL-2008
not						none (unary)	Single bit	Single bit	Any version
not						none (unary)	1D array	1D array	Any version

The pairs are formed in left-to-right order and not by numeric index. The result gets the same range as the left operand.

```
"110" and "011"  --> "010"   Using array × array version
```

When there is no left operand, the right operand is a 1D array, and the output is a single-bit value, the operators are called *reduction* operators. These operators work as an *N*-input logic gate whose inputs are the elements of the array and whose output is a single-bit value. Reduction operators are unary operators because they take a single argument. Here are some examples:

```
and "110"  --> '0'      or "000"  --> '0'      xor "010"  --> '1'
and "111"  --> '1'      or "001"  --> '1'      xor "011"  --> '0'
```

An important remark about reduction operators is that they have a different precedence level than binary logical operators. The binary versions have the second lowest precedence, whereas the unary versions have the highest possible precedence.

Mixing different logical operators within an expression is disallowed unless parentheses are used. The parentheses effectively turn part of an expression into a subexpression, which is evaluated independently and then replaced with a single value. Furthermore, only associative logical operators can be chained without parentheses. This provides some safety against mistakes caused by forgetting that sequences with multiple nand/nor operators are order-sensitive. Note that the same precaution was not taken with the arithmetic subtraction and division operators: these two operators can be chained in an expression, even though this could yield different results depending on the order of evaluation.

A final point to mention about logical operators is that the binary and, nand, or, and nor for types bit and boolean are so-called *short-circuit operators*. This means that in some cases, the right-hand side operator does not need to be evaluated to provide the result. For instance, if the left operand in an and operation is '0', then the result will be '0' no matter the value of the right operand, and therefore the right operand is not evaluated. Because this could produce surprising results, especially if the operand is an expression with side effects, it is best not to write code that relies on this behavior.

8.1.3 Relational Operators

Relational operators compare two operands, testing them for equality, inequality, and relative order. In VHDL, the term *inequality* is different from its conventional mathematical meaning. Here, it means the logical complement of the equality operation. The mathematical inequality symbols <, >, ≤, and ≥ are called *ordering operators* in VHDL.

Relational operators can be classified into two groups. *Ordinary* relational operators are the conventional comparison operators found in other programming languages. They compare two operands, which may be two scalars (single-element values) or two 1D arrays, and produce a boolean result. *Matching* relational operators were introduced in VHDL-2008. They also compare two scalars or two arrays, but their result type is the same as their operands. For example, the result of a matching equality operation between two std_ulogic_vectors would be an std_ulogic value. Table 8.3 shows the relational operators available in VHDL.

Matching relational operators were introduced in VHDL to address certain limitations or inconveniences of the traditional relational operators. The new operators have several advantages. First, they understand the *don't care* symbol '-' when used with standard logic types. This allows us to write with a single comparison what would require several comparisons using ordinary operators. Second, the matching ordering operators compare values based on their binary integer values rather than letter by letter. Third, they propagate the original operand types in an expression, rather than forcing a conversion to boolean in the middle of the expression. This allows us to combine several operators in a single expression and still assign the result directly to an object of a bit or standard logic type.

The following example illustrates two of the above cases: the use of *don't care* symbols and assigning the expression result directly to a signal of a logic type. Prior to VHDL-2008, if we wanted to make a comparison using only certain bits of an operand,

Table 8.3 Relational operators

Operator	Operation	Result Type	VHDL Version
=	Equality	boolean	Any version
/=	Inequality	boolean	Any version
< <= > >=	Ordering	boolean	Any version
?=	Matching equality	Operand type or array element type	VHDL-2008
?/=	Matching inequality	Operand type or array element type	VHDL-2008
?< ?<= ?> ?>=	Matching ordering	Operand type or array element type	VHDL-2008

we had to use array slices. Also, because the result of the operation was a boolean, it took a conditional signal assignment (using the keyword when) to convert the result back to a logic value. The use of a matching operator solves both issues:

```
-- Match any address from 0x070000 to 0x07ffff:
ram_sel <= '1' when address(23 downto 16) = x"07" else '0'; -- Before VHDL-2008
ram_sel <= (address ?= x"07----");   -- After VHDL-2008
```

When performing a comparison between two std_ulogic operands using the matching equality operator (?=), the only possible results are 'U', 'X', '0', or '1'. The rules and the output table are summarized in figure 8.1.

When the matching equality operator ?= is applied to two std_logic_vectors or std_ulogic_vectors, the result is determined by creating an intermediate array and then applying the reduction operator and to this array. The intermediate array is determined by applying the ?= operator pairwise between the corresponding elements of the two input arrays. The end result will be '1' if and only if all elements match, according to the rules of figure 8.1.

One last point that may cause some surprise is how the predefined ordering operators work with arrays. Every time we create a scalar type (a numeric type or an enumeration), VHDL provides for free all the ordinary relational operators. For example, in an enumeration type, a < b is defined to be true if a is to the left of b in the type declaration.

When we create an array type whose element type is a scalar, VHDL also provides for free the ordinary relational operators. However, the behavior of these operators may or may not be what we want. For example, arrays of character-based elements (such as bit_vector or std_logic_vector) are first left-aligned and then compared as if they were strings, letter by letter. This may lead to some unexpected results, especially when the arrays have different sizes:

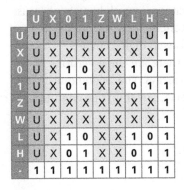

	U	X	0	1	Z	W	L	H	-
U	U	U	U	U	U	U	U	U	1
X	U	X	X	X	X	X	X	X	1
0	U	X	1	0	X	X	1	0	1
1	U	X	0	1	X	X	0	1	1
Z	U	X	X	X	X	X	X	X	1
W	U	X	X	X	X	X	X	X	1
L	U	X	1	0	X	X	1	0	1
H	U	X	0	1	X	X	0	1	1
-	1	1	1	1	1	1	1	1	1

Rules for the *matching equality* operator (?=) used with std_ulogic operands:

- A '-' matches any other symbol and returns '1'.
- A 'U' with any other symbol (except '-') returns 'U'.
- In the remaining cases, if any operand is 'X', 'W', or 'Z', the result is 'X'.
- The remaining symbols ('0', '1', 'L', 'H') are matched according to their equivalent logic levels and return a '1' or a '0'.

Figure 8.1
Rules and result table for the matching equality operator ?= with std_ulogic operands.

```
variable address: std_logic_vector(23 downto 0);
...
address := x"100000";
if (address > x"ff") ...  -- false with predefined ">" operator
if (address > x"ff") ...  -- true with overloaded ">" from numeric_std_unsigned
```

Note that this problem will only be noticed if we try to use the predefined ordering operators with types that do not have an inherent numeric meaning, such as std_logic_vector. If we use the package numeric_std_unsigned, then the predefined operators will be overloaded with their correct mathematical meaning. In any case, a better solution is to use the types signed and unsigned from package numeric_std, which were planned from the beginning for use in arithmetic operations.

8.1.4 Shift Operators

Shift operators are available for 1D arrays of boolean, bit, or standard logic types. The left operand is the array, and the right operand is an integer that specifies the shift amount. The return value is an array with the same subtype and range as the array operand. For simple shifts and rotates, the operators result in cleaner code than the alternatives using loops or manual concatenation:

```
reg := reg(REG_WIDTH-2 downto 0) & '0'; -- Using concatenation
reg := reg sll 1;                       -- Using shift operator
```

8.1.5 Adding Operators

Adding operators perform addition (+), subtraction (-), or concatenation (&). Addition and subtraction are predefined for any numeric type and have their conventional mathematical meaning. The concatenation operator (&) joins arrays or scalars end to end and returns an array. It can be used with any 1D array type, including user-defined types. The features provided by the concatenation operator are worth knowing because they can help us write code that is clearer and more concise. It is also well suited for low-level bit manipulation and other common operations in digital circuits.

For each array type, including user-defined types, there are three versions of the concatenation operator. The first one concatenates two arrays. The second one concatenates one array and one individual element, or vice versa. The third one concatenates two individual elements into an array.

```
"000" & "111"    -- "000111"    Array & Array --> Array
"000" & '1'      -- "0001"      Array & Element --> Array
'0' & "111"      -- "0111"      Element & Array --> Array
'0' & '1'        -- "01"        Element & Element --> Array
```

Concatenation can be useful to implement hardware structures such as FIFOs, buffers, and shift-registers. Compared to shift operators, an advantage of using concatenation is that we can shift the existing data and insert a new value into a register using a single operation:

```
-- Shift old values to the left and shift in a new value
shift_reg <= shift_reg(30 downto 0) & data_in;
```

Concatenation is not only for bits. Whenever we create an array type, the concatenation operator is implicitly defined for it. The array elements may be of any type, including integers, records, and even other arrays. In the following example, we create a record type (sample_type) and a type for arrays of that record (sample_array_type). Then we use concatenation to shift the entire array of records as if it were a shift register:

```
type sample_type is record
    voltage: float32;
    timestamp: timestamp_type;
end record;

type sample_array_type is array (natural range <>) of sample_type;

signal samples: sample_array_type(0 to 15);
...
new_sample <= (analog_reading, timestamp);

-- Shift old values to the right and shift in a new value
samples <= new_sample & samples(0 to 14);
```

Concatenation can be used to join multiple array slices into a single array. This can be used, for instance, to group several objects into a single source value in an assignment, helping reduce the number of statements:

```
pixel_value(31 downto 24) <= alpha;
pixel_value(23 downto 16) <= red;
pixel_value(15 downto 8) <= green;
pixel_value(7 downto 0) <= blue;

-- Same as the four assignments above
pixel_value <= alpha & red & green & blue;
```

Note that it is not possible to use concatenation on the left side of an assignment because the concatenation operator is a function, and as such it returns a value to be used in an expression. It does not make sense to have a plain value on the left side of an assignment. For a similar construct that can be used as the target in an assignment, use aggregates (see section 8.2.2).

Another common use for concatenation is to join two strings:

```
-- Build the full pathname for stimuli file
constant STIMULI_FILE_PATHNAME: string := TESTBENCH_ROOT & "stimuli.dat";
...
-- Create log message dumping contents of the data bus
write(output, LOG_HEADER & "data_bus: " & to_string(data_bus));
```

Concatenation is a versatile operation, but it is not the best tool for every job. For simple shifts and rotates, the shift operators are more concise. When the same name appears multiple times in a concatenation, an aggregate may be a better choice. Finally, long sequences of concatenations may be a sign that the code needs to be better thought out. The following example is a case where other methods would be more appropriate than concatenation:

```
-- Do not do this! Use numeric_std.resize or an aggregate instead
operand_ext := operand(7) & operand(7) & operand(7) & operand(7) & operand;

operand_ext := (operand, others => operand(7)); -- Ok, using an aggregate
operand_ext := resize(operand, 12);             -- Ok, using numeric_std.resize
```

8.1.6 Operator Overloading

Because operators are actually functions, they can be overloaded like any subprogram. The basics of overloading were introduced in section 7.2.3. Here we demonstrate how overloading is used with operators and introduce some characteristics that are unique to this kind of overloading.

Operator overloading allows us to define or redefine the behavior of VHDL operators. This includes changing the behavior of operators for standard types, overriding the default behavior provided for custom data types, or creating new behavior for custom types. Needless to say, changing the behavior of predefined or standard types is not recommended. Our readers have the right to assume that what they know about VHDL is also true for our code.

Providing new behavior or overriding the existing operations is especially useful for user-defined types and abstract data types (ADTs are introduced in section 13.4). Suppose we created a type for complex numbers based on a synthesizable floating-point type. The underlying data structure for this ADT could be a record:

```
type synth_complex_type is record
    re: float32;   -- Real part
    img: float32;  -- Imaginary part
end record;
```

To use objects of this type more naturally in expressions, we could overload the arithmetic operators. Here is how this could be done for the addition operator:

```
function "+"(left_op, right_op: synth_complex_type)
    return synth_complex_type is
begin
    return (left_op.re + right_op.re, left_op.img + right_op.img);
end;
```

The function defining the operator behavior is declared like any subprogram; the only visible difference is that the operator symbol must appear within double quotes. The parameter list should have a number of arguments that is compatible with the operator profile (one for unary or two for binary). Then the overloaded operator can be used in the code just like the predefined ones:

```
variable a, b, sum: synth_complex_type;
...
sum := a + b;
```

One thing to note in this example is that only the operator *symbol* was overloaded. We did not create a second version of an operation that already existed, which would require the use of overloading rules for disambiguation. The term *operator overloading* is used with both meanings: when there are overloaded subprograms and when only the operator symbol is overloaded.

There is one caveat when overloading the equality operator. For the *case* statement and for some internal uses by the simulator, the comparison is always made using the predefined equality operator, even if an overloaded version exists. This may cause unexpected results. Suppose we overload the equality operator for the float32 type to allow for some margin in the comparison:

```
function "="(left, right: float32) return boolean is begin
    -- Return true if values are close enough
    return (abs(left-right) <= 0.001);
end;
```

This could lead to diferent results depending on whether a comparison is made with an *if* or a *case* statement. In the following example, when we compare two floats using an *if*, the overloaded operator is used, and the result of the comparison is true. When we use a *case*, the implictly defined operator is used, and the comparison result is false.

```
constant F_100_0: float32 := to_float(100.0);
constant F_100_001: float32 := to_float(100.001);
...
-- Compare the two values using an 'if' statement:
if F_100_0 = F_100_001 then
    print("if ---> values are equal");
else
    print("if ---> values are different");
end if;

-- Compare the two values using a 'case' statement:
case F_100_0 is
    when F_100_001 => print("case -> values are equal");
    when others => print("case -> values are different");
end case;

-- Output:
--    if ---> values are equal
--    case -> values are different
```

Another difference in behavior is that overloaded versions of the logical operators and, nand, or, and nor are not evaluated in the short-circuit manner described in section 8.1.2. For overloaded operators, all operands are evaluated before being passed as arguments to the function.

Finally, because the list of available operators is set in the VHDL grammar, it is not possible to create new operators. If we want to implement new behavior, we must use one of the available operators or a function named with a regular identifier.

8.2 Operands

As seen in the previous sections, expressions are composed of operators and operands. Operators are functions identified by special symbols or keywords. Operands are the values used by operators in expressions. Many different constructs can be used as operands: objects, attributes, literals, aggregates, function calls, qualified expressions, type casts, allocators, and expressions enclosed in parentheses. This section reviews the different kinds of operands available in VHDL.

8.2.1 Literals

A *literal* is a value written directly in the code. VHDL allows several kinds of literals—namely, numeric literals, character literals, string literals, and bit-string literals. These categories were introduced in section 5.4. There is no need to know the categories by name, but it is a good idea to look at the examples in table 8.4 to know what forms are available for use in our code. The different kinds of literals have unique shapes because they need to be identified early in the analysis process (during lexical analysis) before any other context is available.

Numeric literals provide values of integer, floating-point, or physical types. The compiler uses the format to distinguish among these literals: a decimal point identifies floating-point literals, a unit identifies physical literals, and plain integer numbers identify integer literals. When we write a number without a decimal point in the code, it can only be used where an integer value is acceptable.

Numerical literals are interpreted in base 10 by default, but we can write values in other bases using based literals. For instance, we could write the decimal value 51 as 16#33# in hexadecimal base or 2#0011_0011# in binary. The allowed bases are from 2 to 16. Note that although the literal is written in binary or hexadecimal, it is still an integer value, so it would be an error to assign it to an object that is not of an integer type.

In VHDL, string literals are commonly used with two different purposes: to denote text or vectors of bits. However, the compiler is not allowed to look inside a string to determine its type—it must use only the context in which the string appears. This fact is important to understand the difference between a string literal and a bit-string literal. They are just two different ways to write a string in VHDL; visually in the code, the only difference is that a bit-string literal has a base specifier before the opening quote. When the source code is processed, bit-string literals get expanded automatically to regular strings, following certain rules that take into account the base specifier. In the simplest form, the base specifier is a single character (b, d, o, or h) denoting the binary, decimal,

Table 8.4 Kinds of literals in VHDL

Base Category	Subcategory	Examples
Numeric literal	Integer, decimal literal	51, 33, 33e1
Numeric literal	Integer, based literal	16#33#, 2#0011_0011#
Numeric literal	Real, decimal literal	33.0, 33.0e1
Numeric literal	Real, based literal	16#33.0#, 2#0011_0011.0011#
Numeric literal	Physical literal	15 ns, 15.0 ns, 15.0e3 ns
Enumeration literal	Identifier	true, ESC, warning, failure
Enumeration literal	Character literal	'a', ' ', '0', '1', '½', '@', '''
String literal	String literal	"1100", "abc", "a", "0", "1", "123"
Bit-string literal	Bit-string literal	b"1100", X"7FFF", x"7fff", 16d"123"
Null literal	Null literal	null

octal, or hexadecimal base. The following examples show how three bit-string literals are expanded to the same string value:

```
d"123"       -- Expands to "1111011"
b"111_1011"  -- Expands to "1111011"
7x"7b"       -- Expands to "1111011"
```

The default number of bits in the expanded string depends on the base specifier. For binary, it has the same number of bits as the bit-string literal; for octal, it has three times the number of digits; for hex, four times the number of digits; and for decimal, the minimum number of bits needed to represent the value. It is also possible to specify the size of the expanded string explicitly by writing the number of bits in front of the base specifier. The resulting string will be padded or truncated accordingly, including sign extension if a signed base specifier (sx, so, or sb) is used. The unsigned base specifiers ux, uo, and ub are equivalent to the simple base specifiers x, o, and b. The following examples demonstrate the use of signed and unsigned base specifiers and fixed-length bit-string literals:

```
8d"123"       -- Expands to "01111011"
8b"111_1011"  -- Expands to "01111011"
8ub"111_1011" -- Expands to "01111011" (unsigned, no sign extension)
8sb"111_1011" -- Expands to "11111011" (signed, perform sign extension)
x"7b"         -- Expands to "01111011"
ux"7b"        -- Expands to "01111011" (unsigned, no sign extension)
8sx"b"        -- Expands to "11111011" (signed, perform sign extension)
```

Underscores are also allowed; they are stripped out automatically before the string is expanded:

```
x"07ff_ffff"   -- Expands to "00000111111111111111111111111111"
b"1100_0011"   -- Expands to "11000011"
d"150_000_000" -- Expands to "1000111100001101000110000000"
```

An important note about underscores is that they are not allowed in ordinary string literals that get assigned to objects of bit vector types because the enumeration types bit and std_ulogic do not include the underscore character. If we want to use underscores, then we must use a bit-string literal:

```
variable slv: std_logic_vector(7 downto 0);
...
slv := "1100_0011";    -- Error: Character '_' not in enumeration type bit
slv := b"1100_0011";   -- Ok, expands to "11000011", then gets assigned to slv
```

Bit-string literals written with the decimal base specifier are a good choice when we want to assign a value to a bit vector and still keep it easily readable:

```
constant HORIZONTAL_RESOLUTION: coord_type := d"800"; -- more readable than "1100100000"
constant PLAYER_X_POS: coord_type := d"640";          -- more readable than "1010000000"
```

8.2.2 Aggregates

An *aggregate* is a VHDL construct that combines separate values into a composite value—an array or a record. We can use an aggregate wherever a composite value is needed. Two common cases are to initialize an array or a record or provide the source value in an assignment to a composite object. Aggregates appear frequently in VHDL, so it is important to know how to identify them in the code and use them properly. As we will see in the examples, aggregates can be a great help in writing less verbose code.

The basic syntax of an aggregate is a list of comma-separated expressions enclosed in parentheses. We can write an aggregate using two different notations. The first one uses the order of the expressions to define the place of each element in the resulting composite object—the first expression will be the leftmost value in the target object, and so on. This notation is called *positional association*:

```
-- Agregate syntax using positional association:
(expression_1, expression_2, ...);
```

The following aggregate uses positional association. In the example, it is used to set the value of an entire array at once:

```
variable expected_outputs: real_vector(0 to 3);
...
-- An agregate using positional association:
expected_outputs := (1.0, 2.0, 9.0, 16.0);
```

The second association scheme is called *named association*, and it allows us to state explicitly the position of each expression in the resulting composite value:

```
(choices_1 => expression_1 , choices_2 => expression_2, ...  , others => expression_n )
```
                                            optional   optional

We can rewrite the previous example using named association. Here, we explicitly state an index for each value:

```
-- Right-hand side of assignment is an aggregate using named association:
expected_outputs := (0 => 1.0, 1 => 2.0, 2 => 9.0, 3 => 16.0);
```

Named association provides a lot of flexibility when writing an aggregate. Because the indices are explicitly given, we can write the expressions in any order. The `choices` field uses the same syntax as the choices in a `case` statement (see section 10.3). There are four possible ways to write a `choices` field: using a single value (as in the previous example), using a list of values separated by vertical bars (|), using a range, or using the keyword `others`. In an aggregate, the keyword `others` means "any remaining elements." This notation is useful when many elements in the array have the same value.

Here are some examples of array aggregates using named association:

```
signal status_register: std_logic_vector(31 downto 0);
-- Same as: status_register <= b"00000000_00000000_00000000_00000000";
status_register <= (others => '0');


variable pascals_triangle: integer_vector(0 to 14);
-- Same as: pascals_triangle := (1, 1, 1, 1, 2, 1, 1, 3, 3, 1, 1, 4, 6, 4, 1);
pascals_triangle := (4 => 2, 7 to 8 => 3, 11|13 => 4, 12 => 6, others => 1);


variable float_sequence: real_vector(0 to 7);
-- Same as: float_sequence := (1.0, 2.0, 2.0, 3.0, 3.0, 3.0, 0.0, 0.0);
float_sequence := (0 => 1.0, 1|2 => 2.0, 3 to 5 => 3.0, others => 0.0);


signal byte: std_logic_vector(7 downto 0);
-- Same as: byte <= "11110000";
byte <= (7 downto 4 => "1111", 3 downto 0 => '0');
```

This last example shows that when a range is associated with an array (the bit-string literal `"1111"` in the example), the entire array is copied to the specified range. However, when a range is associated with a scalar value (the character literal `'0'` in the example), this value is repeated through the entire range.

There are certain restrictions regarding the use of named and positional association in aggregates. The first one is that it is not possible to mix the two in a single array aggregate (however, this is allowed in record aggregates). Moreover, if the keyword *others* is used, then it must appear last. Therefore, the following would be illegal:

```
byte <= ('1', 6 downto 0 => '0');   -- Error: mixed named/positional associations
byte <= (others => '0', 7 => '1'); -- Error: 'others' must come last
```

Another restriction is that when the keyword *others* is used, it must be possible for the compiler to infer the size and range of the aggregate. This can be done in a few different ways. For instance, the range can be inferred from the target object in an assignment:

```
variable byte: std_logic_vector(7 downto 0);
byte := (others => '0'); -- Ok, aggregate range is determined from object 'byte'
```

When the range cannot be inferred from an object, we can use a qualified expression:

```
subtype byte_type is std_logic_vector(7 downto 0);
if byte = byte_type'(others => '0') then -- Ok, range determined from 'byte_type'
```

Finally, in the prior examples, we could replace the keyword others with an explicit range, using the range of the target object obtained with an attribute:

```
if byte = (byte'range => '0') then ... -- Ok, range is given in the aggregate
```

Now let us look at some record aggregates, which can also use positional or named association. For the record type rgb_color_type defined next, these are some examples of record aggregates:

```
type rgb_color_type is record
    red, green, blue: integer range 0 to 255;
end record;

-- A record aggregate using positional association:
constant OXFORD_BLUE: rgb_color_type := (6, 13, 69);

-- A record aggregate using named association:
constant AQUAMARINE: rgb_color_type := (red => 97, green => 232, blue => 164);

-- A record aggregate using only the keyword others:
constant BLACK: rgb_color_type := (others => 0);

-- A record aggregate using named association and the keyword others:
constant NAVY_BLUE: rgb_color_type := (blue => 128, others => 0);

-- Mixed named and positional associations
-- (this is valid only for record aggregates, not for array aggregates):
constant SCARLET: rgb_color_type := (255, green => 33, others => 164);
```

In all the examples seen so far, for both arrays and records, the aggregates used only literal values. However, it is possible to use any expression in an aggregate. This includes signals, variables, function calls, and even other aggregates:

```
-- Concatenate the literal value "001000" with signals rs, rt, and immediate:
addi_instruction <= ("001000", rs, rt, immediate);
```

```
-- Calculate the 4 basic operations for a and b and put results in a record:
calculator_results := (add => a+b, sub => a-b, mul => a*b, div => a/b);
```

Since VHDL-2008, it is possible to mix array slices with individual elements in an array aggregate:

```
signal sign_bit: std_ulogic;
signal nibble: std_ulogic_vector(3 downto 0);
signal byte: std_ulogic_vector(7 downto 0);
...
byte <= (sign_bit, '0', nibble, "00");   -- Legal only in VHDL-2008
```

There is one case in which it is not possible to use positional association: when the aggregate has a single element. To explain this restriction, we need to understand the syntax of parentheses and expressions. Just like in mathematical formulas, it is possible to surround a value or an expression with as many pairs of parentheses as we like. For example, the literal number 3 and the expression (3) have the same meaning in the code. Therefore, it is not possible for the compiler to treat this last form as an aggregate. The solution is to always use named association when we want to write an aggregate with a single element:

```
variable single_value: integer;
variable integer_array: integer_vector(0 to 0);
...
single_value := 3;              -- Ok, literal is the integer 3
integer_array:= (3);            -- Error, literal is still the integer 3
integer_array:= ((3));          -- Error, literal is still the integer 3
integer_array:= (0 => 3);       -- Ok, literal is the array ( 3 )
integer_array:= (others => 3);  -- Ok, literal is the array ( 3 )
```

Aggregates as Targets in Assignments   In all the examples so far, aggregates appeared only on the right-hand side of an assignment. However, we can use aggregates as assignment targets as well. In this case, the aggregate must be composed exclusively of variable names (in the case of a variable assignment) or signal names (in the case of a signal assignment). This can be used to collapse several assignments into a single assignment or to split an array into separate objects.

Suppose we need to split a word representing an instruction into the several fields that compose an instruction in a CPU. Given the following declarations,

```
signal instruction: std_logic_vector(31 downto 0);
signal opcode: std_logic_vector(5 downto 0);
signal register_s: std_logic_vector(4 downto 0);
```

```
signal register_t: std_logic_vector(4 downto 0);
signal immediate: std_logic_vector(15 downto 0);
```

one way to break down an instruction into the basic fields specified earlier would be to use several assignments:

```
opcode <= instruction(31 downto 26);
register_s <= instruction(25 downto 21);
register_t <= instruction(20 downto 16);
immediate <= instruction(15 downto 0);
```

However, using an aggregate as target, we can do it in a single assignment:

```
(opcode, register_s, register_t, immediate) <= instruction;
```

As another example, suppose we need to extract the carry bit from a sum. For this task, we could take advantage of the VHDL-2008 syntax enhancements that allow mixing scalar values, arrays, and array slices in a single aggregate. Given the following declarations,

```
signal a, b, sum: unsigned(WIDTH-1 downto 0);
signal carry_out: std_ulogic;
```

we could extract the carry bit from a sum by extending each operand by one bit and then assigning the result to an aggregate:

```
(carry_out, sum) <= ('0' & a) + ('0' & b);
```

8.2.3 Names

A name is a handle for an item declared in the code. In VHDL, anything that we declare and name is called a *named entity*. Here, the term *entity* is used in its general sense and should not be confused with other uses in VHDL. A named entity can be almost anything, including an object (a constant, variable, signal, or file), an architecture, a subprogram, a type, a package, or a design entity. VHDL uses several kinds of names; here we survey the ones that are most used in practice.

A *simple name* is the identifier given to a named entity at its declaration or an alias for this identifier. The rules for creating valid identifiers in VHDL are simple:

1. Identifiers can only use letters, digits, and underscores.
2. Identifiers are not case sensitive.
3. Identifiers must begin with a letter.
4. An underscore is not allowed at the end of an identifier or adjacent to another underscore.

A *selected name* is a name written using the dot notation `prefix.suffix`. It is used to specify elements in a record or to access named entities that are nested inside other entities. In the following example, the prefix `instruction` is the name of a record object, and the suffix `src_reg` denotes a record element:

```
-- Denotes element named 'src_reg' inside record object named 'instruction':
instruction.src_reg
```

Another typical use for a selected name is to refer to named entities declared inside a package. For example, suppose we want to use procedure `finish`, declared in package env, which is compiled into library `std`. We can start a selected name using as a prefix a name that is directly visible, such as library `std` (which is always visible by default). Then we use as a first suffix the name of a library unit (package env) and as a second suffix a name from inside the package (procedure `finish`):

```
std.env.finish;  -- library 'std', package 'env', procedure 'finish'
```

In a selected name, it is possible to use the suffix `all` to denote all named entities inside the entity denoted by the prefix. If the prefix is a design library, then the suffix `all` denotes all primary units in the library. If the prefix is a package, then the suffix denotes all named entities declared immediately within the package. As an example, we can use a selected name in a *use* clause to make all named instances from a package directly visible in our code. In the situation of the previous example, we could write:

```
use std.env.all;
```

Then we could call the procedure by its simple name:

```
finish;
```

An *indexed name* denotes an element of an array. It consists of a prefix that specifies an array, followed by an expression enclosed in parentheses:

```
type matrix_type is array (natural range <>, natural range <>) of real;
variable matrix: matrix_type(1 to 3, 1 to 3); -- matrix is a simple name
...
matrix(1, 1) -- matrix(1, 1) is an indexed name
```

A *slice name* denotes a subarray: a 1D array that is a contiguous sequence of elements from another 1D array. The subarray has the same class as the original array; for instance, if the original array is a signal, then the subarray is also a signal.

```
variable instruction: std_ulogic_vector(31 downto 0); -- instruction is a simple name
...
instruction(15 downto 0);  -- instruction(15 downto 0) is a slice name
```

The kinds of names we have seen so far can also use a function call as a prefix. This means that we can chain the parentheses of a slice or indexed name with the parentheses of a function call, as the following example demonstrates. In the declaration of constant three_ones, a call to the function ones(4) is chained with the slice range (1 to 3), so that only the first three elements are assigned to the constant.

```
-- Return an integer vector with N elements, all set to 1.
-- Ex: ones(3) --> (1, 1, 1)
function ones(N: integer) return integer_vector is begin
    return (1 to N => 1);
end;

-- Create a vector with (1, 1, 1, 1):
constant four_ones: integer_vector(1 to 4) := ones(4);

-- Create a vector with (1, 1, 1, 1), and then take a
-- 3-element slice from that vector:
constant three_ones: integer_vector(1 to 3) := ones(4)(1 to 3);

-- Result:
--   four_ones = (1, 1, 1, 1)
--   three_ones = (1, 1, 1)
```

External names allow us to reach objects belonging to other units in the design hierarchy, sidestepping the normal scope and visibility rules. We can use an external name to refer to any constant, signal, or shared variable in a design. To reach an object, we must provide a path specifying where it is located in the design hierarchy besides the object class and subtype. An external name has the form:

```
<< object_class path: subtype >>
```

The class must be signal, variable, or constant, and the path may be an absolute path (starting at the root of the hierarchy) or a relative path (starting from the current design unit). Absolute paths start with a dot, followed by the name of the top-level entity in the hierarchy, and then by labels of instantiated components. The tool will trust us during analysis that this information is correct and that the object exits. During elaboration, this information will be verified.

External names are used in testbenches for *white-box testing*—a kind of test where we can read or control the internal state of a design. As an example, suppose we need to write a test for an FSM. Because the current state of an FSM is usually hidden inside a design unit, the only way to test it would be indirectly, by interpretation of its outputs. Alternatively, we could read the state directly using an external name:

```
assert << signal duv.main_controller.current_state: state_type >> = coin_return
    report "Unexpected FSM state";
```

External names are a powerful tool; however, they must be used judiciously because they go against good design principles such as modularity and information hiding. They also introduce dependencies between the test and the internal structure of a design, which should be private.

A final point to note about names is that the most important thing in a name is choosing it well. Choosing good names is one of the most important tasks of a programmer. Chapter 16 presents guidelines and recommendations to help you choose great names for use in your designs.

8.2.4 Aliases

An *alias* is an alternative name for an existing named entity or for part of a named entity. Aliases can be helpful in making the code more readable and concise. They are also efficient in simulation because they are simply a handle to objects or entities that already exist.

The simplest way to declare an alias has the following format:

```
alias new_name is original_name;
```

This simply declares another name for an existing named entity, maintaining all the characteristics of the original entity. A common use case is to provide a shorter name for an external name. As seen in the previous topic, external names allow us to refer to objects that exist inside other units in the design hierarchy. Because they must include the object class, type, and path, such names tend to be long and cumbersome:

```
assert << signal duv.main_controller.current_state: state_type >> = coin_return
    report "Unexpected FSM state";
```

We can use an alias to provide a shorter handle for the object:

```
alias duv_state is << signal duv.main_controller.current_state: state_type >>;
...
assert duv_state = coin_return report "Unexpected FSM state";
```

Another common use is to create an alias for part of a composite object. In this case, the original name will be an indexed name or a slice name. This can be used for convenience or to document the meaning of a specific bit or field. In many cases, we can move information from a comment to an alias, making the code more self-documenting. In the following example, an alias named sign_bit is declared for the most significant bit of signal alu_result:

```vhdl
signal alu_result: unsigned(31 downto 0);
alias sign_bit is alu_result(31);
```

As another example, in the following code, an alias is created for each field of a CPU instruction:

```vhdl
-- Break down an instruction using aliases for its four basic fields
signal instruction: std_ulogic_vector(31 downto 0);
alias opcode is instruction(31 downto 26);
alias register_s is instruction(25 downto 21);
alias register_t is instruction(20 downto 16);
alias immediate is instruction(15 downto 0);
```

Another use for an alias is to provide a different view of an object, as if it had a different subtype or other constraints. This requires a different syntax, in which we specify a subtype in the alias declaration:

```vhdl
alias new_name: new_subtype is original_name;
```

This can be used, for instance, to normalize the range of an array so that it has known bounds and a known direction. In a subprogram, it may be difficult to work with array parameters if they can have any range. A possible solution is to normalize the arguments so they have a uniform indexing scheme. This can be done using aliases. In the following example, the parameters vector and mask are normalized to the same ascending range, starting from one:

```vhdl
function apply_bitmask(vector: integer_vector; bitmask: bit_vector)
    return integer_vector
is
    -- Create aliases for the operands so they can be indexed from 1 to N
    alias normalized_vector: integer_vector(1 to vector'length) is vector;
    alias normalized_bitmask: bit_vector(1 to bitmask'length) is bitmask;
    variable masked_vector: integer_vector(1 to maximum(vector'length, bitmask'length));
begin
```

```
    assert vector'length = bitmask'length
        report "length mismatch in apply_bitmask" severity failure;

    -- Now we can index the array elements without risking an access out of bounds...
    for i in masked_vector'range loop
        masked_vector(i) := normalized_vector(i) when normalized_bitmask(i) else 0;
    end loop;

    return masked_vector;
end;
...
apply_bitmask( (1, 2, 3, 4, 5), "10101" );  -- (1, 0, 3, 0, 5)
```

All the previous examples showed *object* aliases, which are aliases to constants, signals, variables, or files. The other kinds of aliases are called *nonobject* aliases. In VHDL, almost any named entity can be aliased. In fact, only labels, loop parameters, and generate parameters cannot be aliased.

Nonobject aliases are used similarly to object aliases, except for subprograms and enumeration literals, which require a signature in the alias declaration. When we write an alias for a subprogram, we must provide a signature describing the types of its parameters and, for a function, its return value. A *signature* is a VHDL construct that lists within square brackets the types of the parameters and return value of a subprogram. The simplified syntax for a subprogram alias is:

```
alias new_name is original_name [type_1, type_2, ... return type_n];
```

For example, suppose we want to create an alias for each version of subprogram to_integer defined in numeric_std. The motivation is that the package declares two overloaded versions, one for signed and the other for unsigned parameters. Because there are two possible meanings, we cannot call the function with a string literal as argument:

```
int := to_integer("11111111"); -- Error: Subprogram "to_integer" is ambiguous
```

However, we can create two aliases with different signatures:

```
alias int_from_sig is to_integer[signed return integer];
alias int_from_uns is to_integer[unsigned return integer];
```

Then each aliased subprogram can be used unambiguously with a string literal:

```
int := int_from_sig("11111111");
int := int_from_uns("11111111");
```

8.2.5 Other Kinds of Operands

We close this section with an overview of the other kinds of operands that can be used in expressions.

A *function call* is an operation that invokes the execution of a function, obtains a return value, and puts this value in place of the original call. A function call is a kind of expression. Therefore, it cannot exist by itself in the code; it can only appear as part of a statement or declaration.

A *qualified expression* is a notation that allows us to explicitly state the type of an expression. Sometimes, especially when using literals, an expression can have more than one interpretation, each one yielding a value of a different type. This may cause trouble for the compiler as it tries to choose one of the possible meanings. In these cases, we can give the compiler a hint by using a qualified expression: we write the intended type for the expression, followed by a tick (') symbol and the expression enclosed within parentheses. In the following example, a qualified expression is used to disambiguate between the signed and unsigned versions of a subprogram:

```
int := to_integer( "11110000" ); -- Error: subprogram "to_integer" is ambiguous
int := to_integer( signed'("11110000") );    -- Ok, use signed version
int := to_integer( unsigned'("11110000") );  -- Ok, use unsigned version
```

For more details about the use of qualified expressions, see *Type Qualification* in section 13.3.

A *type cast* or *type conversion operation* performs type conversions between closely related types, such as 1D arrays created from the same element type. The syntax of a type cast is the intended type name followed by an expression enclosed within parentheses:

```
variable value_slv: std_logic_vector(31 downto 0);
variable value_uns: unsigned(31 downto 0);
...
-- A type cast from unsigned to std_logic_vector:
value_slv := std_logic_vector(value_uns);
```

More details about the use of type casting are found in section 13.3.

An *allocator* is a special construct that creates an object when evaluated and returns a value of an access type that can be used to manipulate the new object. The value returned by an allocator must be stored in a variable so that the object can be accessed later; it is not possible to use the return value for any other operation. The syntax for an allocator is the keyword new followed by a type name or a qualified expression:

```
new integer  -- Create a new object of type integer with the default value for the type
new integer'(10)  -- Create a new object of type integer initialized with the value 10
```

Allocators and access types are discussed in section 12.2.

Finally, the last kind of operand that can be used in an expression is another expression within parentheses. This may seem trivial, but this feature allows us to create arbitrarily complex expressions using operators that otherwise could not be mixed, such as the logical operators and and or. To use an expression as an operand, it must be enclosed in parentheses. This effectively turns the operand into a subexpression, which will be evaluated independently and replaced with a single value.

8.3   Attributes

Attributes are a mechanism to retrieve special information about named entities. We can think of attributes as special functions whose first argument is always a named entity and whose return type may be a constant value, a signal, a type, a range, or information from the simulator. For example, the attribute 'event can be applied to a signal and returns information from the simulator. Attribute 'range can be applied to an array and returns a range. Attribute 'left can be applied to a type and returns a value. None of these would be possible with ordinary functions.

Attributes are used in the code by writing a name followed by the tick (') symbol and the identifier for the attribute:

```
named_entity'attribute_identifier
```

For attributes that require an argument, the identifier may be followed by an expression within parentheses:

```
named_entity'attribute_identifier(argument)
```

Here are two examples of how attributes are used in the code:

```
-- Result is true if clock value has changed in the current simulation cycle:
clock'event

type bitmap is array (1 to 16, 1 to 32) of color;

-- Result is 32 (length of the second array dimension):
bitmap'length(2)
```

There are at least three ways to classify attributes: by return value, by the kind of named entity to which they can be applied, and as predefined or user-defined.

Attributes can be applied to all named entities. In practice, however, the most important cases are predefined attributes that can be applied to scalar types, array types, array objects, and signals. These categories are discussed in this section.

As for the kind of return information, an attribute can return a constant value, type, subtype, range, function, or signal. Attributes that return a value can be used as constants in the code. Attributes that return a range are useful for creating generic code, such as loops that iterate over all the elements of an array regardless of its size. Attributes that return a signal cause the creation of implicit signals managed by the simulator. We will see examples of all those cases when we discuss each attribute in the next topics. First, however, we must discuss the difference between predefined and user-defined attributes.

Predefined and User-Defined Attributes   Another way in which attributes can be classified is as predefined or user-defined. *Predefined attributes* are specified in the VHDL LRM and are available for use in any model, with the provision that not all predefined attributes are synthesizable. The predefined attributes available in VHDL are presented in the next topics of this section and are listed in tables 8.5 to 8.8.

User-defined attributes are constants that can be associated with named entities to provide additional information about them. We can create user-defined attributes of any type and associate them with any entity; however, their main use is to provide information about a model to a VHDL tool. We should check the tool documentation for the attributes it understands. In the following example, the attribute ram_init_ file informs to the synthesis tool the name of a file with the initial contents for a RAM. The named entity associated with this attribute is an array signal called ram:

```
type memory_type is array (0 to 1023) of unsigned(31 downto 0);
signal ram: memory_type;
attribute ram_init_file: string;
attribute ram_init_file of ram: signal is "ram_contents.rif";
```

The category of user-defined attributes includes a subcategory, *synthesis-specific* attributes—standardized attributes that help a synthesis tool in interpreting and implementing a VHDL design. Synthesis-specific attributes are specified in IEEE Std 1076.6[1] and are covered in section 19.6.1.

The next topics introduce the different categories of predefined attributes, organized by the kind of named entity to which they can be applied.

Predefined Attributes of Scalar Types   Certain attributes are applied directly to the name of a type or subtype. Such attributes provide information about the type (such

Table 8.5 Predefined attributes of types

Attribute	Applied to	Return Type	Return Value
T'**left**	Scalar type	*T*	Leftmost value of *T*
T'**right**	Scalar type	*T*	Rightmost value of *T*
T'**low**	Scalar type	*T*	Smallest value of *T*
T'**high**	Scalar type	*T*	Largest value of *T*
T'**ascending**	Scalar type	boolean	True if range is ascending
T'**image**(*v*)	Scalar type	string	String representation of *v*
T'**value**(*s*)	Scalar type	T	Value of *T* represented by *s*
T'**pos**(*v*)	Discrete or physical type	integer	Position of v in *T*
T'**val**(*i*)	Discrete or physical type	*T*	Value of *T* at position *i*
T'**succ**(*v*)	Discrete or physical type	*T*	Value of *T* at position *i*+1
T'**pred**(*v*)	Discrete or physical type	*T*	Value of *T* at position *i*-1
T'**leftof**(*v*)	Discrete or physical type	*T*	Value of *T* one position to the left of *v*
T'**rightof**(*v*)	Discrete or physical type	*T*	Value of *T* one position to the right of *v*
T'**base**	Any subtype	Type	The base type of *T*

as its limits and direction) or about a value belonging to that type (such as its position in an enumeration or textual representation). Note that these attributes cannot be applied to *objects* of the type; they are always applied to the type itself. The only case in which they can return information about an object is when the attribute accepts a value within parentheses, which can be an object.

Table 8.5 summarizes the attributes that can be applied to types and subtypes. In the table, *T* is a type name, *v* is a value of that type, *i* is an integer, and *s* is a string. Note that some of the attributes can be applied to any kind of scalar types (numeric and enumeration types), whereas others can be applied only to discrete types (enumeration and integer types) or physical types.

Here are some usage examples of the predefined attributes of scalar types. The attributes are applied to the enumeration types `quaternary_digit` and `countdown`, defined as follows:

```
type quaternary_digit is (zero, one, two, three);
subtype countdown is quaternary_digit range two downto zero;
```

```
quaternary_digit'left       --> zero     The leftmost value of the type
quaternary_digit'right      --> three    The rightmost value of the type
countdown'left              --> two      The leftmost value of the subtype
countdown'right             --> zero     The rightmost value of the subtype
quaternary_digit'low        --> zero     Lowest value (leftmost for ascending)
quaternary_digit'high       --> three    Highest value (rightmost for ascending)
```

```
quaternary_digit'ascending      --> true      Range is specified with 'to'
countdown'ascending             --> false     Range is specified with 'downto'

quaternary_digit'image(two)     --> "two"     String representation of value two
quaternary_digit'value("two")   --> two       Value with string representation = "two"

quaternary_digit'pos(one)       --> 1         Position of value one in the type
countdown'pos(one)              --> 1         Position of value one in the base type
quaternary_digit'val(1)         --> one       Value at position 1 in the type
countdown'val(1)                --> one       Value at position 1 in the subtype

quaternary_digit'succ(zero)     --> one       Value that comes after zero in the type
quaternary_digit'pred(three)    --> two       Value that comes before three in type
quaternary_digit'succ(three)    --> Error     'succ of value three is illegal

quaternary_digit'leftof(one)    --> zero      Value to the left of one in the type
quaternary_digit'rightof(one)   --> two       Value to the right of one in the type
countdown'leftof(one)           --> two       Value to the left of one in the subtype
countdown'rightof(one)          --> zero      Value to the right of one in subtype
```

Attributes of scalar types are useful for writing type conversion functions and generic code without resorting to constants or duplicating information. For example, the attributes 'left, 'right, 'low, and 'high can be used to test the limits of a type without resorting to explicitly declared constants. Attributes 'pos and 'val can be used to convert from a type to an integer and vice versa. The attribute 'pos returns the index of an enumeration value in its type declaration, starting at zero for the leftmost element.

One thing to note is that the attributes 'pos and 'val refer to the position of a value in the base type, not in the subtype. In the prior example, the value of the attribute 'pos(one) returns 1 for both the type and subtype, even though quaternary_digit is defined from zero to three and countdown from two down to zero.

Finally, although the attribute 'image can be used to convert a value to a string, a better approach is to use the function to_string, which is automatically defined for every scalar type that we create. This function does not require us to write the type name, making the code less verbose, and it is available for most types commonly used in VHDL models, making the code more uniform.

Predefined Attributes of Array Types and Objects   Certain attributes can be applied directly to an array type or array object. Most of these attributes provide information about the dimensions of an array. Table 8.6 summarizes such attributes. In the table, A means an array type, object, or slice, and d is an integer corresponding to one of the array dimensions (starting from one). If d is omitted, then it is assumed to be one, and the attribute can be used without the parentheses.

Table 8.6 Predefined attributes of array types and array objects

Attribute	Applied to	Return Type	Return Value
A'**left**(d)	Array type or object	Type of dimension d	Left bound of dimension d
A'**right**(d)	Array type or object	Type of dimension d	Right bound of dimension d
A'**low**(d)	Array type or object	Type of dimension d	Lower bound of dimension d
A'**high**(d)	Array type or object	Type of dimension d	Upper bound of dimension d
A'**range**(d)	Array type or object	Range of dimension d	Range of dimension d
A'**reverse_range**(d)	Array type or object	Range of dimension d	Reverse range of dimension d
A'**length**(d)	Array type or object	integer	Size (# of elements) of dimension d
A'**ascending**(d)	Array type or object	boolean	True if range of dimension d is ascending
A'**element**	Array type or object	A subtype	Subtype of elements of A

Here are some usage examples of the attributes of array types and objects:

```
type matrix_type is array (positive range <>, positive range <>) of integer;
variable matrix: matrix_type(3 to 5, 1 to 2);
```

```
matrix'left         --> 3
matrix'right        --> 5
matrix'left(1)      --> 3 (same as matrix'left)
matrix'left(2)      --> 1
matrix'length       --> 3 (same as matrix'length(1))
matrix'length(1)    --> 3
matrix'length(2)    --> 2
matrix'range(1)     --> 3 to 5
matrix'range(2)     --> 1 to 2
```

Array attributes are important for writing generic code using arrays. The most common case is to write a loop that iterates through all the elements of an array:

```
variable integer_seq: integer_vector(1 to 10);
...
for i in integer_seq'range loop
    integer_seq(i) := calculate_some_value_based_on_i(i);
end loop;
```

This is especially useful when writing subprograms that take arrays as parameters. In the subprogram declaration, we can leave the parameter unbounded and then use array attributes in the subprogram body to perform calculations and to check the array

Table 8.7 Predefined attributes of signals

Attribute	Applied to	Return Type	Return Value
S'delayed(t)	Any signal	Base type of S	A signal equivalent to S but delayed t time units
S'stable(t)	Any signal	boolean	A signal that is true if S had no events in the last t time units
S'quiet(t)	Any signal	boolean	A signal that is true if S had no transactions in the last t time units
S'transaction	Any signal	bit	A signal of type bit that toggles on every transaction of S
S'event	Any signal	boolean	True if S had an event in the current simulation cycle
S'active	Any signal	boolean	True if S had a transaction in the current simulation cycle
S'last_event	Any signal	time	Time interval since last event on S
S'last_active	Any signal	time	Time interval since last transaction on S
S'last_value	Any signal	Base type of S	Previous value of S just before its last event
S'driving	Any signal	boolean	False if driver in the surrounding process is disconnected
S'driving_value	Any signal	Base type of S	Value contributed to S by the surrounding process

limits. The subprogram will be usable with arrays of any dimension because the array parameters will get their sizes from the actual objects used as arguments in the subprogram call.

As a final remark, array attributes never depend on the contents or value of an array. Even when applied to an object, the return values are constants that depend on the type and subtype only.

Predefined Attributes of Signals Certain attributes are applied directly to signals. Contrary to the other predefined attributes, signal attributes return dynamic information that depends on the current value of an object. Signal attributes have several uses, including to provide synchronization between processes, monitor changes in a signal, and describe clock edges. Table 8.7 summarizes the attributes that can be applied to a signal. In the table, S is any signal, and t is an optional time value (assumed to be zero if omitted).

Signal attributes can be divided into those that return a value and those that return a signal. The attributes that return a signal are 'delayed, 'stable, 'quiet, and 'transaction. When one of these attributes is used in the code, the simulator creates a new implicit signal that is updated automatically. These attributes can be used as any other signals existing in the code (e.g., in the sensitivity list of a process).

The attribute 'transaction produces a signal of type bit that toggles every time a transaction occurs on the original signal. This is useful in situations where we want to trigger the execution of a process whenever an assignment is made, and not only when the signal value changes. If we use a signal name in the sensitivity list, then the process is triggered only when the signal value changes. In contrast, if signal_name'transaction is used, the process is executed whenever the signal gets assigned to, even when its value does not change. This can be useful in behavioral models and testbenches because it allows a single signal to be used to transport data and as a control mechanism. This kind of use is demonstrated in the examples of chapter 20. A final note about the 'transaction attribute is that its actual value should not be used because its initial value is not specified by the language.

Signal attributes that return a value can be used as regular values in the code. The value returned by the attribute can change in the course of a simulation or with the value of the original signal.

The attribute 'driving_value has a particular use. In versions of VHDL prior to 2008, it was not allowed to read the value of an *out*-mode port inside an architecture. However, there are no restrictions to the use of an attribute on those ports. Because 'driving_value returns the value that a process assigns to a signal, it can be used for this same purpose, and we can obtain the signal value without reading from the port explicitly:

```
process (clock, reset) begin
    if reset = '1' then
        count <= (others => '0');
    elsif rising_edge(clock) then
        --count <= count + 1;                -- Illegal prior to VHDL-2008
        count <= count'driving_value + 1;   -- Legal since VHDL-93
    end if;
end process;
```

The only signal attributes that are expressly supported for synthesis are 'event and 'stable, which can be used to denote clock edges. If 'stable is used, then the optional delay argument is ignored. In any case, for better portability and readability, it is recommended to use the predefined functions rising_edge(S) and falling_edge(S) to denote signal edges.

Predefined Attributes of Named Entities   This category of predefined attributes provides textual information about the name and path of a named entity in the design hierarchy. This information can be useful during debugging and in testbenches. Table 8.8 summarizes such attributes.

Table 8.8 Predefined attributes of named entities

Attribute	Applied to	Return Type	Return Value
E'simple_name	Named entity	string	The simple name of the named entity *E*
E'instance_name	Named entity	string	Hierarchical path of *E* in the elaborated design, including the names of instantiated design entities
E'path_name	Named entity	string	Hierarchical path of *E* in the elaborated design, excluding the names of instantiated design entities

Here is an example of the strings provided by such attributes:

```
-- Top-level entity where the entity containing signal 'result' is instantiated
entity alu is
end;


architecture behavioral of alu is
begin
    adder_inst: entity work.adder;
end;

-- Instantiated entity containing signal 'result'
entity adder is
end;


architecture rtl of adder is
    signal result: bit_vector(31 downto 0);
begin
    process begin                      -- Examples of strings returned by the attributes:
        report result'simple_name;   --> "result"
        report result'path_name;     --> ":alu:adder_inst:result"
        report result'instance_name;--> ":alu(behavioral):adder_inst@adder(rtl):result"
        wait;
    end process;
end;
```

The main use for these constructs is to help locate error messages in designs where a named entity is instantiated multiple times. If the entity is instantiated, say, inside a *for-generate* statement, then each path will include the number of the generate parameter.

8.4   Guidelines and Recommendations for Operators, Operands, and Attributes

Prefer Human-Readable Literals for Numeric Values   Even if an object is a vector of bits (including `bit_vector`, `std_logic_vector`, unsigned, and others), we do not need to specify its value using binary literals. Consider the example:

```
variable player_x_pos: unsigned(10 downto 0);
...
player_x_pos := "01010000000";
```

In this example, although the data type is a vector of bits, the object has a numeric meaning. Therefore, we can make the code more readable if we use literals that are easily recognizable by humans. Adding a comment with the decimal value would help, but this is a form of duplicate information, and comments are never guaranteed to be up to date with the code. A better approach is to use a bit-string literal with a decimal base specifier:

```
player_x_pos := 11d"640";
```

This makes the value more evident, but it still has one problem: the length of the binary value is hardcoded in the literal value, which may cause problems if the size needs to change in the future. A better solution would be to use a length-independent value. One alternative is to use an integer type instead of a vector-based type; then we could simply write the integer literal 640 in the assignment. Another possibility, without changing the object type, is to use a conversion function:

```
player_x_pos := to_unsigned(640, player_x_pos'length);
```

or

```
player_x_pos := to_unsigned(640, player_x_pos);
```

These two alternatives are equivalent. In the second example, the object player_x_pos is used inside the routine only for the purpose of determining the width of the return value. In any case, all the alternatives shown are more readable than the binary string literal.

Use Aggregates to Make the Code Less Verbose   As seen in section 8.2.2, there are many opportunities to use aggregates to make the code less verbose. Here we show two of the most common cases. The first is to collapse multiple assignments into a single statement by using an aggregate as the target of the assignment:

```
upper_byte <= word(15 downto 8);     -- One assignment to the upper byte
lower_byte <= word(7 downto 0);      -- Another assignment to the lower byte


(upper_byte, lower_byte) <= word;  -- Same as the two assignments above
```

The second case is to use an aggregate as the source value in the assignment:

```
tx_buffer(0) <= START_BIT;            -- One assignment to the LSB
tx_buffer(8 downto 1) <= tx_word;     -- One assignment to the data bits
tx_buffer(9) <= parity;               -- One assignment to the MSB


tx_buffer <= (START_BIT, tx_word, parity); -- Same as the three assignments above
```

In the previous assignment, we could have used concatenation as well. However, only an aggregate is possible when we want to use named association or the keyword others.

Avoid Explicit Comparisons with True or False   If the object in an equality or inequality test is of type boolean, then there is no need to explicitly write the values true or false in the expression. The code will be cleaner and will read better in English if the comparisons are made implicitly:

```
if pll_locked = true then ...       -- Bad, unnecessary comparison with 'true'
if pll_locked then ...              -- Good, boolean value used directly


while end_of_file = false loop ...  -- Bad, explicit comparison with 'false'
while not end_of_file loop ...      -- Good, boolean value used directly
```

Organize Numeric Tests to Make Them Read More Naturally   We can use the fact that it is easy to swap the operands in an ordering operator to make the code easier to read. When comparing an object against a constant value, prefer putting the constant value on the right.[2] The expressions will read more naturally because this is how we make this kind of comparison in English. For example, you should be able to read this sentence and imagine the corresponding operation easily:

"If the count value is less than the maximum possible value, then ..."

However, trying to imagine the following version would probably cause us to do a double take:

"If the maximum possible value is greater than the count value, then..."

In the code, this second sentence translates as:

```
if COUNT_MAX > count then ...   -- Reads a little unnaturally
```

While the first sentence translates as:

```
if count < COUNT_MAX then ...   -- Good, puts the changing value on the left
```

However, when we want to compare the same object against two limits to check whether it is inside a given range, it is better to put the changing value in the middle[3]:

```
if (TEMP_MIN <= current_temperature) and (current_temperature <= TEMP_MAX) then ...
```

Putting the constant values at the extremities makes it clearer that they act as boundaries for the value in the middle.

Use the Reduction Operators and Array-Scalar Operators for Less Verbose Code   The reduction operators and array-scalar operators introduced in VHDL-2008 can make the code less verbose in certain situations. Reduction operators work as a big logic gate whose inputs are the elements of an array. In the following code, the designer has modeled an 8-input *nor* gate:

```
zero_flag := not (
    alu_result(7) or alu_result(6) or alu_result(5) or alu_result(4) or
    alu_result(3) or alu_result(2) or alu_result(1) or alu_result(0)
);
```

Using a reduction operator, the same code could be written as:

```
zero_flag := nor alu_result;
```

Array-scalar operators repeat the same operation between each element of an array and a single-bit value. Before VHDL-2008, this was a cumbersome operation. Here are two possible ways to do it:

```
-- Create an aggregate with the same size as array operand and do a bitwise and
and_array_bit <= (array_op'range => bit_op) and array_op;

-- Test the single-bit operand and return either the original array or zeros
and_array_bit <= array_op when bit_op = '1' else (others => '0');
```

In VHDL-2008, we can apply the logical operators directly between a scalar and an array:

```
and_array_bit <= bit_op and array_op;
```

Use the Matching Relational Operators for Less Verbose Code   There are at least two ways in which the matching relational operators can help reduce verbosity. One is by allowing comparisons using the *don't care* value:

```
chip_select <= (address ?= x"07----"); -- Matches all values with top byte = 0x07
```

The other is by propagating logic values at the output of each operator, instead of forcing a conversion to boolean. When the ordinary (nonmatching) relational operators are used in an expression, they always return boolean values, which must be converted back to a logic level to be assigned to a signal. This can be done with a conditional signal assignment:

```
warn_led <= '1' when (temperature >= TEMP_MAX and warning_enabled = '1') else '0';
```

With matching operators, the original type (bit or std_ulogic) is preserved through the entire expression, and no conversion is needed:

```
warn_led <= (temperature ?>= TEMP_MAX) and warning_enabled;
```

Use Array Attributes to Create Flexible Code   Subprograms that work with array parameters do not need to specify a fixed range for their arguments. Most of the times, it is possible to use array attributes to write completely generic code.

The following function overloads the logical operator and to work with the type mv14 (defined in the example of listing 12.4). It uses the attributes 'length, 'subtype, and 'range to accept vectors with any range and direction.

```
function "and"(left_op: mv14; right_op: mv14_vector) return mv14_vector is
    -- Use attribute 'length to create array with the same size as argument
    variable right_op_normalized: mv14_vector(1 to right_op'length) := right_op;

    -- Use attribute 'subtype to avoid duplicating code from above line
    variable result: right_op_normalized'subtype;
begin
    -- Use attribute 'range to iterate through all elements of the array
    for i in result'range loop
        result(i) := left_op and right_op_normalized(i);
    end loop;

    return result;
end;
```

Part III   Statements

9 Concurrent Statements

Chapter 7 introduced the two kinds of statements in VHDL: concurrent and sequential. This chapter takes a closer look at concurrent statements, the main language mechanism to support the description of hardware structures and components that operate in parallel. A special emphasis is given to the *process* statement, one of the most versatile constructs in VHDL.

9.1 Introduction

A deeper knowledge of concurrent statements is essential for understanding what happens in a simulation and how VHDL code can model real hardware. Unlike a program in a general purpose programming language, which provides a sequence of instructions to be executed in order by a processor, a VHDL description is more accurately described as a network of communicating processing elements. These elements execute independently from one another and communicate via an elaborate mechanism provided by the language. In VHDL, the processing elements are concurrent statements, and their communication channels are signals. Concurrent statements can exist only in the statement part of an architecture, as shown in figure 7.1.

Concurrent statements have three main characteristics. First, they execute side by side, conceptually in parallel. Second, they execute continuously, as if in a loop. Third, they have no predefined order of execution among them. These three characteristics are similar to the behavior of real hardware, making concurrent statements appropriate for modeling digital circuits.

Concurrent statements fall into one of four categories: those that are a process, those that are a shorthand notation for a process, those that instantiate subcomponents, and those that control the conditional or iterative inclusion of statements into an architecture. Statements that are processes or equivalent to a process execute a continuous loop: they wait for a condition to happen, read their inputs, update their outputs, and then wait again. Statements that instantiate subcomponents are effective only at the beginning of a simulation, in a phase called elaboration; their purpose is to copy the description of the subcomponent into the instantiating architecture. Statements that control the inclusion of other statements in the architecture are also active only during elaboration, when they are replaced by zero or more copies of a group of concurrent statements of the other forms (elaboration was introduced in section 5.5).

Some designers find concurrent statements easier to grasp because they behave like real hardware. However, it is neither practical nor possible to describe all kinds of behavior using only concurrent statements. Nevertheless, concurrent statements are the first level of organization in a model and must be mastered by all VHDL designers. This chapter presents all the concurrent statements available in VHDL and provides guidelines for their use.

9.2 The *process* Statement

A *process* is a delimited region of sequential code inside an architecture body. Although the statements inside a process are sequential, the process is a concurrent statement because it executes in parallel with all other processes and concurrent statements in an architecture.

Processes are the most fundamental and versatile construct for describing behavior in VHDL. Even when the same behavior could be modeled with other concurrent statements, using processes has some advantages. Certain kinds of behavior are easier to describe algorithmically, using sequential statements. In many cases, processes are also more efficient from a simulation point of view. One example of these two advantages is the use of variables, which are only allowed in sequential code. Variables are conceptually simpler than signals because they are always updated immediately; they are also more efficient in simulation because they do not generate events and keep only their current value.

Figure 9.1 shows the main parts of a *process* statement. Because processes are used everywhere in VHDL, you are probably already familiar with the *process* statement. In any case, you may be surprised to see that so many of its parts are optional.

The first optional part of a *process* statement is its label. Labeling a process has two main advantages: it is a good way to document our intent, and it helps identify processes during a simulation. The downside is that unless the label is meaningful and accurate, it only adds clutter to the code. As is the case with all optional elements in the code, we should use our judgment: if they really help you in your coding process, then use them. However, keep a critical eye on them and ask yourself from time to time which labels are useful and which ones are not. Do not feel obliged to add a label for each and every process if it does not provide useful information. We can use the same reasoning for the *is* keyword, which is optional and exists only for consistency with other constructs.

The sensitivity list is also optional, in the sense that a process should contain either a sensitivity list or *wait* statements, but not both (sensitivity lists are covered in section

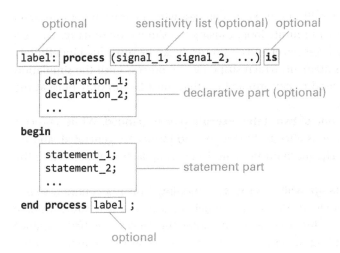

optional   sensitivity list (optional) optional

```
label: process (signal_1, signal_2, ...) is
        declaration_1;
        declaration_2;          declarative part (optional)
        ...
begin
        statement_1;
        statement_2;            statement part
        ...
end process label ;
```

optional

Figure 9.1
Parts of a *process* statement.

9.2.2). Finally, the declarative part can be used to introduce elements that are used only inside the process and do not need to be visible to the rest of the architecture, such as variables, constants, types, and local subprograms.

9.2.1   The Ins and Outs of VHDL Processes

Because the *process* statement is highly flexible and has so many peculiarities, it is a common source of mistakes in VHDL, especially among beginners. This section sheds some light on the most confusing topics about processes.

Every process starts executing during the initialization phase of a simulation. When a simulation starts, each process is run once until it suspends. This is true for all processes in the code, regardless of whether they have a sensitivity list or *wait* statements.

Once activated, a process keeps running until it executes a *wait* statement. The only way to suspend a process is by executing a *wait* statement. Note, however, that a process with a sensitivity list has an implicit *wait* statement immediately before the *end process* keywords. This implicit *wait* statement suspends the process until an event occurs in one of the signals in the sensitivity list. When the process resumes, it will loop around and continue executing from the first statement at the top.

Statements in a process are executed in an infinite loop. After the last statement in a process is executed, the process begins again immediately from the first statement. There is no pause, and the process does not suspend unless it has a sensitivity list, in which case it has an implicit *wait* statement just before the *end process* keywords.

This has two important ramifications. First, a process without any kind of *wait* statement will execute forever in an infinite loop, causing the simulation to hang. The second implication is that a process never ends. At most, it can be suspended indefinitely by an unconditional *wait* statement, which suspends the process until the simulation finishes. This trick can be used to perform actions that must be run only once at the beginning of a simulation.

A processes is always in one of two states: executing or suspended. While a process is executing, simulation time is effectively stopped, no events are generated, and no signal values get updated. The mechanism behind signals updates is elaborated in the next paragraph.

Signal values do not change while a process is running. This is a common source of mistakes among newcomers to VHDL. If a signal is assigned a new value inside a process, then it will assume this new value only after the process has had a chance to suspend. In fact, all pending signal updates will be applied at the same time in the simulation, after all processes that were active in the current simulation cycle have suspended. This ensures that all processes observe the same values for each signal during a simulation cycle, making the simulation results deterministic. For more details about the simulation cycle, see section 5.6.4.

A process can declare variables, which are local to the process. The only place where variables can be declared in VHDL is the declarative part of a process or subprogram (with the exception of shared variables, covered in section 14.5). Variables declared in a process can only be used in the statement part of the corresponding process. Such variables may be used as auxiliary values in calculations or to keep the part of the system state that is only relevant inside the process.

Variables declared in a process are initialized only once when the simulation begins. This means that when the process loops back to the first statement, the values of all variables are kept. In other words, the process declarative part is executed only once. For this reason, variables can be used to keep the state of a system and can infer storage elements such as flip-flops and latches. Contrast this with variables in a subprogram: because they are initialized on each call, variables in a subprogram cannot be used to model system state.

A process cannot declare signals. Signals, unlike variables, are intended for interprocess communication. Even if a signal could be declared locally to a process, it would not be visible by other processes and therefore would not be useful.

Although processes cannot declare new signals, they can access existing ones. A process can read and assign values to all signals visible at the architecture level. However, assigning values to the same signal from different processes takes additional care and requires the use of resolved signals.

If a process makes multiple assignments to a signal, only the last assignment before the process suspends is effective. This is another common source of confusion. Because simulation time is effectively stopped while a process is executing, assignments to signals never take place immediately; rather, they are always scheduled for a future time. If an assignment does not specify a time delay, the new value is scheduled for the next delta cycle. This is always the case in synthesis, where time delays are ignored. This mechanism is explained with more detail in section 11.3. Delta cycles are explained in section 5.6.4.

If multiple assignments occur before the process has had a chance to suspend, they will all be scheduled for the same instant, and each assignment will effectively overwrite the previous ones. To illustrate this, consider the following example. The same signal, called integer_signal, is assigned three values in sequence (lines 2–4). Because simulation time does not advance between the assignments, each assignment overwrites the previous one.

```
1       process begin
2           integer_signal <= 1;
3           integer_signal <= 2;
4           integer_signal <= 3;
5
6           if some_condition then
7               integer_signal <= 4;
8           end if;
9
10          wait for 1 ms;
11      end process;
```

To test your understanding, try to predict the values assumed by integer_signal each time the process suspends, based on the current value of some_condition (which can be true or false). You should conclude that integer_signal is updated every millisecond; it gets the value 3 when some_condition is false and 4 when it is true.

A process should contain a sensitivity list or wait statements but not both. A process with a sensitivity list is shorthand for a group of sequential statements that are executed whenever a signal in the sensitivity list changes. To achieve this effect, a process with a sensitivity list cannot contain explicit *wait* statements; otherwise, the process could suspend at different places, and different sequences of statements would be executed when it resumed. Therefore, when you write a process using a sensitivity list, remember that *wait* statements are not allowed.

Technically, we could write a process without a sensitivity list and without *wait* statements. However, such a process would be useless because it would execute indefinitely without ever giving other processes a chance to execute. In fact, it would hang the simulator.

Processes can model combinational or sequential logic. The fact that variables and signals keep their values between activations of a process can be exploited to model storage elements. The choice of whether a processes implements combinational or sequential logic is determined by the nature of the sequential statements inside the process body and its sensitivity list. This will be explained in more detail in the next section, when we discuss the process sensitivity list.

The order of execution among processes is random and should be irrelevant. A simulator is free to execute concurrent statements in parallel or in any chosen order. The VHDL language does not define the order in which such statements must be executed. In fact, the VHDL specification states that any code that depends on a particular order of execution of concurrent statements is erroneous.[1]

Conceptually, all processes execute concurrently. Keep in mind that processes, like real hardware components, exist and execute independently of each other. One process cannot influence what happens inside another process except through proper communication channels. Do not assume any timing relationship between processes, such as their relative order of execution. Any timing dependency between processes must be explicitly coded by the designer.

9.2.2   The Sensitivity List

A *sensitivity list* is a list of signals to be monitored for changes while a process is suspended. When one of these signals changes, the process is reactivated. This is a useful construct for modeling combinational logic, where the outputs must be reevaluated whenever an input changes. It is also important for modeling sequential logic, where the circuit state is updated on the command of a synchronism signal.

The sensitivity list gives important cues about the nature of a process. When skimming through the code, the sensitivity list helps infer the process goals and tells which signals control and influence the process. By looking at the signals in a sensitivity list, we can tell whether the process is intended to model sequential or combinational logic. Because the sensitivity list always appears at the top of a process, this information is presented in a prominent place in the code.

There are two ways to specify when a process gets suspended and reactivated: using a sensitivity list or using *wait* statements. The *wait* statement is a sequential statement that causes the containing process to suspend, and it is covered in detail in section 10.5. Figure 9.2 shows the equivalence between the two alternatives. The key rule is that a process with a sensitivity list is equivalent to a process with a single *wait on* statement located immediately before the *end process* keywords. The signals in the sensitivity list are the same ones that would be used in the *wait on* statement. In other words, the sensitivity list acts as an implicit *wait* statement that suspends the process until there is a change in one of the listed signals.

```
process (a, b) begin              process begin
   y <= a and b;                     y <= a and b;
end process;                         wait on a, b;
                                  end process;
```

Figure 9.2
Equivalence between a process sensitivity list and a *wait on* statement.

In most cases, the choice between these two alternatives is a matter of style. A process with a sensitivity list can always be converted to a process with a *wait* statement. However, the opposite is not always true: for instance, a process with multiple *wait* statements cannot be rewritten using a sensitivity list. Moreover, a process may contain a sensitivity list or *wait* statements but not both.

Choosing to use a sensitivity list offers some advantages: the process is a little more compact, the monitored signals are in an eminent place, and sensitivity lists are easily understood by developers and synthesis tools. As will be discussed in section 10.5.1, synthesis tools may have problems with multiple *wait* statements in a process. The behavior of a process with a sensitivity list is also more restricted and predictable: because it cannot contain explicit *wait* statements, a process with a sensitivity list always runs from the first to the last statement when activated.

In contrast, a process with explicit *wait* statements is more flexible because it can suspend from different places and resume under different conditions. Also, an explicit *wait* statement at the end of a process makes it evident that the process always suspends at that point. This communicates more clearly that the process body will be executed at least once in the initialization phase before the process suspends.

Among designers, there seems to be a clear preference for writing processes with a sensitivity list in RTL code. For simulation and testbenches, both styles are common. Because of the advantages described earlier, it may be a good idea to use a sensitivity list when possible. Nevertheless, it is important to know both alternatives and the equivalency between them.

We now turn our attention to the contents of the sensitivity list: what signals to include in it based on the kind of logic that we intend to model.

Sensitivity List for Combinational Logic   A combinational logic circuit is one whose outputs depend only on the current input values and not on any internal state or previous input history. This kind of circuit has no memory or feedback loops, and the value of its outputs is given as a logic function of the inputs.

In combinational circuits, a change in an input can potentially cause a change in any output depending on the implemented logic function. Therefore, whenever an input changes, this function must be recomputed and the outputs updated accordingly.

To ensure that the function is reevaluated whenever necessary, the sensitivity list must include all the inputs to a combinational logic block. Otherwise it would be possible for an input to change without triggering the corresponding update on the outputs.

To ensure that the sensitivity list includes all the required signals, we have two choices: one is to create and update the list manually. The other, available since VHDL-2008, is to use the keyword *all* instead of specifying individual signal names. Using the word *all* means that the process will be sensitive to changes in all signals that are read within its sequential statements.

To illustrate these choices, figure 9.3 shows two equivalent processes intended to model combinational logic. The code shown is part of a finite state machine (FSM) next state logic, a typical combinational circuit. The first process (figure 9.3a) uses a manu-ally created sensitivity list, including all the signals read in the process. Besides being tedious work, this is also error-prone: it is easy to include a signal that should not be in the list or overlook a necessary signal. Furthermore, we can easily forget to update the list when we modify the statements inside the process. The second process (figure 9.3b) has only the reserved word *all* in its sensitivity list. This approach makes the code easier to write, easier to maintain, and more likely to be correct. In general, it is a good idea to use the keyword *all* whenever possible.

So far, we have seen why the sensitivity list of combinational processes must include all signals read inside the process. But what happens if we break this rule? The most

```
next_state_logic: process (current_state, nickel_in, dime_in, quarter_in, coin_return)
begin
    case current_state is
        when accepting_coins =>
            ...
    end case;
end process;
```

(a) A manually written sensitivity list.

```
next_state_logic: process (all)
begin
    case current_state is
        when accepting_coins =>
            ...
    end case;
end process;
```

(b) With the reserved word *all* (VHDL-2008 only).

Figure 9.3
Two ways to write the sensitivity list for a combinational process.

likely result is that the simulations will not match the synthesized circuit. To illustrate this, consider the following process, which is intended to implement a simple boolean function but has an incomplete sensitivity list:

```
process (a) begin
    y <= (a and b) or c;
end process;
```

The author's original intention for the synthesized circuit can be inferred from the logic expression. However, strictly speaking, this code specifies that the output should not change when signals b or c change because only signal a is in the sensitivity list.

This poses a problem for the synthesizer. Should it interrupt the synthesis process, declaring the code incorrect? Note that the code in the example is, after all, valid VHDL code. Should it assume that there is a mistake in the sensitivity list or an error in the statements inside the process? Faced with this situation, most synthesizers will issue a warning that the sensitivity list is incomplete and proceed with the synthesis as if all the required signals were in the list.

The simulator, in contrast, is a pure software application, and it is not limited to what can be implemented as physical hardware. Therefore, the simulator can reproduce the behavior specified in the code and execute the assignment only when signal a changes. This causes a mismatch between simulation results and the synthesized circuit.

Now that we have seen that the sensitivity list must include all signals read within the process, is this enough to guarantee that the modeled circuit will be combinational? Definitely not. The nature of the modeled circuit ultimately depends on the statements inside the process. Figure 9.4 shows three sequential circuits as counterexamples, all modeled with processes using the keyword *all* in their sensitivity lists.

In figure 9.4a, the test for the rising edge of clock ensures that the value of q is updated only on this condition and that it is kept otherwise. In figure 9.4b, testing the enable signal has the same effect. Finally, figure 9.4c shows that it is possible to

```
process (all) begin                process (all) begin                process (all) begin
    if rising_edge(clock) then         if enable then                     q <= r nor q_bar;
        q <= d;                            q <= d;                        q_bar <= s nor q;
    end if;                            end if;                         end process;
end process;                       end process;
```

(a) A D-type, edge-triggered flip-flop.   (b) A D-type latch.   (c) An SR latch.

Figure 9.4
Sequential circuits modeled with the keyword *all* in the sensitivity list.

model state-holding circuits even with simple assignments that are executed in every run of the process. Clearly, the keyword *all* is not enough to guarantee that a process is combinational.

Sensitivity List for Sequential Logic A sequential logic circuit is one whose outputs depend on the internal system state, which is preserved in memory elements such as flip-flops or latches. The storage elements are usually controlled by special signals such as clock and reset. In a sequential circuit, the outputs cannot be predicted by looking only at the current input values.

Before we look into the details of sequential circuits, we should be careful not to mix up sequential *statements*, which are VHDL statements that can be used in sequential code (such as *if*, *for*, and *wait*), with sequential *circuits*, which are digital circuits containing storage elements. There is no implication between the two: sequential circuits can be modeled with both kinds of statements (concurrent or sequential), and sequential statements can be used to model both kinds of circuits (combinational or sequential). However, the combination of sequential circuits modeled with concurrent statements in unusual.

Sequential circuits are further divided into *synchronous* and *asynchronous*. In synchronous circuits, state changes are allowed only at discrete times, dictated by the system synchronism pulse or *clock*. In asynchronous circuits, changes in the system state can happen at any time in response to changes on the inputs. Today, the vast majority of digital devices is built with synchronous circuits, which have better timing predictability and are easier to design and verify. Unless you are an experienced designer, properly trained in asynchronous techniques, it is a good idea to stick with synchronous designs. Section 19.3.2 has more information about the synchronous design paradigm. In most of this book, whenever we mention sequential logic, we are referring to synchronous digital circuits.

The terms *synchronous* and *asynchronous* do not need to refer to an entire circuit. To understand this, we must recognize that synchronicity is not a self-standing attribute. If we say that an event is synchronous, then it has to be synchronous to some other event. A signal can only be synchronous with respect to another signal—typically, one of the system clocks. Within the same circuit, it is possible that some signals are synchronous to the system clock whereas others are not. A common combination is to have storage elements that register new values only on a clock edge but can be cleared immediately in response to a reset signal. This kind of storage element is available in many hardware technologies used with VHDL.

To model such elements, the code must accurately represent their behavior, which is often described in a process with a sensitivity list. The process must be activated whenever an event happens that could potentially change the system state. There are two kinds of such events: clock edges and asynchronous control events.

```
1   process (clock, reset) begin          1   process (clock) begin
2       if reset then                     2       if rising_edge(clock) then
3           count <= 0;                    3           if reset then
4       elsif rising_edge(clock) then      4               count <= 0;
5           if load then                   5           elsif load then
6               count <= load_value;       6               count <= load_value;
7           elsif enable then              7           elsif enable then
8               count <= count + 1;        8               count <= count + 1;
9           end if;                        9           end if;
10      end if;                            10      end if;
11  end process;                          11  end process;
```

(a) Synchronous counter with   (b) Synchronous counter with
    asynchronous reset.   synchronous reset.

Figure 9.5
A process modeling a sequential circuit (a counter) with synchronous and asynchronous assignments.

The clock edge is used to update the value of a storage element at precise moments in time. To model this behavior, the clock needs to be in the sensitivity list, and the assignment must be inside a test for the clock edge. As for asynchronous signals, because they must change the stored value immediately, they must also be in the sensitivity list; however, in the process statements, they must be outside any test for a clock edge.

The upshot is that any control signal dictating when the state is allowed to change must be in the sensitivity list. But what about the signals that determine the actual value written to the storage elements: should they also be in the sensitivity list? The answer is no, for the following reason. Signals that are the target of synchronous assignments are the outputs of flip-flops. Because the value of a flip-flop does not change except in a clock edge, there is no need to re-execute the process whenever one of such signals changes. Adding these signals in a sensitivity list would be superfluous and inefficient. The synthesis tool knows that the equations used in assignments under control of a clock edge should model a block of combinational logic and will infer the right kind of hardware. For details, see section 19.4.

To make these concepts clearer, let us look at an example. Figure 9.5 shows a process describing a sequential circuit, an up-counter with synchronous load and enable. The figure shows two different versions: figure 9.5a uses asynchronous reset, and figure 9.5b uses synchronous reset.

The clock signal is mandatory in the sensitivity list of both cases. The reset signal, however, requires closer attention. We need to determine whether it is working as an asynchronous control signal or used only in synchronous assignments, in which case it deserves no special treatment. In the process of figure 9.5a, the reset signal is tested in line 2, outside the test for the clock edge. It is, therefore, asynchronous to this clock and needs to be in the sensitivity list. In the process of figure 9.5b, in contrast, the reset

Table 9.1 Rules and recommendations for sensitivity lists of combinational or sequential logic

In processes modeling <u>combinational</u> logic, the sensitivity list ...	In processes modeling <u>sequential</u> logic, the sensitivity list ...
• should not include synchronism signals, such as clock and reset • should include all signals that are read within the process • is not enough to guarantee that a process is combinational	• should include the synchronism signals (clock, reset) • should include any asynchronous control signals (set, reset, enable) • should not include any other signals

In both cases, the sensitivity list ...

- should not contain unnecessary signals, for efficiency reasons
- should include all the necessary signals, or simulation results may differ from the synthesized circuit

signal is read inside the test for the clock edge. Therefore, it works as any other synchronous input that influences the next value of the counter and should not be included in the sensitivity list.

As for the assignments that define the new count value stored in the register, they are all synchronous, and therefore their inputs should not be in in the sensitivity list.

Putting It All Together   One of the biggest problems with writing sensitivity lists is that as long as they contain valid signal names, they are legal VHDL code. Therefore, if we add a totally unrelated signal to a sensitivity list, or if we forget to add an important signal, then the compiler may give us a warning, but it will not generate an error. The end result will be inconsistent behavior between simulation and synthesis, or even erroneous behavior of the compiled circuit.

To make the designer's job easier, table 9.1 summarizes the most important rules to keep in mind while writing a sensitivity list.

9.2.3   Equivalence between Processes and Other Concurrent Statements
Another way to see the *process* statement is as a basic template from which other concurrent statements are built. Most of the commonly used concurrent statements can be seen as a shorthand notation for typical processes.

To illustrate this equivalence, consider the architecture with two concurrent statements shown in listing 9.1. The statement in line 13 is a *concurrent signal assignment* (covered in section 11.1), and the statement in line 15 is a *concurrent procedure call* (covered in section 9.3). Both statements are located inside the architecture body, which is a region of concurrent code. Therefore, both statements are concurrent and execute in parallel.

Listing 9.1 A combinational circuit modeled with two concurrent statements

```
1    entity adder_subtractor is
2        port (
3            operand_a: in unsigned(31 downto 0);
4            operand_b: in unsigned(31 downto 0);
5            sum: out unsigned(31 downto 0);
6            difference: out unsigned(31 downto 0)
7        );
8    end;
9
10   architecture concurrent_statements of adder_subtractor is
11   begin
12
13       sum <= operand_a + operand_b;
14
15       subtract(operand_a, operand_b, difference);
16
17   end;
```

Because both statements are concurrent, there is no predefined order of execution between the two. But how does a simulator know when each statement should be executed? One way is to think that each kind of concurrent statement has its own rules for when the statement should be executed. For instance, for a concurrent assignment, we know that the statement is executed whenever one of the signals on the right side of the assignment changes.

Another approach, also used in the Language Reference Manual (LRM) that specifies the VHDL language, is to describe the concurrent statement in terms of an equivalent process. To create this process, the concurrent statement is converted to a sequential statement and then wrapped in a *process* statement. The process can be written with a sensitivity list or with a *wait* statement at the end; the LRM uses this second approach.

To illustrate this equivalence, listing 9.2 shows an architecture equivalent to the one from listing 9.1. In this new architecture, each concurrent statement was converted to an equivalent processes with a sensitivity list. The concurrent signal assignment from line 13 was converted to the process in lines 21–23, and the concurrent procedure call from line 15 was translated to the process in lines 25–26. In both cases, the sensitivity list is composed of all the signals read in the statement: in the concurrent assignment, the list includes all signals read in the expression; in the procedure call, the list includes all signals of mode *in* or *inout* used as parameters in the subprogram call.

Listing 9.2 Architecture for the `adder_subtractor` using processes with a sensitivity list

```
18    architecture equivalent_processes_with_sensitivity_list of adder_subtractor is
19    begin
20
21        process (operand_a, operand_b) begin
22            sum <= operand_a + operand_b;
23        end process;
24
25        process (operand_a, operand_b) begin
26            subtract(operand_a, operand_b, difference);
27        end process;
28
29    end;
```

Finally, remember that a process with a sensitivity list is equivalent to a process with a *wait* statement at the end. Therefore, the architectures of listings 9.1 and 9.2 are also equivalent to the architecture shown in listing 9.3.

Listing 9.3 Equivalent architecture for the `adder_subtractor` using processes with a *wait* statement

```
30    architecture equivalent_processes_with_wait_statements of adder_subtractor is
31    begin
32
33        process begin
34            sum <= operand_a + operand_b;
35            wait on operand_a, operand_b;
36        end process;
37
38        process begin
39            subtract(operand_a, operand_b, difference);
40            wait on operand_a, operand_b;
41        end process;
42
43    end;
```

In practice, there is not much reason to convert a concurrent statement to an equivalent process, which would always be longer than the original statement. However, it is important to be aware of this possibility because sometimes (as in the LRM) statements are defined in terms of their equivalent processes. In other occasions, looking at a concurrent statement as a kind of process may help determine more clearly when it will be executed.

9.2.4 Guidelines and Recommendations for Writing Processes

Now that we have covered the important topics and cleared up many of the common misconceptions about processes, we can discuss how to write processes that make our code more organized, easier to understand, and more likely to be correct.

Treat Each Process as a Module   Processes are one of the most important constructs for implementing and organizing behavior in an architecture. In many senses, a process in an architecture is comparable to a module in a design hierarchy. When planning how to distribute functionality across processes, it pays to keep in mind the principle of modularity and the guidelines for architecting modular systems presented in section 4.1.3. Here we review the most important recommendations about modularity that are applicable to process statements.

The first recommendation is to give each process a clear responsibility. Start by having a clear idea of what the process is supposed to do; writing this down as a comment in the source code may help keep the process focused. Do not group unrelated functionality into a process—it is better to have several processes that are easy to understand than a single process that no one can make sense of. Use the principle of cohesion to decide what should be in the process: after defining a main purpose, check that each and every statement supports this central goal. Anything that is not strictly related should be moved elsewhere.

Another recommendation is to keep each module to a manageable size. Short processes reduce the amount of information we need to keep in our heads at a time, making the code easier to understand, debug, and maintain. It is difficult to give hard and fast rules about how long a process should be, but most of the recommendations about the length of subprograms are also valid for processes; see section 15.4 for details. As a rule of thumb, if your processes are too long to fit in one screen, then they are probably on the long side. Look for ways to make them shorter, such as moving parts of the process to separate routines.

Although the *process* statement does not have an explicit interface, we should have a clear idea of its inputs and outputs. In a sense, VHDL processes are like routines where all input and output happen through global data (which would be a terrible interface for a routine!) because a process has no input or output parameters but can manipulate objects declared in the architecture. However, just because a process can access such objects, it does not mean it should. If a task can be done completely inside the process without accessing other elements from the architecture, then do it locally. Use variables whenever possible to keep state and to calculate values that are only relevant inside the process.

Finally, an architecture will be easier to reason about if it minimizes the number of connections between processes. When you split a process, try to do it in a way that minimizes the number of signals needed to communicate between them. Use the number of signals as a yardstick to assess how tightly coupled the processes are to each other and the enclosing architecture. Refer to the principle of loose coupling introduced in section 4.4 for the details.

Make Each Process Purely Combinational or Purely Sequential Before you start writing the code for a process, give it a clear responsibility. Then think about what kind of logic the process is supposed to model—combinational or sequential. The kind of logic defines several characteristics of a process, such as what should go in the sensitivity list and whether the process is allowed to keep an internal state.

There is much to say about modeling sequential and combination logic. For details about the sensitivity list, see "Sensitivity List for Combinational Logic" and "Sensitivity List for Sequential Logic" in section 9.2.2. For more about writing synthesizable code for both kinds of logic, see section 19.4. The following paragraphs are only the essence of what you should keep in mind at all times.

If the process is supposed to model *combinational* logic, then the sensitivity list should include all signals read within the process, or it should be the reserved word *all*. The alternative with the keyword *all* is cleaner and less error-prone, and it should be used if your tools support it (this is a VHDL-2008 feature). Clock and reset signals should not be in the sensitivity list, and they probably should not be used in the process at all. Furthermore, be careful not to read any signal or variable before assigning a value to it, in any possible execution path—this would imply unwanted storage elements.

If the process is to model *sequential* logic, then only synchronism signals (such as the system clock) or asynchronous control signals (such as an asynchronous reset) should be in the sensitivity list. As for the assignment statements inside the process, if the sequential logic is synchronous, then all assignments should happen inside a region of code that checks for a clock edge. To keep the process focused, check that the ultimate goal of every assignment and expression is to compute the values to be stored in state-holding elements. Any other expression or statement is probably noncohesive and should be moved to a different process.

If a Declaration Is Used Only in the Process, Then Make It Local to the Process The process declarative part may contain several kinds of declarations, such as constants, variables, aliases, types, subtypes, and subprograms. If one of these items is used only within a process, then there is no need to declare it in the architecture, where it would be visible to all other concurrent statements. In such cases, it is good practice to make each declaration as localized as possible. In this way, the declarations will not get in the way of a reader trying to get the bigger picture of the architecture; moreover, each declaration will be closer to where it is used, improving readability.

Prefer Variables to Signals for Keeping State That Is Local to a Process This is a special case of the previous recommendation (keeping declarations local to a process), but it deserves closer examination because of its subtleties and because choosing between signals and variables is a common design decision.

When a process needs to hold data values, the designer has two choices: keep the data in variables declared inside the process, or keep the values in signals declared in the architecture. As usual, it is good practice to make each declaration as localized as possible, to avoid introducing more names than a reader needs to know and to keep the declarations closer to where they are used. However, for signals and variables, the place where the object is declared defines its class: if the object is declared in an architecture it must be a signal, whereas if the object is declared in an process it must be a variable. Unless there is a specific reason for using a signal, prefer a variable for modeling state-holding elements in a process. As a bonus, variables are significantly more efficient than signals in simulations. For a list of guidelines to assist in choosing between signals and variables, see section 14.4.

Consider Rewriting the Process as One or More Concurrent Statements   For all their virtues, processes are not always the most compact way to implement simple behavior. Sometimes it is possible to model the same behavior in a more compact form using one or more concurrent statements. The following example shows a process modeling a multiplexer:

```
process (a, b, sel) begin
    if sel = '0' then
        y <= a;
    else
        y <= b;
    end if;
end process;
```

The same behavior can be modeled with a concurrent statement called a *conditional signal assignment*:

```
y <= a when (sel = '0') else b;
```

Being familiar with the concurrent statements introduced in this chapter and with the assignment statements introduced in chapter 11 will allow you to write shorter code in many situations.

9.3   The Concurrent Procedure Call Statement

Procedures and *functions* are the two kinds of subprograms available in VHDL. Both are discussed in detail in chapter 15. This section focuses on one aspect of procedures: a

```
entity adder is                              procedure add(
    port (                                       a, b: in integer;
        a, b: in integer;                        signal sum: out integer
        sum: out integer                     ) is begin
    );                                           sum <= a + b;
end;                                         end;

architecture behavior of adder is
begin
    sum <= a + b;
end;
```

(a) As an entity (**adder**).   (b) As a procedure (**add**).

Figure 9.6
Two alternatives for modeling an **adder**.

procedure can also be used like a concurrent statement, when it is called from a concurrent code region.

Most developers are familiar with the idea of using procedures to organize sequential code, by encapsulating a sequence of statements in a subprogram and then calling it from a region of sequential code. The procedure is invoked when the procedure call statement is reached in the normal execution flow. However, a lesser known fact is that procedures can be called from concurrent code as well. The following examples illustrate this case.

Suppose we need to model a block of combinational logic (illustrated here with an adder) as part of a larger VHDL model. There are at least two alternatives. The first is to create a separate entity to implement the logic function (figure 9.6a), and the second is to use a procedure to the same effect (figure 9.6b).

The two implementations can be used similarly in concurrent code, as shown in the architecture of figure 9.7. The *component instantiation statement* in line 3 instantiates the entity from figure 9.6a, whereas the *concurrent procedure call* in line 4 invokes the procedure from figure 9.6b. In both cases, the component declaration and the package must be made visible with *use* clauses (not shown in the code).

One thing to note in passing is that the entity was named as a noun (*adder*), whereas the procedure was named as an action (*add*). This is in line with the recommended naming conventions presented in chapter 16. As a suggestion, try to pay attention to how different VHDL constructs are named in the examples in this book, and see whether you can figure out other general rules.

How do we know when a concurrent procedure call will be executed? Like most concurrent statements, a concurrent procedure call can be written as an equivalent process. This process has an implicit sensitivity list, created from all the signals used as

```
1   architecture behavior of dual_adder is
2   begin
3       adder_component: adder port map (addend, augend, sum_1);
4       adder_procedure: add(addend, augend, sum_2);
5   end;
```

Figure 9.7
Usage example of entity adder and procedure add.

```
architecture sensitivity_list of adder is        architecture wait_statement of adder is
begin                                             begin
    process (addend, augend) begin                   process begin
        add(addend, augend, sum_2);                      add(addend, augend, sum_2);
    end process;                                         wait on addend, augend;
end;                                                  end process;
                                                  end;
```

(a) With a sensitivity list.   (b) With a *wait* statement.

Figure 9.8
Equivalent processes for the concurrent procedure call.

parameters of mode *in* or *inout* in the procedure call. Figure 9.8 shows two equivalent processes for the concurrent procedure call from line 4 of figure 9.7.

From any of the equivalent processes, it is easy to see that the procedure is invoked whenever there is a change in one of the two signals, addend or augend. The two processes shown in figure 9.8 are equivalent, but the version from figure 9.8b makes it a little clearer that the procedure is executed at least once when the simulation begins.

Finally, note that the sensitivity list is constructed from the actual parameters used in the call. This means that if one of the actual parameters is an expression involving several signals, the procedure will be invoked when any of the involved signals changes.

The main advantage of using a concurrent procedure call instead of a design entity is the reduced code overhead. A design entity requires both an entity declaration and an architecture body, and many coding conventions require that an entity–architecture pair be put in a separate file. Procedures, in contrast, can be grouped in a package containing related functionality. The downside is that, in general, procedures cannot keep internal state between calls. Therefore, unless the state is kept outside the procedure, a concurrent procedure call can only be used to model combinational logic.

9.4   The Component Instantiation Statement

The component instantiation statement is the basis for all structural design in VHDL, as it allows the use of subcomponents defined in other design units. A component instantiation is different from other statements in that it is not "executed" during a simulation. Instead, it is used in the analysis and elaboration phases to flesh out the hierarchy of a design.

There are two basic types of component instantiation statements, defined by the kind of unit being instantiated: it is possible to instantiate a *declared component* or a *compiled library unit*. In the latter case, the library unit can be a *design entity* or a *configuration*. The diagram in figure 9.9 illustrates these possibilities.

In the instantiation of a *declared component*, the code that instantiates a unit must provide a description of its interface. This description is called a *component declaration*, and it may be located in the declarative part of an architecture or in a package. Because the instantiating code contains all the information required to use the subcomponent in the design, it is not necessary to compile the instantiated unit beforehand. In other words, the unit is instantiated as a black box.

In the instantiation of a *compiled library unit*, the instantiating code does not provide a description of the subcomponent's interface; it just uses it. For this reason, the correctness of the code cannot be verified in isolation. Therefore, the instantiated unit must be analyzed first and put into a library. Because the unit is used without being declared first in the code, this kind of instantiation is often called *direct instantiation* of an entity or a configuration.

Unfortunately, the LRM makes things unnecessarily confusing by calling the instantiation scheme on the left side of figure 9.9 an *"instantiation of a component"* (which clashes with the generic name of the statement, *component instantiation statement*) and the instantiation scheme on the right an *"instantiation of a design entity"* (which sounds like it excludes the instantiation of a configuration). This correspondence with the names used in the LRM is presented here for completeness, but we avoid those terms in our discussions to prevent any ambiguity.

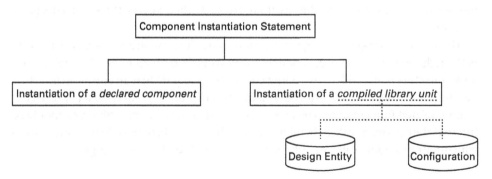

Figure 9.9
Kinds of component instantiation statements.

Figure 9.10 demonstrates the basic syntax of the three forms of component instantiation. Number 1 shows the instantiation of a declared component. Note that this is the only case where the specification of the kind of unit (*component*, in this case) is optional. Numbers 2 and 3 demonstrate the direct instantiation of a design entity and a configuration, respectively. In these two cases, because the unit is fetched from a design library, it is necessary to specify the library name unless a *use* clause has been used.

Figure 9.11 shows in a source code example the differences between the instantiation of a declared component (figure 9.11c) and the instantiation of a library unit—in this case, a design entity (figure 9.11b). Because the syntax for instantiating a design entity is essentially the same as for a configuration, figure 9.11b shows only the first form.

Figure 9.11a shows the unit we would like to instantiate (adder), which is the same in both cases. In figure 9.11c, the unit is instantiated as a declared component. In this case, the component interface is described by a *component declaration* in the architecture (starting with component adder is...). Because the interface is described inside the instantiating file, the source file component_instantiation_example.vhd can be compiled independently from any other file in the design.

In figure 9.11b, the adder unit is instantiated as a design entity. Note that it is not necessary to describe the interface of the instantiated unit; however, it is necessary to specify its complete name, including the library where it has been compiled. Because the code does not describe the interface of the instantiated unit completely, the file adder.vhd must be compiled into library work before entity_instantiation.vhd can be compiled.

Now that we have seen the different forms of the component instantiation statement, which one should we use? First, let us discuss the advantages and disadvantages

The three forms of the component instantiation statement:
1 Instantiation of a declared component
2 Direct instantiation of a design entity
3 Direct instantiation of a configuration

Simplified syntax:

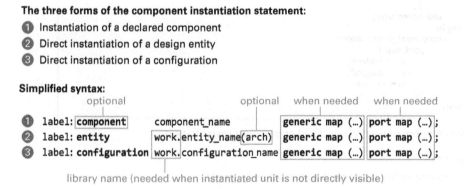

Figure 9.10
Basic syntax of the three forms of the component instantiation statement.

```
-- File adder.vhd

entity adder is
    port (
        a, b: in integer;
        sum: out integer
    );
end;

architecture rtl of adder is
begin
    sum <= a + b;
end;
```

(a) Subcomponent to be
    instantiated.

```
-- File entity_instantiation.vhd

entity entity_instantiation is
    port (
        addend, augend: in integer;
        sum: out integer
    );
end;


architecture example of entity_instantiation is
begin
    entity_inst: entity work.adder
        port map(
            a => addend,
            b => augend,
            sum => sum
        );
end;
```

(b) Subcomponent used via direct instantiation
    of a compiled library unit (in the example,
    a design entity).

```
-- File component_instantiation.vhd

entity component_instantiation is
    port (
        addend, augend: in integer;
        sum_1: out integer
    );
end;

architecture example of component_instantiation is
    component adder is
        port (
            a, b: in integer;              ------ component declaration
            sum: out integer
        );
    end component;
begin
    component_inst: adder
        port map (
            a => addend,
            b => augend,
            sum => sum
        );
end;
```

(c) Subcomponent used via declared component instantiation.

Figure 9.11
Usage example of the component instantiation statement.

of a declared component instantiation (figure 9.11b). On the positive side, it allows us to compile the instantiating file in isolation, so this may prevent unnecessary compilations. In fact, the instantiated unit does not even need to exist. This approach is also useful if we want to defer to a later time the choice of which architecture to use (e.g., when there is a need to simulate different implementations of a same entity).

On the downside, the component declaration is duplicate code, created with copy-and-paste. As with any code duplication, it may force us to update several places in the code when we want to make a change. Therefore, if you do not have a strong reason to use this method, it is probably better to instantiate the design unit directly.

Now let us turn to the advantages and disadvantages of directly instantiating a design unit (figure 9.11c), which are more or less the complement of those for a component instantiation. On the positive side, there is no code duplication. On the negative side, the instantiated unit must be compiled and put in the library before the instantiating unit can be analyzed. Also, the entity must be determined at coding time, and this binding is hardcoded in the statement (as the name work.adder in the example).

Direct instantiation does not offer the flexibility of declared component instantiation; however, it simplifies things by completely eliminating one level of indirection between the instantiating and instantiated units. For most low- or medium-complexity designs, it is possible for a designer to completely ignore the existence of components and configurations—a common source of confusion among VHDL learners. By using only the simpler version of the statement, the direct instantiation of design entities, designers can put away the complexities of dealing with components and configuration until a later time, when they are involved with more complex projects. Therefore, it is a good idea to use direct instantiation (as in figure 9.11c) whenever possible.

9.5   The *generate* Statement

The *generate* statement allows the conditional or iterative inclusion of concurrent statements in a VHDL description. With conditional inclusion, it is possible to create models that exhibit different behaviors depending on configurable parameters. With iterative instantiation, it is possible to replicate part of a design and create regular hardware structures such as memories, processing arrays, and arithmetic circuits.

The *generate* statement comes in three flavors: the *if-generate* and *case-generate* versions are used for conditional inclusion of statements, and the *for-generate* version is used for iterative inclusion. We will see the three forms in turn; their basic syntax is shown in figure 9.12.

if-generate   The *if-generate* statement (figure 9.12a) conditionally includes or excludes from the code one or more groups of concurrent statements. In the simplest form, an *if-generate* has only one condition. When this condition is true, the concurrent

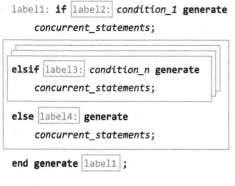

(a) *if-generate* statement.

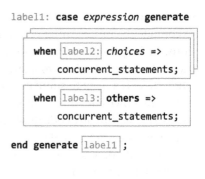

(b) *case-generate* statement.

```
label1: for generate_parameter in generate_range generate
    concurrent_statements;
end generate label1 ;
```

(c) *for-generate* statement.

Figure 9.12
Simplified syntax of the three forms of the *wait* statement (boxed parts are optional).

statements are copied to the architecture. When it is false, no statements are copied to the architecture.

The *if-generate* statement may have *elsif* clauses, each followed by a boolean expression. The *elsif* expressions are evaluated in sequence. When the first condition that evaluates to true is found, the corresponding concurrent statements are copied to the architecture, and the remaining clauses are ignored. The statement also accepts an *else* clause, whose statements are copied to the description if none of the conditions is true.

Because *generate* statements must choose which statements (if any) will be included in the code for elaboration, the condition expression must be evaluated at compile time. Therefore, it is not possible to use signals or variables in the expression. If you need to use a condition whose value can change dynamically, then use a sequential *if* statement inside a process. In fact, mixing up the concurrent *if-generate* statement with the sequential *if* statement (covered in section 10.2) is a common mistake. To clear this up, the following paragraphs elaborate on the differences between them.

The *sequential if statement* selects for execution a region of sequential code based on the value of a condition expression. This statement can be used in models intended for synthesis or simulation, and the expression can be static or dynamic (an expression containing signals or variables is dynamic). During a simulation, the expression is evaluated whenever the *if* statement is reached; if the condition is true, the underlying statements are executed. During synthesis, the result depends on whether the expression value is constant at compile time. If the value is constant, then the compiler may ignore the irrelevant cases (such as the *else* clause when the *if* expression is true). If the value can change dynamically, then the synthesized hardware must contemplate all the possibilities, maybe including a selector circuit whose control input is given by the expression value.

The *concurrent if-generate statement*, in contrast, selects for inclusion in the code a group of concurrent statements based on the value of a condition expression. This statement is relatively simpler than the sequential *if* because the condition must always be evaluated at compile time. When the condition is true, the statements are included in the code for elaboration; otherwise no statements are included. Therefore, only one of the *if*, *elsif*, or *else* clauses is synthesized, and the compiler does not generate unnecessary hardware.

In practice, the *if-generate* statement can be used for several different purposes. One example is to generate different logic depending on a configuration option (passed, for instance, via a generic constant). In the following example, two different circuits can be generated. When the constant CPU_HALT_ENABLED is true, signal halt is connected to an input port; otherwise it has a fixed value of '0'.

```
halt_signal_generate: if CPU_HALT_ENABLED generate
    halt <= halt_request_input;
else generate
    halt <= '0';
end generate;
```

Those familiar with C or C++ may find this similar to the #ifdef preprocessor directive. In fact, a common question among C/C++ developers coming to VHDL is how to achieve the same effect of an #ifdef preprocessor directive. In most cases, an *if-generate* is the closest alternative.

The *if-generate* statement can also be used to instantiate different subcomponents depending on a configuration option. In the following example, the value of the generic constant USE_ARP_V2 is used to select between a version of the component called arp_v2 (when the condition is true) or another version called arp_v1 (when the condition is false).

```
arp_generate: if multi_entry_arp: (USE_ARP_V2) generate
    arp_responder: entity work.arp_v2 port map (...);
else single_entry_arp: generate
    arp_responder: entity work.arp_v1 port map (...);
end generate;
```

The previous example illustrates the use of optional labels in the alternatives (immediately after the *if, elsif, else,* or *when* keywords). It also illustrates how the excessive use of labels can make the code look cluttered, so remember to use your judgment in each case and ask yourself whether the labels are more helpful or harmful.

Another common place for an *if-generate* is nested inside a *for-generate* statement to account for deviations in a regular structure. In such cases, the condition of the *if-generate* can use the generate parameter (the loop index) of the *for-generate* statement to create different descriptions for the first and last elements in an array of similar hardware elements. We will see an example of this use when we discuss the *for-generate* statement.

Finally, note that versions of VHDL prior to VHDL-2008 do not allow *elsif* or *else* clauses in *if-generate* statements. In those cases, the solution is to use several *if-generate* statements with different conditions.

case-generate   VHDL-2008 introduced a new version of the *generate* statement. The *case-generate* statement (figure 9.12b) is similar to an *if-generate* in that it provides a mechanism for conditional inclusion of concurrent statements in the code. However, its syntax is similar to the sequential *case* statement (presented in section 10.3).

The *case-generate* statement evaluates one expression, which is then compared with several alternatives. When an alternative matches the expression, the underlying concurrent statements are included in the source code, and all other alternatives are ignored.

The following example shows how a *case-generate* can be used to choose between several concurrent signal assignments based on the value of the generic constant DEGREE.

```
type degree_type is (degree_4, degree_8, degree_16, degree_32);
...
polynomial: case DEGREE generate
    when degree_4 =>
        lfsr_input <= lfsr_state(3) xor lfsr_state(4);
    when degree_8 =>
        lfsr_input <= lfsr_state(4) xor lfsr_state(5) xor lfsr_state(6) xor lfsr_state(8);
    when degree_16 =>
        lfsr_input <= lfsr_state(11) xor lfsr_state(13) xor lfsr_state(14) xor lfsr_state(16);
    when degree_32 =>
        lfsr_input <= lfsr_state(1) xor lfsr_state(2) xor lfsr_state(22) xor lfsr_state(32);
end generate;
```

As in the sequential *case* statement, a *case-generate* can be more compact than a series of *if-generate* statements. It is also more regular: the reader knows that all alternatives are simple comparisons against the same expression value. However, an *if-generate* statement with *elsif* clauses is more flexible because it can use independent expressions for the *elsif* clauses.

for-generate   The *for-generate* statement (figure 9.12c) allows the iterative replication of concurrent statements. In hardware, there is often a need to replicate part of a design or instantiate the same subcomponent a number of times. Examples include arithmetic circuits, multicore processors, and memory arrays. A *for-generate* can be used in all those cases. The concurrent statements within the generate body are replicated once for each value of the generate parameter (the "loop index").

A common use for a *for-generate* statement is to replicate a component instantiation statement, using the generate parameter to configure the connections to the subcomponents. Figure 9.13 shows an example of this situation. Suppose we wish to replicate a full-adder circuit (figure 9.13a) a number of times to implement an *N*-bit ripple-carry adder (figure 9.13b shows an example for a 4-bit adder). If we ignore the irregularities at the circuit edges, then we could write a *for-generate* to instantiate four copies of the full-adder block:

```
instantiate_adders: for i in 0 to 3 generate
    adder_bit: entity work.full_adder
        port map (
            a => a(i), b => b(i), carry_in => carry_out_vector(i-1),
            sum => sum(i), carry_out => carry_out_vector(i)
        );
end generate;
```

The trick for using the *for-generate* in such cases is to have all ports of the instantiated components connected to signals of vector types. Then, to make the connections, we must write the port mappings in a generic way that works for all values of the generate parameter. To make this possible, the vector indices must be written as expressions, using the value of the generate parameter.

In practice, it is unlikely that all instantiations can be made exactly the same way. Usually, the first or last element must be connected or configured differently. To account for such irregularities, we have a few choices. The first is to use a nested *if-generate* inside the *for-generate* statement. If we compare the generate parameter with the first and last values in the generate range, then we can write different component instantiation statements for these special cases. Listing 9.4 demonstrates this approach in the complete VHDL code for the ripple-carry adder of figure 9.13b.

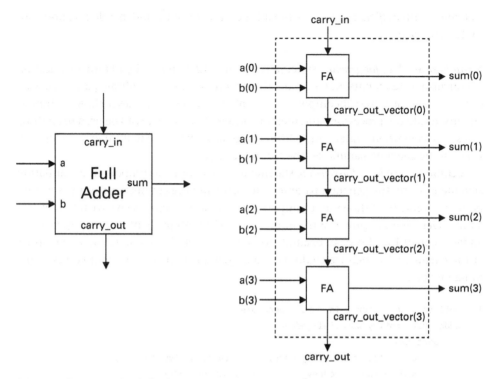

(a) Basic element (1-bit full adder).   (b) Replicated structure (4-bit adder).

Figure 9.13
Replication of a 1-bit, full-adder circuit to create a 4-bit adder.

Listing 9.4 Complete code for the ripple-carry adder of figure 9.13b

```
1    library ieee;
2    use ieee.std_logic_1164.all;
3
4    entity adder is
5       generic (
6          WIDTH: integer := 4
7       );
8       port (
9          a, b: in std_logic_vector(WIDTH-1 downto 0);
10         carry_in: in std_logic;
11         sum: out std_logic_vector(WIDTH-1 downto 0);
12         carry_out: out std_logic
13      );
14   end;
```

```
15
16    architecture structural of adder is
17        signal carry_out_vector: std_logic_vector(WIDTH-1 downto 0);
18
19    begin
20        instantiate_adders: for i in 0 to WIDTH-1 generate
21            first_bit: if i = 0 generate
22                adder_bit: entity work.full_adder
23                    port map (
24                        a => a(i), b => b(i), carry_in => carry_in,
25                        sum => sum(i), carry_out => carry_out_vector(i)
26                    );
27            end generate;
28
29            other_bits: if i /= 0 generate
30                adder_bit: entity work.full_adder
31                    port map (
32                        a => a(i), b => b(i), carry_in => carry_out_vector(i-1),
33                        sum => sum(i), carry_out => carry_out_vector(i)
34                    );
35            end generate;
36        end generate;
37
38        carry_out <= carry_out_vector(WIDTH-1);
39    end;
```

There are a few remarks about this version of the adder code. First, it could have been written with an *else* clause instead of a second *if-generate*, but in this case, the code would be legal only in VHDL-2008. Second, the component instantiation inside the first *if-generate* (lines 22–26) could be left outside the loop, in which case the *for-generate* could start at one instead of zero and no *if-generates* would be necessary.

A third remark is that this approach still creates a lot of duplicated code. Of course, the code is not exactly the same inside both generate bodies, but they are similar enough to cause us trouble if something needs to change. To prevent this problem, we could use a slightly different approach. Instead of working around the edge conditions by treating them as special cases, we could arrange the internal signals inside the architecture so that the component instantiation statement can be the same for all instances. This was not possible in the example of listing 9.4 because we wanted to connect the carry_in of the first full adder to the carry_in port of the ripple-carry adder entity. We could make things simpler by declaring two distinct vectors: one for connecting to all the carry_in inputs of the subcomponents, and another for connecting to all carry_out outputs. Then we would simply connect the two using a signal assignment. Listing 9.5 illustrates this approach.

Listing 9.5 Use of a single component instantiation statement by declaring separate carry_in and carry_out vectors connected with a signal assignment

```
1    architecture structural of adder is
2        signal carry_in_vector: std_logic_vector(0 to WIDTH-1);
3        signal carry_out_vector: std_logic_vector(0 to WIDTH-1);
4
5    begin
6        instantiate_adders: for i in 0 to WIDTH-1 generate
7            adder_bit: entity work.full_adder
8                port map(
9                    a => a(i), b => b(i), carry_in => carry_in_vector(i),
10                   sum => sum(i), carry_out => carry_out_vector(i)
11                );
12       end generate;
13
14       carry_in_vector <= (carry_in, carry_out_vector(0 to WIDTH-2));
15       carry_out <= carry_out_vector(WIDTH-1);
16   end;
```

The assignment in line 14 makes the connections between the carry_in inputs and the carry_out outputs. Because array assignments are always made in left-to-right order, carry_in_vector(0) receives the value of the input port carry_in, carry_in_vector(1) receives the value of carry_out_vector(0), and so on. This connection could have been made with a *for-generate*, but the solution shown here is much more compact. This approach also simplifies the expressions that produce the vector indices because no calculations are necessary; the generate parameter can be used directly as the index of all vector signals.

As a recommendation, whenever you see in your code a sequence of duplicate or similar concurrent statements, ask yourself whether a *for-generate* statement could be used instead. Whenever you see individual assignments made to elements of a vector, ask yourself whether an array assignment or a *for-generate* could help.

9.5.1 General Remarks about *generate* Statements

We conclude this section with some general remarks about *generate* statements. First, statements in the body of a *generate* statement should be concurrent, so control structures such as *if*, *case*, and *loop* are not allowed. If you find yourself wanting to use these structures, then remember that the *generate* statement instantiates a concurrent region of code, for which there is no predefined order of execution. Therefore, it would not make sense to use sequential statements.

Second, the condition expression for all *generate* statements must be determined at compile time (more specifically, during elaboration). Consequently, it cannot use objects whose values change dynamically, such as signals and variables. The most commonly used objects in *generate* expressions are generic constants.

Third, any of the three forms can have an optional declarative region, where it is possible to declare signals, types, and other elements allowed in an architecture declarative part. Use this region to declare any objects that are used only inside the *generate* statement body. Note that it is possible to declare different object instances for each alternative in a *generate* statement. Another point to note is that it is possible to declare signals local to a *generate* statement. This can be useful when making connections that are entirely enclosed within the statement. The declarations will be closer to where they are used, and the architecture will not be cluttered with signals that are used only by part of the statements.

Fourth, be careful not to make your code too hard to read. The mixture of nested conditional and iterative statements, along with several different versions of code on the same screen, can be daunting. Consider implementing the concurrent region as a subprogram or a separate design entity. Units that must be replicated are natural candidates for implementation as proper modules.

Remember to choose well the name and type of the generics used in the expressions. If you want to enable or disable particular features, use a boolean, and not an integer with values 0 and 1. If a generic parameter only allows a few numeric values, then consider using an enumeration type (if the values are sparse) or an integer subtype (if the values are contiguous).

Finally, always keep in mind the differences between a *generate* statement and the sequential control structure with the same name (*if-generate* vs. sequential *if* statement, *case-generate* vs. sequential *case* statement, *for-generate* vs. *for-loop*). Table 9.2 summarizes the key differences between the *generate* statements and sequential control structures.

9.6 Other Concurrent Statements

We close this chapter with a brief description of the remaining concurrent statements available in VHDL. Rather than cover all these statements in detail, the goal is to give an idea of when these statements might be useful in our models.

9.6.1 The *block* Statement

The *block* statement offers a way to compartmentalize regions of concurrent code in a design. Blocks have two main uses in VHDL. The first is to organize an architecture, grouping concurrent statements that form a distinct part of a design. The second is to disconnect signal drivers. We will see each use in turn.

The first use for a *block* statement is to delineate regions of concurrent code in a model. A block can have its own ports, generics, and declarations. The declarations are the same as those allowed in an architecture, such as signals, constants, types, and

Table 9.2 Key differences between *generate* statements and sequential control structures

	generate Statements	Sequential Control Structures
Statements	*if-generate*, *case-generate*, *for-generate*	*if*, *case*, *loop* (including *for*, *while*, and simple loops)
Allowed in	concurrent code regions (architecture body)	sequential code regions (processes or subprograms)
Enclosed statements (statements allowed in the body)	concurrent statements	sequential statements
Main use	to select concurrent statements for inclusion in the code (during elaboration)	to select sequential statements for execution; to control the execution flow of the code
Condition expressions	must be static (constant at compile time/elaboration time)	may be static or dynamic
Objects allowed in the condition expression	only constants and generics	variables, signals, constants, generics
Usage examples	to generate different versions of an entity; to replicate regular hardware structures	to implement algorithmic descriptions

subprograms. In other words, a block is almost like a nested entity, whose code is written inline inside an architecture. There is, however, an important difference: the declarations made inside the block are not visible to the enclosing architecture, whereas any declarations made in the architecture are visible inside the block.

The following example shows a *block* statement used to group the signals and operations related to a multiply-and-accumulate (MAC) unit inside a larger signal processing architecture. Signals used only inside the block are declared locally, helping keep the rest of the architecture code cleaner.

```
architecture rtl of dsp is
    ...
begin

    mac_unit: block
        port (a, b: in float32; mac_out: out float32);
        port map (a => x, b => y, mac_out => acc);
        signal multiplier_out: float32;
        signal adder_out: float32;
    begin
        multiplier_out <= a * b;
        adder_out <= multiplier_out + mac_out;
```

```
        process (clock, reset) begin
            if reset then
                mac_out <= (others => '0');
            elsif rising_edge(clock) then
                mac_out <= adder_out;
            end if;
        end process;
    end block;


    ...

end;
```

The second use for the *block* statement is to disconnect signal drivers. To understand this use, we must introduce some VHDL terms such as guarded signals, guarded blocks, and guarded signal assignments. *Guarded signals* are a special kind of signal whose drivers can be turned off (or *disconnected*) and then stop contributing values to the signal's resolution function. *Guarded blocks* contain a special *guard expression*, which controls the execution of *guarded signal assignments*. When the guard expression changes to false, all guarded assignments in the block have their drivers disconnected. When it changes to true, the drivers are reconnected.

The following example shows how guarded blocks and assignments can be used to model a system bus where different blocks have write access at different times. Signal data_bus is declared as a guarded signal, which can be seen from the *bus* keyword used in its declaration. Blocks b1 and b2 are guarded blocks. Each guard expression checks whether the value of signal address is within a certain range. When the value of one of the expressions turns to false, the corresponding driver is disconnected and vice versa.

```
architecture behavioral of system_bus is
    signal data_bus: std_logic_vector(31 downto 0) bus;
    ...
begin

    b1: block (x"0000" <= address and address < x"0010") begin
        data_bus <= guarded peripheral_1_data_out;
    end block;

    b2: block (x"0010" <= address and address < x"0020") begin
        data_bus <= guarded peripheral_2_data_out;
    end block;


    ...

end;
```

```
architecture ... begin                    architecture ... begin

    sum <= operand_a + operand_b;             process begin
                                                  sum <= operand_a + operand_b;
    ...                                           wait on operand_a, operand_b;
                                              end process;
end architecture;
                                              ...

                                          end architecture;
```

(a) A concurrent signal assignment statement.   (b) Equivalent process.

Figure 9.14
Two ways to perform a concurrent signal assignment.

Guarded blocks can be used in models where different parts of the design may provide sources to a signal, one at a time. The previous example shows one of the most typical uses, modeling a system bus. Keep in mind, however, that guarded blocks are generally not supported for synthesis.

Guarded blocks and guarded assignments have several inconveniences, and in general are best avoided. To name a few problems, guarded blocks are generally not supported for synthesis, most designers are not familiar with the construct, and the resulting code may be harder to read because the guard expression and subordinate assignments may be located far apart in the code. Also the standardization of the *std_ logic_1164* multivalue logic system introduced the high-impedance value 'Z', which can be used to model buses in a way that is much more familiar to hardware designers.

Despite its limited use in VHDL designs, the *block* statement has an important role. In the Language Reference Manual, many VHDL statements are described or specified in terms of an equivalent *block* statement. It is, therefore, an important construct for language designers and the developers of VHDL tools.

9.6.2   The Concurrent Signal Assignment Statement

A concurrent signal assignment statement is equivalent to a process that assigns values to a signal. Concurrent signal assignments are one of the most common statements in VHDL. Assignments are covered in detail in chapter 11; here we touch only briefly on the subject, to complete the introduction of all the concurrent statements available in VHDL.

We saw an example of a concurrent signal assignment when we discussed the equivalence between concurrent statements and processes (figure 9.14). There are simple, conditional, and selected versions of the concurrent signal assignment statement (figure 9.15). All three forms are discussed in detail in chapter 11.

```
target <= expression;
```

```
target <=
   value_1 when condition_1 else
   value_2 when condition_2 else
   default_value;
```

```
with expression select
   target <=
      value_1 when choice_1,
      value_2 when choice_2,
      default_value when others;
```

(a) A simple assignment.   (b) A conditional assignment.   (c) A selected assignment.

Figure 9.15
The three kinds of concurrent signal assignment.

9.6.3   The Concurrent Assertion Statement

Assertions are statements that allow VHDL code to check itself as it is simulated, compiled, or synthesized. The *assert* statement is unique in that it is not used to implement functional behavior; rather its primary goal is to assist the developers while the code is being written or verified.

The *assert* statement checks a condition that is expected to be true. If the condition is in fact true, then nothing happens. If the condition is false, then an *assertion violation* happens. The effect of a violation depends on the tools processing the code: a simulator may report a message and interrupt the simulation, whereas a synthesizer may abort the synthesis process.

Just like a signal assignment, there are concurrent and sequential versions of the assert statement. Both have the same syntax. The sequential version is used in sequential code and checks the condition when the statement is reached in the normal execution flow, from top to bottom. The concurrent assertion checks the condition whenever there is a change in one of the signals involved in the expression. In the following example, if address is between 10000000h and 7FFFFFFFh, then an assertion violation occurs, and a message including the offending address will be printed.

```
assert (address < x"1000_0000") or (address >= x"8000_0000")
   report "Invalid access at address " & to_string(address)
   severity warning;
```

In general, sequential assertions are used more often than concurrent assertions because they can be wrapped in *if* statements that check for specific conditions before making an assertion (e.g., the rising edge of a clock). Sequential assert statements are discussed in section 10.6.

10   Sequential Statements

Sequential statements are used to implement algorithms in sequential regions of code—processes and subprograms. Most of the sequential statements available in VHDL are similar to the statements found in any imperative programming language. Although this helps us write algorithms using familiar statements and constructs, there is more to learn before we can use them to create effective VHDL models. Besides the peculiarities of sequential statements in VHDL, we need to know where they should be used and how they can model real hardware.

Compared with concurrent code, sequential code is usually simpler to understand, especially unconditional statements that are always executed from top to bottom. However, few algorithms can be implemented in a strictly linear fashion. In a sequential region of code, the order of execution can be changed with conditional statements called *control structures*. Because such structures can make the code highly complicated, they play a major role in determining its complexity. This chapter presents the sequential statements available in VHDL, with a special emphasis on control structures. Sequential assignment statements are discussed in the next chapter.

10.1   Control Structures

Sequential statements are executed from top to bottom in the order in which they appear in the source code. However, most interesting algorithms require variations in this order of execution. Sometimes a series of statements must be run only when a condition is met; at other times we would like to repeat a sequence of statements iteratively.

Statements that allow us to deviate from linear execution are called *control flow statements* or *control structures*. This name comes from the fact that they change the *flow of control* of a program—the order in which statements are executed. In VHDL, because we cannot control the execution order of concurrent statements directly, it only makes sense to speak of control flow in sequential code. For this reason, control flow statements such as *if*, *case*, and *loop* can only be used in sequential code regions.

While coding, we should pay close attention to control structures because they greatly contribute to the code's complexity. In most complexity metrics, a straight sequence of statements has the lowest possible complexity, no matter how long it is. Complexity increases proportionally to the number of conditional or iterative statements in the code. Control flow complexity has been correlated with poor reliability and a higher incidence of bugs[1], so any effort to make control structures simpler and more readable has a big payoff in the quality of our code.

There are three basic kinds of control structures in VHDL: *if*, *case*, and *loop*. We will see each one in turn after we discuss a typical hazard common to all control structures.

10.1.1   The Problems with Deep Nesting

Because the order of execution in a region of sequential code is well defined, it should be easy to follow. We should be able to read one instruction at a time, figuring out what it does and why it is relevant to the task at hand.

Just like in a maze, things start to get complicated when we meet a bifurcation. When we come across an *if* statement, we must mentally evaluate its condition. If it is true, then we must keep in mind that we are entering a region where the condition was met and continue reading. Each condition is a new level that must be kept in our "mental stack." When we meet an *end if* clause, we need to "pop" the most recent condition and recall the previous state. Unfortunately, our brain is not as efficient as a computer in this kind of task.

The use of control structures inside other control structures is called *nesting*. Understanding deeply nested code is highly taxing on our brains: a study by Ed Yourdon, Noam Chomsky, and Gerald Weinberg suggests that few people can understand more than three levels of nested *if* statements[1]. The Institute for Software Quality (IfSQ) recommends that conditional statements should have a maximum depth of four.[2] Besides taking up space in the reader's mental stack, nesting has other disadvantages: it makes the code harder to read and write because of the long line lengths and excessive indentation. It also makes the compiler's job harder and can inhibit some synthesis optimizations.[3]

There are two general approaches to reducing nesting: reimplement the code using simpler control logic or extract part of the code to a separate routine. We will see more specific guidelines in the following sections when we discuss each kind of control structure in detail.

10.2   The *if* Statement

The *if* statement enables or disables the execution of other statements based on the value of an expression. When an *if* statement is executed, its expression gets evaluated. If the expression result is *true*, then the statements in the *if clause* are executed. If

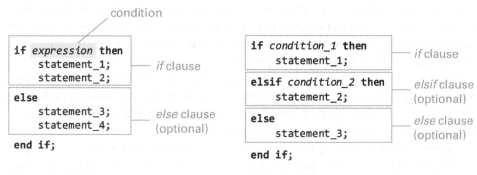

(a) A simple *if-else* statement.

(b) An *if-elsif-else* statement.

Figure 10.1
Parts of an *if* statement.

the expression is *false*, the statements in the (optional) *else clause* are executed (figure 10.1a).

Because the *if* statement is a control flow instruction, it can only be used in regions where there is an explicit order of execution—namely, in sequential code. This is a common source of confusion among VHDL beginners. However, once we understand that there is no predefined order of execution in a region of concurrent code, it makes no sense to use an *if* statement in such regions. When would the *if* be executed? Which other statements would be executed before or after the *if* statement? Once our intentions are clear, it is easy to find a concurrent statement that can accomplish the intended goal (e.g., a selected signal assignment or an *if-generate*).

The *if* statement can get quite complex if all optional parts are used (figure 10.1b) or if it is nested within other *if* statements. Whenever we use conditional statements, we should pay close attention to readability. The next topic presents guidelines to improve the readability of *if* statements and some general advice about their use.

10.2.1 Guidelines for the *if* Statement

Choose the Order of the if and else Clauses Consciously
When we write an *if-else* statement, we can choose which statements to put in the *if* clause and which to put in the *else* clause. We can freely swap them around without changing the original behavior, provided that we negate the condition expression. If putting one of the two blocks first can make our code more readable, then we should use this to our benefit. Here are a few guidelines.

In an *if-else* statement, usually one of the paths is considered the "normal" case, and the other is a deviating case. The normal case may be the one that happens more often or corresponds to the positive value of the condition expression, or the one that processes valid data instead of handling an error condition. In such cases, it is a good idea to write the *if-else* statement so that the normal case comes first. This minimizes the amount of code a reader needs to process when skimming through the code.

In other cases, one of the clauses is short and simple, and the other is more complex. You may choose to put the simple case first to get it out of the way sooner. This has the extra advantage of making both the *if* and *else* keywords visible on the same screen, making it easier to understand the overall logic.[4]

Finally, consider ordering the clauses in a way that the positive case comes first so the condition expression does not have to be negated. We can combine this recommendation with the previous two tips: first choose the case that should appear at the top, then rename the variables or rewrite the condition as needed. For instance, instead of testing if not processing_error, we could rewrite the condition as if processing_ok.

Write the Nominal Path First, Then Add the Unusual Cases   When coding an algorithm, it is usually a good idea to write the nominal path first, ignoring the branches and exceptions that will later become the *else* clauses.[5] Make sure that the nominal path is clear and correct—it will be the back bone holding the algorithm together. Then add the exceptions and deviations. Writing an algorithm in this way ensures that it has a clear normal path and improves readability.

Replace Complicated Tests with a Boolean Variable   Each *if* clause adds a new possible path through the code, increasing its complexity. Although we cannot escape the algorithm's inherent complexity, we can ensure that the statements are no more complicated than necessary by making the conditions unmistakably clear.

A long condition expression is usually composed of two or more subconditions. A good way to make a condition more readable is to find the pieces that have some particular meaning and move them to explaining variables. This is best explained with an example. Suppose you are skimming through a section of code and you run into the following *if* statement:

```
if (sync_counter = 63 and (data_in_sync(7 downto 0) = "01000111" or
    data_in_sync(7 downto 0) = "00100111") and (current_state =
    acquiring_lock  or current_state = recovering_lock) )
then
    locked <= '1';
    ...
end if;
```

It would take a while to decipher this condition and decide whether the contents of the *if* statement are relevant to your current task. Now, suppose the author of the code used a few explaining variables (`waiting_lock`, `last_sync_bit`, and `preamble_detected`), as shown next, to compose the condition:

```
alias current_byte is data_in_sync(7 downto 0);
...
waiting_lock := (current_state = acquiring_lock) or (current_state = recovering_lock);
last_sync_bit := (sync_counter = 63);
preamble_detected := (current_byte = "01000111") or (current_byte = "00100111");

if waiting_lock and last_sync_bit and preamble_detected then
    locked <= '1';
    ...
end if;
```

In this second example, it is much simpler to decide whether the code inside the *if* statement is relevant to our task.

Sometimes, like in the prior example, it is enough to break the expression into a few meaningful terms and then combine them in the condition. At other times, it is better to put the entire expression in a variable. Note that this technique is not restricted to variables; signals and aliases can be used as well.

Replace Complicated Tests with a Boolean Function If the condition in an *if* statement performs a meaningful test on one or more data objects, then we can extract it to a dedicated function. Consider the following example:

```
-- Check whether the two rectangles intersect.
if not (
    (rectangle_1.bottom < rectangle_2.top) or
    (rectangle_1.top > rectangle_2.bottom) or
    (rectangle_1.left > rectangle_2.right) or
    (rectangle_1.right < rectangle_2.left)
) then
    ...
end if;
```

The expression in the *if* statement checks whether the two rectangles intersect. This is a meaningful operation performed on two data objects, so we could extract it to a function:

```
function rectangles_intersect(rect_1, rect_2: rectangle_type)
   return boolean is ...
```

In this way, the *if* statement becomes much simpler and more readable:

```
if rectangles_intersect(rectangle_1, rectangle_2) then
   ...
end if;
```

As a bonus, we can get rid of the (now) redundant comment before the statement.

Use the Condition Operator "??" to Clean up the Expression   In other programming languages, automatic type conversion allows us to use expressions of types other than boolean in *if* statements. If an expression returns an integer, then the compiler converts it to *false* when the value is zero and *true* otherwise.

VHDL, in contrast, requires that the expression produce a value of type boolean. Prior to VHDL-2008, this caused excessive verbosity because the data in a condition expression is often of a logic type (*bit* or *std_logic*). Thus, we had to write:

```
if chip_sel = '1' and enable = '1' and write = '1' then ...
```

In one of the many improvements aimed at reducing verbosity, VHDL-2008 introduced the *condition operator*, "??", that converts from other types to boolean. Thus, we can rewrite the previous expression as:

```
if ??(chip_sel and enable and write) then ...
```

This helps clean up the condition expression a little. However, VHDL-2008 introduced another nice feature: it applies the condition operator automatically, whenever possible, in conditional statements such as *if*, *while*, or *when* (see section 8.1.1 for the details). Thus, we could simply write:

```
if chip_sel and enable and write then ...
```

The condition operator is predefined for the logic types *bit* and *std_logic*, but we can write overloaded versions for our custom data types.

Consider Using a Conditional Assignment Instead of Trivial *if* Statements   Sometimes the only action performed in an *if-else* statement is to select one of two possible values and assign it to a data object:

```
if error_detected then
    next_state <= locking_error;
else
    next_state <= locked;
end if;
```

In such cases, replacing the *if* statement with a conditional assignment makes the code more concise:

```
next_state <= locking_error when error_detected else locked;
```

In VHDL-2008, conditional assignments can be used with either signals or variables and in concurrent or sequential code.

Convert Nested ifs to a Chain of if-elsif Statements   To reduce nesting, consider converting a nested sequence of *if* statements into a flat chain of *if-elsif* statements. In the example of figure 10.2, a structure with three *if* statements and two levels of nesting (figure 10.2a) was reduced to a single *if-elsif* chain (figure 10.2b).

Strictly speaking, the *if-elsif* chain is still a nested control structure. However, it reads as a series of conditions at the same depth or level because we can read the conditions sequentially until one of them is met. Because there is no need to keep the previous conditions in our mental stack as we read the *elsif* clauses, the structure in figure 10.2b is easier to follow. However, this does not always make the code more readable; you should experiment with the two alternatives and choose the one that works best in each case.

```
if (a) then                          if (a and b) then
    if (b) then                          action_when_a_b;
        action_when_a_b;             elsif (a and not b) then
    else                                 action_when_a_not_b;
        action_when_a_not_b;         elsif (not a and b) then
    end if;                              action_when_not_a_b;
else                                 else
    if (b) then                          action_when_not_a_not_b;
        action_when_not_a_b;         end if;
    else
        action_when_not_a_not_b;
    end if;
end if;
```

(a) Nested *if-then-else* statements.   (b) A flat chain of *if-elsif* statements.

Figure 10.2
Converting nested *if*s to a chain of *if-elsif* statements.

```
process (all) begin          process (all) begin          process (all) begin
   if sel = "01" then            if sel = "01"  then          output <= 0;
      output <= mem(0);             output <= mem(0);         if sel = "01" then
   elsif sel = "10" then         elsif sel = "10" then           output <= mem(0);
      output <= mem(1);             output <= mem(1);         elsif sel = "10" then
   end if;                       else                            output <= mem(1);
end process;                        output <= 0;              end if;
                                 end if;                   end process;
                              end process;
```

(a) A latch is inferred because   (b) Latch prevented with   (c) Latch prevented with a
    there are values of *sel* for   an *else* clause; all possible   default assignment at the
    which *output* is not   paths are covered.   top of the process.
    assigned.

Figure 10.3
Latch inference due to incomplete conditionals.

Be Careful with Incomplete Conditionals To generate combinational logic using a process, we must assign a value to every signal and variable controlled by the process, on every possible execution path. If an object is not assigned a value during a run of the process, then it keeps its previous state. In clocked processes, this implies edge-sensitive storage or registers. In nonclocked processes, this implies latches, which are generally undesirable. For details, see "Unintentional Storage Elements (Latches or Registers)" in section 19.4.2.

Figure 10.3a shows a process where a latch is inferred because, depending on the value of sel, it is possible that signal output is not assigned a value. The semantics of VHDL specify that the signal should retain its value, thus implying storage. Because the process is nonclocked, a latch is inferred for output. Figures 10.3b and 10.3c show two possible ways to remove the unwanted latch. This advice is valid for *if* statements as well as any conditional statements in similar situations.

10.3   The *case* Statement

The *case* statement specifies a number of sequences of statements and selects one of them for execution depending on the value of a single expression. Because the choices share a single condition that is evaluated only once, the *case* statement is simpler than other implementation alternatives, such as a chain of *if-elsif* statements.

The *case* statement is formed by a *selector expression* and one or more *alternatives* (figure 10.4a). Each alternative includes one or more *choices* and a sequence of statements. When the *case* statement is executed, the selector expression is evaluated, then the alternative whose choice matches the expression value is selected for execution. Conceptually, this occurs in a single step, rather than going through the choices sequentially.

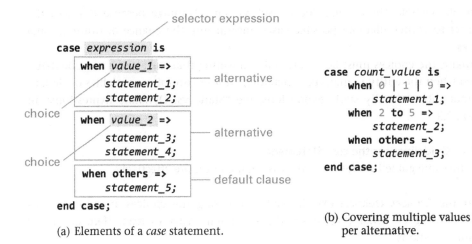

(a) Elements of a *case* statement.

(b) Covering multiple values per alternative.

Figure 10.4
Examples of *case* statements.

Figure 10.4b demonstrates two convenient ways to cover multiple choices in a single alternative: we can list multiple values separated with the vertical bar symbol (|) or use a range when the values are contiguous (2 to 5, in the example).

The *case* statement has several advantages over a chain of *if-elsif* statements. It is easier to read because of its regularity, and the expression does not have to be repeated for every choice, resulting in less clutter. The *case* statement is also easier to understand because its behavior is much more limited than an *if-elsif* statement. The selector expression is evaluated only once when the statement is executed, and all the choices are compared against the same expression. In an *if* statement, in contrast, we must read all the conditions, which may be totally unrelated to one another. Finally, because the expression is evaluated only once, the *case* statement is easier to debug and usually more efficient in simulation.

For all its advantages, the *case* statement has several limitations. The first restriction is that we can only use expressions that produce values of a discrete type or vectors of characters. This sounds like a serious limitation, but in practice, it does not cause much trouble. The allowed types cover most of the interesting cases such as integers, *std_logic_vectors*, and enumerations.

Another restriction is that the alternatives must cover all possible choices. In other words, they must cover all the possible values of the expression return type. If we want to specify behavior for only some of the possible choices, then we must include the default clause *when others* to represent all values not covered in the previous alternatives. If a default clause exists, then it must be the last alternative in the statement. Despite being a restriction, the fact that the alternatives must cover all possible choices may be considered an advantage because it forces the designer to contemplate all possible cases, minimizing the chance of unintentional latches.

Finally, the choices must be locally static, meaning that their value must be determined when the entity is analyzed. Therefore, choices cannot use variables or signals. For more about locally static expressions, see "Static vs. Dynamic Expressions" in section 7.3.

10.3.1 Guidelines for the *case* Statement

Here are some guidelines for using the *case* statement more effectively.

Order the Choices Deliberately Instead of ordering the choices at random, consider choosing an order that makes the statement easier to read. Here are a few recommendations.

Consider writing the most frequent or commonly expected cases first. We can save the reader some time if the alternatives at the top are the most likely or interesting ones.

If the choices have a natural order, such as succeeding states in a finite state machine, then write the alternatives in the same order as they occur during execution. Try to imagine what someone reading the code would like to see next.

Finally, if the *case* statement has a large number of choices that are equally important and do not occur in any predetermined order, then consider ordering them alphabetically or numerically. At least this makes the code easier to navigate.

Keep the Code in the Alternatives Simple and Short The sequence of statements inside each alternative should be simple and short. If you write complex logic inside the alternatives, then the *case* statement will get too long, making it hard to get the big picture of what the statement does. It will be difficult to see the code around the *case* statement, which may contain important initialization steps or default assignments. It will be hard to figure the order in which the alternatives will occur when the code is executed, especially if the statement is controlling a temporal sequence of actions.

If an alternative grows too long, then consider extracting the code inside the alternative to a function or procedure, thus replacing a large number of statements with a single subprogram call. Another option is to break the code into shorter *case* statements. Finally, make sure that the *case* is not trying to perform two or more separate tasks.

Avoid Using the Same Case Statement for Distinct Tasks Sometimes a *case* statement gets long because its alternatives have too many statements. In these situations, it is possible that the statement is doing two or more separate tasks that only share a common selector expression. If this is the case, then separate the code into two or more *case* statements, one for each task.

Use the Matching case Statement ("case?") for *don't care* Inputs   In some cases, we are not interested in all the bits of an input when we make a selection. The *matching case* statement (*"case?"*), introduced in VHDL-2008, allows us to describe such situations:

```
case? data_word is
    when "1---" =>
        leading_zeros_count := 0;
    when "01--" =>
        leading_zeros_count := 1;
    when "001-" =>
        leading_zeros_count := 2;
    when "0001" =>
        leading_zeros_count := 3;
    when "0000" =>
        leading_zeros_count := 4;
    when others =>
        null;
end case?;
```

The main advantage of the matching *case* statement is that it collapses multiple values into a single choice, making the code much more compact in some situations. It is *not* meant for logic optimization; for this purpose, we should use *don't care outputs* in the values of the assignments.

The *don't care* terms ('-') match any value at their corresponding positions. Thus, the first choice in the earlier example ("1---") would match the values "1000", "1111", "1010", and so on. This is different from an ordinary *case* statement, which matches each position literally: in an ordinary *case*, the first choice would be selected only if the input were literally "1---".

As in ordinary *case* statements, there can be no overlap between the choices. In the prior example, it would be illegal to add a choice with the value "11--" because it would overlap with the value "1---" already specified.

Prefer *case* Statements without a Default Clause   Whenever possible, try to cover all the choices explicitly without resorting to a *when others* clause. The use of a default clause could mean that a developer did not consider all the alternatives carefully.[6] Without a default clause, we are forced to specify the actions explicitly for each possible choice.

Certain *case* statements may break under modification due to the default clause. Suppose that your code has many *case* statements whose expressions produce a value of an enumerated type. When you add a new value to the enumeration, two things could happen. If you wrote your *case* statements without a default clause, then the compiler will let you know immediately that the statements do not cover all possible choices. In contrast, if you used a *when others* clause, then the new choice would be silently absorbed—you might have introduced a bug without any warning from the compiler.

If you routinely use default clauses because the type of the expression has too many values, then a useful trick is to declare a new subtype, containing only those values needed in the *case* statement. Then you can qualify the expression in place using the new subtype. This will prevent compiler warnings about uncovered cases.

Do Not Use the Default Clause to Handle a Normal Choice Value   The default clause is the right place to detect errors and unexpected conditions. It is also the place to specify a generic course of action. However, it is not a good idea to use a *when others* clause when the actions inside it could be associated with explicit choice values.

Sometimes we may be tempted to use a *when others* to handle the last choice. However, this weakens the self-documenting nature of the *case* statement. We also lose the compiler checks for uncovered choices and the possibility of using the default clause for error handling. In the cases where you are forced by the compiler to include a *when others* clause, consider using a `null` statement instead of using it to handle a valid choice.

10.4 Loops

Loop is a generic name for any control structure that allows repeated execution of a sequence of statements. The statements are called the loop *body*, and they are executed until a predefined number of iterations is reached, a controlling condition is met, or the loop is exited manually via auxiliary loop control statements.

Loops are useful in both behavioral and RTL code, in simulation or synthesizable code. In behavioral code, loops specify actions to be repeated. In synthesizable code, loops describe algorithmic calculations or replication of hardware structures.

10.4.1 Kinds of Loops

Although VHDL has only one loop statement, it allows three different configurations: a simple (unconditional) loop, a *while* loop that executes until a condition becomes false, and a *for* loop that iterates automatically through a predefined sequence. Figure 10.5 illustrates the three kinds of loops.

Knowing when to use each kind of loop is essential to writing high-quality code. We will see the distinctive characteristics of each one in turn.

```
variable i: integer := 0;     variable i: integer := 0;     for i in 1 to 10 loop
...                           ...                               statements;
loop                          while i < 10 loop              end loop;
    statements;                   statements;
    i := i + 1;                   i := i + 1;
    exit when i = 10;         end loop;
end loop;
```

(a) A simple loop.   (b) A *while* loop.   (c) A *for* loop.

Figure 10.5
Kinds of loops in VHDL. The three examples implement the same behavior.

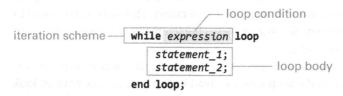

Figure 10.6
Elements of a *while* loop statement.

The *simple* Loop   The *simple* loop statement (figure 10.5a) does not include any kind of automatic iteration control. Left by itself, it would execute indefinitely and would have little use in practice. However, it is mostly used together with auxiliary loop control statements, which allow advancing to the next iteration or exiting the loop instantly. In such cases, although still called a "simple" loop, the control logic can get quite complicated (see section 10.4.2 for the details).

The *while* Loop   The *while* loop (figure 10.6) is a *condition-controlled* loop. The condition is tested before each iteration; if it is true, then the loop body is executed. Otherwise the loop is finished.

Because the condition is reevaluated at the beginning of each iteration, the *while* loop can be used when the total number of iterations is unknown at first. The expression may use variables calculated inside the loop or objects updated elsewhere in the code.

Although the condition is evaluated automatically, we are still responsible for implementing any counters or flags used in the expression. Nevertheless, it is an improvement over the simple loop in terms of organization: the condition appears in an eminent place, and the loop can be used without auxiliary loop control statements.

The *for* Loop   The *for* loop (figure 10.7) repeats a series of statements for a number of times that is known before the first iteration.

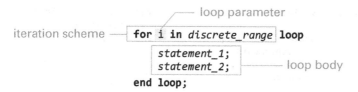

Figure 10.7
Elements of a *for* loop statement.

The *for* loop provides an object, named the *loop parameter*, whose value is controlled automatically during execution. The loop parameter assumes values within the specified range, and it is advanced automatically at each iteration. This value can be used in expressions inside the loop body. Because of these characteristics, the *for* loop falls into the category of *count-controlled* loops.

The *for* loop has two main advantages. First, it puts all the control logic in one place—at the top. This makes the loop easier to read because we do not have to look inside the loop body to understand the iteration logic. Second, all the housekeeping is done automatically; all we have to do is set the loop limits in the beginning. This reduces the risk of mistakes in the loop control logic.

The loop parameter has three important characteristics. First, inside the loop body, it is effectively a constant: it can be read but not modified. Second, the type of the loop parameter is determined from the discrete range specified in the loop statement. It is often numeric, but it does not need to be. For example, for enumeration types, we can use the type name (also called the *type mark*) or the `'range` attribute of an object to specify the loop range. The following code shows an example using a type mark:

```
type pipeline_stage_type is (fetch, decode, execute, memory, writeback);
...
for stage in pipeline_stage_type loop
   report "Current stage: " & to_string(stage);
   ...
end loop;
```

Finally, the loop parameter exists only inside the loop, and it is not necessary to declare it explicitly elsewhere in the code. Because it is implicitly declared, the loop parameter will hide other identifiers with the same name that may exist outside the loop.

10.4.2 Auxiliary Loop Control Statements

Any of the three kinds of loop may use auxiliary control statements to override the normal flow of control defined by their iteration scheme. Loop control statements add a lot of flexibility to the normal iteration scheme; however, they decrease readability and can make the code unnecessarily complicated. There are two auxiliary loop control statements in VHDL: *exit* and *next*. We will see each one in turn.

The *exit* statement finishes the loop immediately, skipping any statements below it and transferring control to the statement immediately after the *end loop* keywords. In the following example, the loop is exited when the value of variable search_value is found in the array stored_values:

```
for i in stored_values'range loop
    if stored_values(i) = search_value then
        match_found := true;
        match_index := i;
        exit;
    end if;
end loop;
```

A common use for the *exit* statement is to test a condition and then exit the loop if it is true. For such cases, the *exit* statement offers a shorthand notation with the keywords *exit-when*:

```
variable instruction: instruction_type;
...
loop
    fetch(instruction);
    exit when instruction = abort;
    execute(instruction);
end loop;
```

In the previous example, the condition expression is evaluated when the *exit* statement is executed. If it is true (i.e., if instruction has the value abort), then the loop is finished immediately.

Sometimes instead of finishing execution of the loop, we just want to advance to the next iteration, skipping the remaining statements in the loop body. In such cases, we can use the *next* statement:

```
for i in entries'range loop
    -- Invalid entries shouldn't contribute to running statistics
    next when entries(i).blank;

    -- Update running statistics
    sum := sum + entries(i).value;
    count := count + 1;
end loop;
```

In the prior example, the *next* statement prevents execution of the two other statements in the loop body when the condition is met (when the current entry is blank).

It is possible to use the *next* and *exit* statements in nested loops. In such cases, the statements refer to the immediately enclosing loop by default. If this is not the intended behavior, then we can label the loops and specify their identifiers after the *next* or *exit* keyword. In the example that follows, the *exit* statements jump out of the outermost loop, and the *next* statement advances to the next iteration of the inner loop:

```
message_loop: for i in messages'range loop
    word_loop: for j in words'range loop
        char_loop: for k in char'range loop
            wait until clock;
            ...
            exit message_loop when framing_error;
        end loop;
        next word_loop when rx_char /= END_OF_WORD;
        ...
        exit message_loop when rx_char = END_OF_MESSAGE;
    end loop;
end loop;
```

Auxiliary loop control statements can make a loop extremely hard to read. The simplest loops have all the control logic in one place. When we use *next* and *exit* statements, we are spreading this control throughout the loop body, forcing us to read the statements inside the loop to understand the iteration logic. When you come across a loop containing several *next* and *exit* statements, think carefully about it and try to rewrite the iteration logic without them. Often we only need to change the condition slightly or enclose part of the *loop* body in an *if* statement. Usually the loop will be clearer without the auxiliary statements.

10.4.3 Loops in Hardware

A synthesizable VHDL model can be used in two widely different settings: verification in a simulator or hardware translation by a synthesis tool. For obvious reasons, we would like the code to exhibit the same behavior in both situations.

This poses an interesting problem: in software, loops are tightly related with the flow of control in a program. In hardware, however, the code is translated to components that work in parallel and are largely independent from one another; there is no centralized entity controlling the entire circuit and choosing which statements to execute next. The synthesis tool has the tricky job of creating a circuit that, without central coordination, exhibits the same behavior seen in the simulator.

If we understand what kind of hardware our loops represent, then we will be making things easier for ourselves and the compiler. The following topics present the four main uses of loops in VHDL.

Parallel Computation In some computations, input and output values are in array form. Such computations are often implemented as a loop. Inside the loop, we use one or more generic expressions to calculate the outputs.

Code written in this fashion is suitable for both synthesis and simulation. In simulation, because signals are updated only at the end of a process, the model behaves as if the values were calculated all at the same time. To obtain the same behavior in hardware, the synthesis tool replicates the hardware structures that implement the expressions. Figure 10.8 shows how a *for* loop can implement a circuit that performs a parallel computation. The statements in the loop (figure 10.8b) implement four multiplications and sums, corresponding to the shaded part of the circuit from figure 10.8a.

Serial Computation Another common use for a loop is to compute an iterative formula. This kind of calculation is characterized by the carryover of state from one iteration to the next. In software, this is easy to accomplish because all instructions in a sequence share and modify the same program state.

In hardware, things are a little different. As we have seen in section 7.1, there is not a one-to-one relationship between VHDL statements and hardware elements in the synthesized circuit. The in-order execution of statements is only a programming model for the developer. Moreover, during the execution of a process, no physical time elapses from the moment it is activated until the moment it suspends. Even if the statements executed include a loop, any computation performed by the process must provide a result immediately. In the cases where a series of actions must be executed over a time period, the designer must create the logic to make it happen, possibly breaking down the computation across multiple clock cycles. This, however, goes beyond the duties of a typical VHDL compiler.

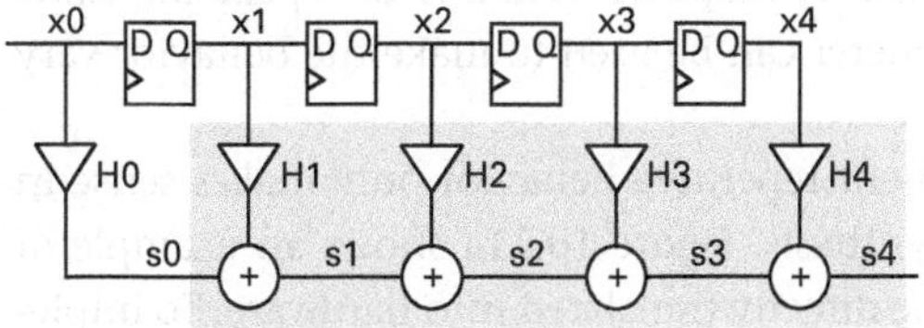

(a) Circuit diagram.

```
for i in 1 to 4 loop
    s(i) <= x(i) * H(i) + s(i-1);
end loop;
```

(b) Implementation using a *for* loop.

Figure 10.8
Example of a parallel computation—an FIR filter with four taps.

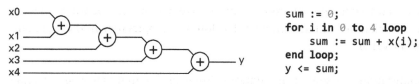

(a) Circuit diagram.   (b) Implementation using a *for* loop.

Figure 10.9
Example of a serial computation—sum of elements in an array.

The net result is that each iteration will originate its own set of hardware elements. In other words, computations performed in the iterations of a loop must exist concurrently, side by side. This is called *loop unrolling* or *spatial replication*, the approach used by a synthesizer to implement iterative formulas. In this way, an iteration has access to the values computed in previous iterations because they exist side by side in the synthesized circuit. The example of figure 10.9 shows how a *for* loop can be used to implement a serial computation—summing all the values in an array.

The point to gain from this example is that certain loops may be implemented as purely combinational hardware. Although the code seems to imply it, there is no need to keep any state or use any storage elements. However, this kind of loop can generate long chains of combinational logic and cause long propagation delays.

Structure and Interconnections   A common need in structural design is to instantiate the same component many times over. This replication is usually implemented with a *for-generate* statement, which is a concurrent statement that also qualifies as a loop. Figure 10.10 shows an example of a component that is replicated four times.

The use of a loop reduces code duplication and also helps in connecting with signals in array form. Instead of writing the array indices literally, their values can be calculated as a function of the loop parameter.

Repetitive Behavior   The last common use of loops in VHDL is to repeat the same action a number of times. The loop parameter can be used to make the behavior vary across iterations.

In general, the use of loops to implement repetitive behavior only makes sense in simulation code and not for hardware synthesis. Figure 10.11a shows an example of behavioral code that cannot, in general, be directly translated into hardware. To implement this behavior in hardware, the designer would need to implement some dedicated control logic, such as the FSM of figure 10.11b.

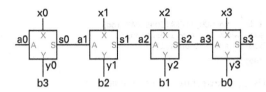

```
connect: for i in 0 to 3 generate
    M(X => x(i), A => a(i),
      S => s(i), Y => y(i));
    a(i) <= 0 when (i = 0) else s(i-1);
    b(3-i) <= y(i);
end generate;
```

(a) Circuit diagram.   (b) Implementation using a *for* loop.

Figure 10.10
Example of hardware replication.

```
loop
    for i in 1 to 3 loop
        red <= '1'; wait until timeout;
        red <= '0'; wait until timeout;
    end loop;

    for i in 1 to 3 loop
        green <= '1'; wait until timeout;
        green <= '0'; wait until timeout;
    end loop;
end loop;
```

(a) Behavioral code for blinking LEDs   (b) An FSM that could reproduce the same
(generally not synthezisable).   behavior in hardware.

Figure 10.11
Not all loops can be translated directly into hardware.

As tools evolve, some synthesizers are already capable of inferring states from a series of *wait* statements that use the same condition expression. Such tools are called *behavioral synthesizers,* and this technique is called *implicit finite sate machine coding.*

In simulation code, loops in VHDL are as flexible as in other programming languages. For testbenches or behavioral models, it is perfectly fine to use loops for implementing repetitive behavior.

So What Kinds of Loops Are Synthesizable?   The answer to this question has changed over time. Once it was possible to say that only *for* loops with constant limits were synthesizable. Today, however, many synthesis tools accept all three examples shown in figure 10.5. If this answer is really important to you, then check your synthesis tool's manual—or, even better, write some sample code and test it. You might be surprised with what synthesis tools support these days.

Still, one common limitation is that to be synthesizable, a *for* loop must have a static range. This means, for instance, that we cannot use variables or signals to specify the loop limits (although a static attribute of a signal, such as 'range or 'length, would

```
sum := 0;
for i in 1 to num_iterations loop
    sum := sum + inputs(i);
end loop;
```

(a) If the synthesis tool supports
    dynamic ranges.

```
sum := 0;
for i in 1 to NUM_ITERATIONS_MAX loop
    exit when i > num_iterations;
    sum := sum + inputs(i);
end loop;
```

(b) Using an *exit* statement to replace
    a dynamic range.

```
sum := 0;
for i in 1 to NUM_ITERATIONS_MAX loop
    if i <= num_iterations then
        sum := sum + inputs(i);
    end if;
end loop;
```

(c) Using an *if* statement to replace
    a dynamic range.

Figure 10.12
Alternative implementations of a *for* loop with a dynamic range.

be fine). In the examples of figure 10.12, num_iterations is a signal of an integer type, and it sets the number of iterations to be performed in the loop. If your synthesizer supports dynamic ranges in *for* loops, then you can use the solution in figure 10.12a. Otherwise try one of the choices in figure 10.12b or 10.12c.

10.4.4 Guidelines for Using Loops

Choose the Right Kind of Loop How do we choose the right kind of loop for each situation? Because the *for* loop controls the iteration automatically, we should use it whenever possible to minimize the chance of coding mistakes. Use it whenever the number of iterations is known in advance—in other words, whenever it can be determined at compile time.

If the number of iterations cannot be determined before the loop starts executing, then use a *while* loop. Try to limit the test to a single place, in the condition expression that is part of the statement. If possible, avoid auxiliary loop control statements.

The simple loop is the right choice only for infinite loops, or for loops that finish via an *exit* statement and whose stop condition cannot be rewritten to use a *while* loop. For example, sometimes you may want the action in the loop body to be executed at least once. In such cases, a possible alternative is to use a simple loop, with an *exit-when* as the last statement. However, a better solution is to rewrite the control logic, so that the condition can be checked at the beginning, and then use a *while* loop.

```
for i in 0 to 31 loop
    a(i) <= '0';                    a <= (others => '0');
end loop;
```

(a) Needless use of a *for* loop.   (b) Using an array aggregate with an *others* choice.

Figure 10.13
Two ways of setting the elements of an array to zero.

```
for i in 1 to 31 loop
    a(i) <= a(i-1);                 a(31 downto 1) <= a(30 downto 0);
end loop;
```

(a) Needless use of a *for* loop.   (b) Using array slices.

Figure 10.14
Two ways to perform a shift.

Make Sure You Really Need a Loop   Before you decide to use a loop, look for a simpler way to do the job. For example, instead of using a loop to assign the same value to an entire array, we could use an aggregate (figure 10.13).

This technique is not limited to assigning constant values or assigning the same value to all array elements:

```
a <= (0 to 15 => variable_a, 16 to 31 => signal_b);
```

Another place where a loop is *not* needed is to perform shifts and rotations, or other kinds of assignments between array slices (figure 10.14a). In this case, your code will be much more compact and expressive if you use the predefined shift and rotate operators (rol, ror, sll, srl, sla, and sra) or array slices (figure 10.14b).

Get into the habit of asking yourself whether a loop is really necessary; you will probably find many opportunities to simplify your code.

Keep the Loop Control Logic Simple   The simplest way to implement a loop is to put all the control logic in one place, including the declaration and update of the loop control variables. The only kind of loop that allows this is a *for* loop without any auxiliary control statements. Whenever possible, try to write your loops in this format.

In general, prefer loops that have a single exit point. One of the few exceptions to this rule is to reduce nesting (for details, see "Break the iteration early to reduce nesting" later in this section).

If you really need to use loop control statements, then concentrate them in one place, either at the top or bottom of the loop. Always initialize any control variable immediately before the loop.

Treat the Loop Body as a Black Box   From the viewpoint of the surrounding code, the loop body should be treated as a black box.[7] Your loops will be easier to read if you can understand the iteration scheme without looking at the statements inside the loop. The use of auxiliary loop control statements (*next* and *exit*) in the loop body eliminates this possibility.

Besides improving readability, this practice has another advantage: it makes it easy to extract the loop body to its own routine. This can be necessary if the loop gets too long and complicated.

Make the Loop Short Enough to View All at Once   Try to keep your loops small. A loop should not exceed your screen height,[8] although ideally it should be much shorter than that. Any section of code is easier to understand if you can see all of it at once, without vertical scrolling. Jumping up and down forces us to keep pieces of the code in our heads.

The most effective technique to make the loop body shorter is to extract part of it to a separate routine. If the loop's main goal is to assign values to an array, it may be possible to convert the value expression to a function that takes the loop parameter as an input argument, thus reducing the loop body to a simple assignment.

Give Each Loop a Single Purpose   Sometimes a loop gets too long because it is trying to do too much. This happens when the loop body does two distinct tasks that only share the iteration logic. In this case, you can make the code simpler by separating the body into two or more loop statements, one for each task.

Limit the Level of Nesting   The Institute for Software Quality recommends a maximum nesting level of four, not only in loops but for all control structures. Yourdon, Chomsky, and Weinberg recommend a maximum level of three. No matter whose authority you choose, your code gets significantly less readable beyond a few levels of nesting. To reduce nesting, consider moving the inner loops or other control structures to their own routines.

Avoid Literals in the Loop Limits   Loops that use literal values to specify ranges are hard to maintain. In the example of figure 10.15a, if you change the size of array *a*, then you must remember to update all loops that depend on that range. Furthermore, it is hard for someone reading the code to make the connection between a literal value and other elements in the code.

Using a named constant solves part of the problem (figure 10.15b). However, if the number of iterations is determined by the size of an object, then it is best to use the attribute 'range to define the loop limits (figure 10.15c).

```
for i in 0 to 31 loop        for i in 0 to Y_SIZE-1 loop      for i in y'range loop
    a(i) <= i;                   a(i) <= i;                       a(i) <= i;
end loop;                    end loop;                        end loop;
```

(a) Using literals in loop   (b) Using a named constant   (c) Using the 'range
    ranges breaks under   mitigates part of the   attribute solves the
    modification.   problem.   problem completely.

Figure 10.15

Using attributes to avoid literals in loop ranges.

```
for i in 1 to 10 loop             for i in 1 to 10 loop
    statement;                        statement;
    if condition_1 then               next when not condition_1;
        statement_1;                  statement_1;
        if condition_2 then           next when not condition_2;
            statement_2;              statement_2;
            if condition_3 then       next when not condition_3;
                statement_3;          statement_3;
            end if;                end loop;
        end if;
    end if;
end loop;
```

(a) A deeply nested loop.   (b) A flat loop body using *next* statements.

Figure 10.16

The *next* statement can be used to reduce nesting.

Break the Iteration Early to Reduce Nesting   The *next* statement can make the loop harder to read because it spreads the iteration control logic throughout the loop body. There is, however, one case where it can make a loop more readable. In a deeply nested piece of code, such as the one in figure 10.16a, we can use the *next* statement to reduce the level of nesting, as shown in figure 10.16b. Each *next* statement works like an *if* that prevents the remaining statements in the loop from being executed. Whenever an *if* statement extends until the end of the loop, it can be replaced with a *next-when* statement with the test condition negated, reducing one level of nesting and indentation.

Choose Meaningful Parameter Names in Long or Nested Loops   In short loops, it is customary (and even recommended) to use short loop names, such as *i*, *j*, and *k*. In general, such short names should never be used for other kinds of variables, but loop parameters have many distinctive characteristics and deserve their own naming conventions. For a discussion about naming loop parameters, see section 16.5.

However, if the loop is more than several lines long, or if it has more than a couple of levels of nesting, then you should give the loop parameters meaningful names. Consider the following nested loop:

```
for i in X_RESOLUTION'range loop
    for j in Y_RESOLUTION'range loop
        framebuffer(i, j) := BACKGROUND_COLOR;
        for k in 0 to NUM_PLANES-1 loop
            pixel := get_pixel(x => i, y => j, plane => k);
            if pixel /= TRANSPARENT_COLOR then
                framebuffer(i, j) := pixel;
            end if;
        end loop;
    end loop;
end loop;
```

How can we know whether the associations in the highlighted line are correct? The names of the loop parameters give no hint of what they stand for, so we need to go back to the header of each loop and mentally make the connections. Consider the same code giving the loop parameters more meaningful names:

```
for x in X_RESOLUTION'range loop
    for y in Y_RESOLUTION'range loop
        framebuffer(x, y) := BACKGROUND_COLOR;
        for plane_index in 0 to NUM_PLANES-1 loop
            pixel := get_pixel(x => x, y => y, plane => plane_index);
            if pixel /= TRANSPARENT_COLOR then
                framebuffer(x, y) := pixel;
            end if;
        end loop;
    end loop;
end loop;
```

Now the highlighted line is obviously correct. If we had mistakenly exchanged the names in the association list, then we would know it immediately.

10.5 The *wait* Statement

The *wait* statement suspends the execution of a process until a condition is met. The condition may include:

- A **sensitivity list**, causing the process to resume when an event occurs in one of the signals in the list;
- A **condition expression**, causing the process to resume when the expression evaluates to *true*;
- A **time expression**, causing the process to resume when the specified time has elapsed.

The complete *wait* statement has the form:

```
wait on sensitivity_list until condition_expression for time_expression;
```

All three clauses are optional, so it is possible to write:

```
wait;
wait on clock, reset;
wait until clock;
wait for 10 ns;
wait on clock until reset for 100 ms;
```

We will discuss each kind of condition in turn.

wait (unconditional) The unconditional *wait* statement, without any optional clause, suspends the process indefinitely. A possible use for it is to generate stimuli at the beginning of a simulation. In the following example, the process generates a pulse on the *reset* signal and then suspends forever:

```
process begin
    reset <= '0', '1' after 10 ns, '0' after 20 ns;
    wait;
end process;
```

wait on sensitivity_list
The *wait* statement with a sensitivity list causes the process to suspend until an event happens on one of the signals in the list. In the following example, the sensitivity list is composed of the signals clock and reset:

```
wait on clock, reset;
```

When the process reaches this statement, it will suspend. Then whenever clock or reset changes value, the process will resume.

wait until condition_expression The *wait* statement with an *until* clause causes the process to suspend until the given expression becomes true. However, the expression is not checked continuously; it is evaluated only when a signal in the sensitivity list of the *wait* statement changes. The sensitivity list may be specified explicitly with an *on* clause. Consider the following statement:

```
wait on clock until clock = '1' and reset = '0';
```

The expression will be evaluated whenever the value of clock changes; however, the process will resume only if clock is '1' and reset is '0'.

If a sensitivity list is not specified with an *on* clause, then the *wait* statement will have an *implicit sensitivity list*, composed of all signals present in the condition expression. Thus, the following two statements are equivalent:

```
wait on clock, reset until clock = '1' and reset = '0';
wait until clock = '1' and reset = '0';
```

The condition operator is also applied implicitly if the type of the expression is different than boolean, so a third way of writing the same statement is:

```
wait until clock and not reset;
```

The implicit sensitivity list in a *wait until* statement is a common source of confusion among VHDL beginners. Here we explain three of the most common mistakes.

First, because only signals are included in the implicit sensitivity list, this causes a problem when the condition expression uses only variables: the process never resumes because the implicit sensitivity list is empty. Thus, the *wait* statement below would never resume:

```
wait until some_variable = '1';
```

Second, a process always suspends when it reaches a *wait* statement; only then it waits for an event on the sensitivity list. Because of this characteristic, a wait statement like:

```
wait until some_signal = '1';
```

will always suspend, even if the condition is true at the moment when the statement is executed. This surprises some developers who expect the statement to "fall through" in this situation.

Finally, because of the implicit sensitivity list, there is no reason to use the 'event attribute on the sensitivity list or in the condition expression (other than satisfying an old compiler). Thus, instead of writing:

```
wait until clock'event and clock = '1';
```

prefer writing:

```
wait until clock = '1';
```

Or, alternatively, if you want to be even more concise:

```
wait until clock;
```

wait for time_expression The *wait* statement with a *for time_expression* clause suspends the process for the specified time interval. This may be used in a process to model periodic behavior. In the following example, the counter is incremented every 10 nanoseconds:

```
for i in 0 to 255 loop
    count <= i;
    wait for 10 ns;
end loop;
```

The *wait-for* clause can be combined with the two other clauses (*on* and *until*). In such cases, it defines the maximum amount of time during which the process should be suspended. Effectively, it works as a timeout.

Table 10.1 Usage examples for the *wait* statement

Statement	Comments
`wait;`	Execution suspends unconditionally; never resumes.
`wait on some_signal;`	Suspend execution until there is an event on `some_signal`.
`wait until some_signal;`	Suspend execution until there is an event on `some_signal` and its value is true.
`wait on some_variable;`	ERROR: `some_variable` is not a signal.
`wait until some_variable;`	BUG: the process never resumes because the implicit sensitivity list is empty.
`wait on some_signal` `until some_variable;`	OK: process resumes when there is an event on `some_signal` and `some_variable` is true.
`wait for some_time;`	Suspend execution for the given amount of time.
`wait on some_signal` `for some_time;`	Suspend execution until there is an event on `some_signal` or `some_time` elapses—whichever comes first.
`wait until some_signal` `for some_time;`	Suspend execution until `some_signal` becomes true **or** `some_time` elapses—whichever comes first.

`wait on some_signal`
  `until some_signal`
  `for some_time;`
	This is the same as `wait until some_signal for some_time;`

`wait on some_signal`
  `until some_other_signal`
  `for some_time;`
	Suspend until `some_signal` changes and `some_other_signal` is true. Because `some_other_signal` is not in the sensitivity list, it will not cause the process to resume when it changes. Because `some_signal` is not in the condition expression, we could not rely on the implicit sensitivity list.

The timeout clause has no meaning for synthesis. Some synthesizers may give you an error, whereas others will completely ignore it.

Combining the *on*, *until*, and *for* Clauses   Because all three clauses are optional, it is possible to use them in any combination. Table 10.1 shows some usage examples of the *wait* statement with one or more of the optional clauses.

10.5.1   Synthesizable *wait* Statements

The *wait* statement is supported for synthesis but with several limitations. Such restrictions are specified in the *IEEE Standard for VHDL Register Transfer Level (RTL) Synthesis,* IEEE Std 1076.6-2004. This section presents the most relevant restrictions.

First, the *for* clause, which causes a process to suspend for a given time period, is not synthesizable. Because no hardware platform has the capability to produce arbitrary delays, statements with this kind of timing behavior are not synthesizable. To implement delays in hardware, use a specific hardware structure, such as a counter or a FIFO.

Second, if we use several *wait* statements in a process, then all of them should use *the same edge of the same clock* for the process to be synthesizable. To understand this restriction, consider that in a process with multiple *wait* statements, the synthesizer must infer a hardware structure to keep track of the current point of execution. This structure should "pause" at the places specified by the *wait* statements in the process. The approach used by the synthesizer is to create a finite state machine whose transitions occur when the *until* clause becomes true. To keep this FSM synchronous and to prevent it from being too complex, all state registers must share the same clock. If you want to use this style, then check whether your tool offers support for it; not all synthesizers are ready for it yet.

Currently, your best chance to make the synthesizer understand your intention is to use a single *wait* statement, either at the beginning or end of the process. The *wait* statement may have an explicit clock edge, an implicit clock edge, or no clock edge at all. However, in this last case, the standard dictates that an *if* statement following the *wait* statement must have a clock edge in the condition. In general, putting the *wait* statement at the bottom of the process is a good idea because it ensures that signal values are calculated at least once before the process suspends, making simulation results more similar to the real hardware. It also makes the process equivalent to a process with a sensitivity list.

Finally, if your synthesizer has a problem understanding a *wait-on-until* statement, then remember that a *wait on* with an *until* clause is the same as a *wait on,* followed by an *if* statement testing the condition from the *until* clause. Sometimes rewriting the statement using this alternative form helps. Otherwise consider rewriting the process to use a sensitivity list.

10.6   The *assert* Statement

Assertions are statements that allow VHDL code to check itself as it is simulated, compiled, or synthesized. The *assert* statement is unique in that it is not used to implement functional behavior; instead its primary goal is to assist the developers while the code is being created.

The *assert* statement checks a condition that is expected to be true. If the condition is in fact true, then nothing happens. If the condition is false, then an *assertion violation* happens. The effect of a violation depends on the tool processing the code: a simulator may report a message and interrupt the simulation, whereas a synthesizer may abort the synthesis process. In any case, an assertion will never generate extra hardware.

The *assert* statement has the form:

```
assert condition report message severity level;
```

The report and severity clauses are optional, so it is possible to write any of the following combinations:

```
assert (1 <= x) and (x <= 127);
assert x > 0 severity error;
assert x /= 0 report "This should never happen";
assert fractional_value(4 downto 0) = "00000"
    report "Precision loss - value will be truncated"
    severity warning;
```

10.6.1   When to Use an Assertion

Programmers use assertions for several reasons: to confirm that a model is working properly, to ensure that a model has been used correctly, to check assumptions made while writing the code, or as a form of executable documentation. We will see each reason in turn.

To Receive Precise Information When Something Goes Wrong   A programmer may add assertions to the code to be warned when something unexpected happens. Assertions catch errors that a compiler could not, and they give information that is more precise than a black-box testbench could provide. Such testbenches only tell us that an output did not match an expected value, whereas an assertion can pinpoint a precise line of the code and tell us exactly what went wrong. Compared to the typical tests written in testbenches, assertions can cause the code to break earlier and are located closer to the problematic code.

To Ensure That the Code Is Working as Expected   Assertions can be used to ensure that the system is working as expected. If a certain pulse must have a minimum duration, then we could write an assertion to check that. For example, to ensure that pulses of the write_strobe signal have a minimum width of TWS, we could write:

```
wait until falling_edge(write_strobe);

assert write_strobe'delayed'stable(TWS)
   report "minimum write strobe width violation"
   severity error;
```

In the previous example, we need the 'delayed attribute because the *wait* statement resumes the process exactly on a falling edge, and write_strobe would never be considered stable at this moment. Therefore, we need to inspect its immediately preceding value.

To Ensure That a Module Is Working with Valid Inputs   Assertions can be used to ensure that the inputs received by a module are valid. In programming jargon, this is called to *enforce preconditions*.

If a function expects to receive a vector with at least two elements, then write an assertion to check that. If your module would break for certain values of a generic constant, then write an assertion to ensure that the values are valid. This can prevent a model from being used incorrectly and provides the user with valuable debug information.

To Ensure That an Action Was Performed Correctly   Assertions can be used after computations to perform correctness checks and prevent incorrect results from propagating to the rest of the system. In programming terms, this is called to *enforce postconditions*. For example, when writing a function to calculate a square root, you could check that the square of the return value equals the input value.

To Check Assumptions   You can use assertions to check that assumptions made during development still hold as the code is being executed. If you think that something is impossible, then write an assertion to document that assumption. For example, if halfway through a function you believe that a variable's value will always be positive, and if the code that follows depends on that, then write an assertion to ensure that you are not caught off guard when the impossible happens.

To Document the Code   Assertions can be used to document assumptions and expectations. Besides verifying that the data received by a function is correct, an assertion also documents that expectation in the code, serving as a form of executable specification. This is better than documenting the expectation in a comment because the assertion cannot fall out of sync with the code.

To Detect Model Failures in Testbench Code   In testbenches, assertions are used to check the outputs of a design under test (DUT), comparing them against expected values. If an output differs from the expected value, then an assertion is violated. In that case, an error message is logged, and the simulation may be interrupted depending on the severity of the error and on simulator configurations.

10.6.2   When *Not* to Use an Assertion

Sometimes you may find an assertion with a hardcoded value *false*, just to report a message during a simulation. Do not do that. Use the report statement instead:

```
assert false report "Everything normal..." severity note; -- Do not do this!
```

```
report "Everything normal...";   -- Ok; same effect as above.
```

10.6.3   Severity Levels

The severity clause in an assertion statement takes a value of type severity_level. This type is predefined in package *standard* as:

```
type severity_level is (note, warning, error, failure);
```

The VHDL standard recommends that a simulation should be stopped when an assertion of severity *failure* is violated; for any other level, the simulation should continue running. In practice, most simulators allow the user to specify this level. The default severity level for an *assert* statement without a severity clause is *error*, indicating that by default a simulation would not be stopped.

Because the severity level is chosen by the user, a common doubt is which severity level to use in each situation. Table 10.2 shows the typical use cases for each level. These recommendations are based on common industry practices and guidelines provided by the European Space Agency (ESA).[9]

10.7   The *null* Statement

The *null* statement has no effect when executed; it is used to specify a situation in which no action should be taken. It is normally used in *case* statements because we need to cover all possible choices, but an action may not be required in every case.

Table 10.2 Recommended use of severity levels in assertions

Severity Level	When to Use
note	To output debug information when there are no design problems. This level is not very useful for assertions; it is the default level for the *report* statement.
warning	To indicate unexpected conditions that *do not affect the state of the model*, but could affect its behavior. Usually means that the model can continue to run.
error	To indicate unexpected conditions that *affect the state of the model*. Usually associated with unrecoverable error conditions.
failure	To indicate a problem *inside the module* that is reporting the violation. Used in self-checking components to indicate that a bug has been detected and the code needs to be fixed.

In the following example, the *null* statement is used with two different purposes. The first is to make it explicit that when the value of operation is nop, no action should be taken. The second is to specify that for all the remaining values of the enumeration type the *case* statement should have no effect.

```
case operation is
    when add =>
        result := operand_a + operand_b;
    when subtract =>
        result := operand_a - operand_b;
    when nop =>
        null;
    when others =>
        null;
end case;
```

11   Assignment Statements

Assignments are statements that attribute values to data objects. In VHDL, assignments are highly versatile: they can be used to implement combinational or sequential logic, produce outputs or intermediate values, and describe models at the dataflow, RTL, or behavioral level of abstraction. Because of this flexibility, assignments are a frequent source of mistakes in VHDL. To clear up common misunderstandings, this chapter first identifies the kinds of assignments available in VHDL. Then it presents the different constructs that can be used as sources and targets in assignments, to help us write code that is more concise and communicates our intentions more clearly. It closes with a discussion of topics that are useful for testbenches, such as assignments involving timing and delay and force and release assignments.

11.1   Kinds of Assignments in VHDL

At first glance, it may seem like VHDL has many different kinds of assignments, with long and complex names such as "conditional variable assignment" or "concurrent selected signal assignment." In fact, there are only three basic forms of the assignment statement; however, because they can be used in concurrent or sequential code, and with signals or variables, there are many possible combinations. What are generally described as different kinds of assignments are just instances of the three ways to classify an assignment in VHDL: by target object class (signal or variable), by kind of assignment statement (simple, conditional, or selected), or by kind of VHDL code in which it appears (concurrent or sequential).

Figure 11.1 shows the three possible ways to classify an assignment in VHDL. If we study them separately, then we will be able to recognize and use all kinds of assignments. We will see each classification in turn.

11.1.1   Classification by Target Class (Signal vs. Variable)

The first way to categorize an assignment is by its target object's class. In VHDL, the only object classes that can be assigned values are variable and signal. Files cannot be assigned to, and constants have their values set at declaration.

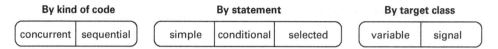

Figure 11.1
The three possible classifications for assignments in VHDL.

```
signal my_signal: integer := 0;          variable my_variable: integer := 0;
signal carry: bit;                        variable carry: bit;
signal sum, op_a, op_b: bit_vector(7 downto 0);   variable sum, op_a, op_b: bit_vector (7 downto 0);
signal vehicle_state: vehicle_state_type;  variable vehicle_state: vehicle_state_type;
...                                       ...
my_signal <= 1;                           my_variable := 1;
(carry, sum) <= ('0' & op_a) + ('0' & op_b);   (carry, sum) := ('0' & op_a) + ('0' & op_b);
vehicle_state <= (                        vehicle_state := (
   gear  => 2,                               gear  => 2,
   speed => 50 mph,                          speed => 50 mph,
   fuel  => 25 gal                           fuel  => 25 gal
);                                        );
```

(a) Signal assignments.   (b) Variable assignments.

Figure 11.2
Examples of signal assignments and variable assignments.

Making a distinction between assignments to signals and variables is important because their behavior is significantly different. Remember that variables and signals are conceptually distinct objects; for a review of their main differences, see "Signals vs. Variables" in section 14.4. The upside of signals and variables being so different from each other is that it is trivial to choose between a signal assignment and a variable assignment: if the target is a variable, then we must use a variable assignment; if the target is a signal, then we must use a signal assignment. The choice is made at the moment we declare the object, not when we write the assignment.

It is easy to tell one kind of assignment from the other in the code. Variable assignments use the variable assignment symbol (:=), whereas signal assignments use the signal assignment symbol (<=). Although the signal assignment symbol coincides with the *less than or equal* operator (<=), it is always possible to determine which one is intended from the context in which they appear in the code.

Figure 11.2 shows some examples of assignments to signals (figure 11.2a) and variables (figure 11.2b). In the source code, the only visible differences between the two are the class of the target object and the symbol used in the assignment statement.

One thing to note in passing is that if an initial value is provided in a signal declaration, then the symbol used is that of a variable assignment (:=). This may look like an inconsistency, but in fact the specification of an initial value has little to do with a signal assignment statement. It does not generate events, and the signal is assumed to have had the same value indefinitely when the simulation begins. We could also reason that the attribution of an initial value is similar to a variable assignment because the signal is initialized instantly, whereas all other signal assignments are scheduled for a future time. The differences in timing behavior between signal and variable assignments are the subject of the next section.

Timing Behavior of Signal and Variable Assignments   The first step in understanding the difference between signal and variable assignments is to understand the differences between signals and variables. Again, refer to "Signals vs. Variables" in section 14.4 if needed. In a nutshell, a variable is an object that holds only its current value, whereas a signal is an object that keeps its past history. Variables are intended for keeping intermediate computation steps or the results of calculations, whereas signals are intended for interprocess communication. For synthesis, both signals and variables can imply storage.

As for their timing behavior, a variable assignment immediately replaces the value of a variable, whereas a signal assignment schedules a new value for the signal at some time in the future. The main consequence of this in sequential code is that the statements following a signal assignment will still see the old value of the signal until the process has a chance to suspend. This is a frequent source of confusion among beginners, so we will look at this in a bit more detail.

The key to understanding this difference is to remember that variables are intended for use *within* a process or subprogram, whereas signals ate intended for communication *between* processes or subprograms. To support those two usages, it follows that the value of a variable must be updated immediately on an assignment, whereas the value of a signal must be held constant during an entire simulation cycle—otherwise the simulation results would be nondeterministic. Each of these statements is discussed separately in turn.

Let us start with signal assignments. A VHDL simulator must give the illusion that all processes and concurrent statements are running simultaneously, in parallel. This implies that the result of a simulation must be independent of the order of execution of all processes and concurrent statements, as seen in section 9.2. However, if a process were allowed to update a signal immediately, the next process or concurrent statement to read this signal would see a different value from what other processes might have seen in the current simulation cycle. This could lead to different results depending on the order of execution of the processes. For this reason, all signal updates are postponed until after all processes have had a chance to execute, guaranteeing that a simulation yields deterministic results.

Now let us turn to variable assignments. A variable is intended for use within a process or subprogram. A typical example is to keep a value representing part of a larger computation, breaking down a complex operation into simpler steps. Another common use is as a loop counter. In both cases, a variable will be useful only if its updated value can be read within the same execution of a process or subprogram. Waiting for a subprogram to return before updating a variable's value would defeat the purpose of using variables in the first place. For this reason, variable assignments are effective immediately, and the updated value is available at the next line of code.

To illustrate these situations, let us see a few examples. The following process calculates the two roots of a quadratic equation in a clocked circuit. The inputs are given by three floating-point signals corresponding to the coefficients: a, b, and c. The outputs are provided in signals root_1 and root_2.

```
1   process (clock)
2       variable discriminant: float32;
3   begin
4       if rising_edge(clock) then
5           discriminant := b**2 - 4 * a * c;
6           root_1 <= (- b + discriminant) / (2 * a);
7           root_2 <= (- b - discriminant) / (2 * a);
8       end if;
9   end process;
```

To simplify the expressions and avoid duplicating code, it is a good idea to calculate the value of the discriminant first and keep it in an intermediate object. In the previous code, a variable was used (line 2). Hence, the first assignment statement is a variable assignment (line 5), and the value gets updated immediately when the statement is executed. Therefore, the subsequent lines referring to the variable (lines 6–7) will read its new value.

Now let us see what happens when we use a signal instead of a variable. The modified code is shown here:

```
1   signal discriminant: float32;
2   ...
3   process (clock) begin
4       if rising_edge(clock) then
5           discriminant <= b**2 - 4 * a * c;
6           root_1 <= (- b + discriminant) / (2 * a);
7           root_2 <= (- b - discriminant) / (2 * a);
8       end if;
9   end process;
```

In this case, the value of the discriminant is calculated in line 5 as well, but because this is a signal assignment, signal discriminant will assume the new value only after the process has suspended. This process has no explicit *wait* statements, so it runs until the last statement before the *end process* keywords. This means that the expressions in lines 6–7 will read the value that discriminant had when the process got activated—its "old" value. From an external perspective (outside the process), the net effect is that the value of the two roots is "delayed" by one clock cycle. This is a frequent source of confusion among VHDL learners; the only way to avoid it is to understand the fundamental difference between signal and variable assignments.

Now let us see an example of how this mechanism of delayed signal updates preserves determinism in a simulation and helps in modeling real hardware. In synchronous circuits, the inputs to all registers are sampled at the same clock edge, and the corresponding VHDL model is a collection of processes sensitive to this event. Because processes are concurrent statements, they do not have a predefined order of execution. Therefore, to preserve determinism, it is imperative that all processes observe the same values for the same signals on an event, independent of their order of execution. As an example, consider the following two processes:

```
process_A: process (clock) begin
    if rising_edge(clock) then
        a <= a + 1;
    end if;
end process;

process_B: process (clock) begin
    if rising_edge(clock) then
        b <= a;
    end if;
end process;
```

In this code, the first process changes the value of signal a, and the second process reads this value. If signal assignments took place immediately, the final value of b would depend on the order of execution of the two processes: if process *A* were executed first, then signal a would be incremented first, and then the incremented value would be assigned to b in the second process. In contrast, if process *B* were executed first, then the original value of signal a (before incrementing) would be copied to b. The value of signal b after executing the two processes would depend on their order of execution. To make this kind of situation deterministic, VHDL "freezes" signal values before executing all processes in a simulation cycle. In this way, no matter which process gets executed first, they all observe the same signal values. Only after all processes have suspended is the simulation cycle allowed to advance.

A final caveat about signal assignments is that when there are multiple assignments to the same signal within a process and the assignments are scheduled for the same time unit, only the last assignment before the process suspends will be effective. This topic is discussed in more detail in section 11.3.

One last thing to note about variable assignments is that they are only allowed in sequential code because variables cannot exist in concurrent code. This naturally makes variables more localized than signals: if a process uses a variable, we know that the variable must have been declared in that process (with the exception of shared variables, discussed in section 14.5). Variable assignments are also conceptually simpler because they do not involve timing, and any new value immediately overwrites the previous one. This saves us the trouble of thinking about drivers, delta delays, and resolution functions. Simplicity and locality are two good reasons to prefer variables over signals when we have this choice.

11.1.2 Classification by Assignment Statement (Simple vs. Conditional vs. Selected)

Sometimes we want an assignment to be executed unconditionally, always evaluating the same expression and assigning the resulting value to a signal. At other times, we want to assign different values depending on one or more conditions. For those cases, VHDL offers three basic forms of the assignment statement: *simple*, *conditional*, and *selected* (figure 11.3). We will see each kind in turn.

Simple Assignment Statement A simple assignment statement (figure 11.3a) always evaluates the same expression and assigns the resulting value to a signal or variable. Of course this does not mean that the value will always be the same; the expression may contain signals and variables, and it may produce a different value each time it is evaluated.

As any other assignment, the target of a simple assignment statement may be a signal or variable. If the target is a variable, then the assignment must be in sequential code, for the simple reason that variables can only exist in sequential code. Simple assignments are often used in combination with control structures to be executed only when a given condition is true.

```
target <= expression;
```
```
target <=
    value_1 when condition_1 else
    value_2 when condition_2 else
    default_value;
```
```
with expression select
    target <=
        value_1 when choice_1,
        value_2 when choice_2,
        default_value when others;
```

(a) A simple assignment. (b) A conditional assignment.   (c) A selected assignment.

Figure 11.3
Assignment statements in VHDL.

Conditional Assignment Statement A conditional assignment statement (figure 11.3b) assigns different values to a data object depending on the value of one or more boolean expressions. This statement evaluates the condition expressions in turn until one is found to be true. Then the corresponding value expression is evaluated, and the result gets assigned to the target object. Note that there are two kinds of expressions involved in a conditional assignment: *value expressions*, which must be of the same type of the target signal, and *condition expressions*, which must be boolean.

A conditional assignment is equivalent to an *if* statement with *elsif* clauses that test the condition expressions and assign one of the possible values to a single data object. Figure 11.4 shows the equivalence between a conditional signal assignment and the corresponding *if* statement. Naturally, if the conditional assignment in figure 11.4a were in a concurrent code region, the *if* statement of figure 11.4b would have to be inside a process.

The main advantage of a conditional assignment in this case is that it provides a compact way of assigning different values to the same object. It is also highly readable when used with simple expressions and boolean variables. However, for complex expressions, the line lengths may get too long; in such cases, it might be better to use the equivalent *if* statement.

Making a parallel with other programming languages, conditional assignments are similar to the conditional ternary operator ? : in languages such as C++ and Java. This operator can be used in expressions and returns one of two values depending on a boolean condition. The main difference is that in VHDL, an assignment is a statement per se and cannot be used as part of an expression. Figure 11.5 illustrates the equivalence between the conditional operator found in other languages and the conditional assignment statement in VHDL.

A final point to note is that a conditional assignment may be used to model level-sensitive or edge-sensitive storage elements. Consider the following D-type flip-flop modeled using a process:

```
target <=
    value_1 when condition_1 else
    value_2 when condition_2 else
    default_value;
```

```
if condition_1 then
    target <= value_1;
elsif condition_2 then
    target <= value_2;
else
    target <= default_value;
end if;
```

(a) A conditional signal assignment.   (b) Equivalent *if* statement.

Figure 11.4
Equivalence between a conditional assignment and an if statement.

```
next_count = (count_direction == up) ? count+1 : count-1;
                        |                    |        |
              boolean condition   value when *true*   value when *false*
```

(a) Conditional operator (C++ or Java).

```
next_count <= count+1 when (count_direction = up) else count-1;
                  |              |                          |
         value when *true*    boolean condition    value when *false*
```

(b) Conditional signal assignment (VHDL).

Figure 11.5
Equivalence between the conditional operator in other languages and the conditional assignment in VHDL.

```
process (clock, reset) begin
    if reset then
        q <= '0';
    elsif rising_edge(clock) then
        q <= d;
    end if;
end process;
```

This code is equivalent to the following concurrent statement:

```
q <= '0' when reset else d when rising_edge(clock);
```

This is probably the most compact way of modeling a flip-flop with an asynchronous reset in VHDL.

Conditional assignments were upgraded in VHDL-2008. In prior versions, the target object had to be a signal, and conditional assignments could only be used in concurrent code. In VHDL-2008, the target object can be a signal or variable, and the conditional assignment may be used in sequential or concurrent code.

Selected Assignment Statement A selected assignment statement (figure 11.6a) assigns different values to a data object depending on a selector expression. The statement includes a number of choices associated with the different values. Each choice is compared with the selector expression in turn; when a choice matches, the associated value is assigned to the target object.

A selected assignment is equivalent to a *case* statement that uses the same selector expression and choices and assigns different values to a single target object.

```
with selector_expression select         case selector_expression is
    target <=                               when choice_1 =>
        value_1 when choice_1,                  target <= value_1;
        value_2 when choice_2,              when choice_2 =>
        default_value when others;              target <= value_2;
                                            when others =>
                                                target <= default_value;
                                        end case;
```

(a) A selected signal assignment.   (b) Equivalent *case* statement.

Figure 11.6
Equivalence between a selected assignment and a case statement.

Figure 11.6 shows the equivalence between a selected signal assignment and the corresponding *case* statement. Again, if the selected assignment in figure 11.6a were in concurrent code, the *case* statement in figure 11.6b would have to be wrapped in a process.

The main advantage of using a selected assignment instead of a *case* statement is that it is more compact. Also, a selected signal assignment can be used in concurrent code, whereas a *case* statement can only be used in sequential code.

In VHDL-2008, the target object can be a signal or variable, and the selected assignment may be used in either sequential or concurrent code. In prior VHDL versions, the target object had to be a signal, and selected assignments could be used only in concurrent code.

11.1.3 Classification by Kind of VHDL Code (Sequential vs. Concurrent)

The last classification of assignment statements is by the kind of VHDL code in which they appear: concurrent or sequential. This distinction is important because it determines when the statement is executed.

Sequential Assignment Statements Sequential assignments, being sequential statements, are executed when they are reached in the normal execution order of the source code, from top to bottom. For this reason, they are often associated with control structures that decide when or whether the assignment should be executed.

Sequential assignments are used in processes or subprograms. The following example shows a combinational process that calculates the value of a signal named sine_table_address based on the value of another signal called phase. The process is activated whenever there is an event on phase. In the process body, the two variable assignments in lines 6 and 7 are executed in sequence. Then one of the signal assignments in lines 10 or 12 is selected for execution.

```
1      process (phase) is
2          variable phase_quadrant: unsigned(1 downto 0);
3          variable phase_offset: unsigned(5 downto 0);
4          variable quadrant_is_odd: std_logic;
5      begin
6          (phase_quadrant, phase_offset) := phase;
7          quadrant_is_odd := not phase_quadrant(0);
8
9          if quadrant_is_odd then
10             sine_table_address <= phase_offset;
11         else
12             sine_table_address <= LUT_ADDRESS_MAX - phase_offset;
13         end if;
14     end process;
```

Note that we used variables instead of signals for the intermediate values phase_ quadrant, phase_offset, and quadrant_is_odd because their updated values are needed immediately after the assignments.

Concurrent Assignment Statements Concurrent assignments exist in concurrent regions of code, and they are executed whenever one of the signals in their implicit sensitivity list changes. This sensitivity list is composed of all the signals referenced in the expressions on the right-hand side of the assignment symbol.

To understand this, remember that, like most concurrent statements, a concurrent sequential assignment is short for a process containing the same statement in sequential code. Consider the following concurrent assignment, which could be located in an architecture body (a concurrent code region). The objects next_count, current_count, and increment are all signals.

```
next_count <= current_count + increment;
```

The expression current_count + increment will be evaluated whenever there is a change in current_count or increment. Then the resulting value will be assigned to next_count. This is equivalent to the following process:

```
process (current_count, increment) begin
    next_count <= current_count + increment;
end process;
```

In the rewritten example, the process is activated whenever a signal in the sensitivity list (current_count or increment) changes. This process has the same behavior as the original concurrent assignment. Because a process with a sensitivity list is equivalent to a process with a *wait* statement sensitive to the same signals, another way to write the same assignment is:

```
process begin
    next_count <= current_count + increment;
    wait on current_count, increment;
end process;
```

As mentioned in "Timing behavior of signal and variable assignments," in section 11.1.1, the target signal (next_count) will be updated only after the process suspends. In other words, during the simulation cycle that caused the process to resume, all other processes will still see the old value of the signal.

Concurrent assignments can be used only with signals for the reason that variables are not allowed in concurrent code regions. Concurrent signal assignments are useful for modeling combinational circuits. They provide a compact and natural way to describe arithmetic and logic equations, which can be interpreted and optimized by a synthesizer, resulting in a cloud of combinational logic.

11.2   Assignment Sources and Targets

Every assignment has two sides: a source, which provides a value, and a target, which takes on that value. VHDL offers several choices for what can be a source or target in an assignment. Knowing the available options is important if we want to write code that is more concise and clearly communicates our intentions. This section surveys the language elements that can be used on both sides of assignments statements.

11.2.1   Assignment Sources

The assignment source provides the value in an assignment and is always an expression. However, in VHDL, many elements qualify as expressions, such as literal values, aggregates, function calls, and the names of signals or variables. To illustrate this, let us see how these elements can be used as sources in assignments.

The simplest source for an assignment is a literal value, which is a constant value written directly in the code:

```
variable items_count: integer range 0 to 15;
variable current_weather: weather_condition_type;
variable tx_data: std_logic_vector(7 downto 0);
...
items_count := 5;          -- An integer literal
current_weather := sunny;  -- An enumeration literal
tx_data := b"11110000";    -- A bitstring literal
```

Other common sources are expressions using operators, the names of signals or variables, and function calls, as shown in the following examples:

```
items_count := items_count + 1;        -- Value given by evaluating an expression
current_weather := yesterdays_weather; -- Value taken from another object
tx_buffer := ("11", others => '0');    -- Value given by an aggregate
average_level := average(samples);     -- Value returned by a function call
```

When the target is an array, two more options are available on the source side: aggregates and array slices. Recall from section 8.2 that an aggregate is a VHDL construct that combines distinct values into a composite object. The syntax of an aggregate is a series of expressions separated by commas and enclosed within parentheses.

The following examples show how aggregates can be used to make the code more compact in common design situations. In lines 2 and 3, an aggregate is used to initialize a constant array in which most of the elements have the same value. In lines 9 and 11, aggregates were used to provide values to arrays whose lengths can be configured through generic constants. The aggregate allows the assignment source to be used with targets of different lengths, making the code self-adjusting. Other examples using aggregates will be shown in the next section.

```
1    -- A lookup table for telling whether an integer number is prime
2    constant IS_PRIME: boolean_vector(0 to 31) :=
3        (2|3|5|7|11|13|17|19|23|29|31 => true, others => false);
4
5    signal rx_buffer: std_logic_vector(RX_BUFFER_LENGTH-1 downto 0);
6    signal tx_buffer: std_logic_vector(TX_BUFFER_LENGTH-1 downto 0);
7    ...
8    -- RX buffer is set to all 0's
9    rx_buffer <= (others => '0');
10   -- TX buffer is set to 100...0011
11   tx_buffer <= (0 to 1 => '1', tx_buffer'high => '1', others => '0');
```

A slice is a subset of consecutive elements in an array. The syntax of an array slice consists of the name of an array object followed by a discrete range within parentheses. The following example shows how slices can be used to separate a vector into four smaller vectors with different meanings:

```
-- Extract the fields from a MIPS I-type instruction
opcode <= instruction(31 downto 26);
rs <= instruction(25 downto 21);
rt <= instruction(20 downto 16);
imm <= instruction(15 downto 0);
```

11.2.2 Assignment Targets

Although several VHDL constructs can be used as a source, only two elements can be the target of assignments in VHDL: a name, corresponding to the whole or a part of a data object, or an aggregate. Most commonly, the target of an assignment is an object name. The following examples show assignments to signals and variables in which the target is an object of a scalar type:

```
carry_out <= (b and carry_in) or (a and carry_in) or (a and b);
current_state <= next_state;
pulse_count := pulse_count + 1;
```

When the target object is an array, the name used as a target can also be a slice. This is shown in the next topic.

Assignments to Arrays and Slices   To make an assignment to a whole array at once, we can specify its name as the target. The following example shows an assignment to an array variable, using as source an aggregate composed of eight literal values:

```
variable fibonacci_sequence: integer_vector(1 to 8);
...
fibonacci_sequence := (0, 1, 1, 2, 3, 5, 8, 13);
```

To make an assignment to part of an array, we can use array slices as targets:

```
tx_buffer(9 downto 2) <= tx_data;
tx_buffer(1 downto 0) <= "01";
```

A common mistake is trying to assign a literal vector, or an array object, to an array element that is a scalar object.

```
tx_buffer(10) <= "1";    -- Error: target type is different from expression type.
tx_buffer(10) <= '1';    -- OK; target and expression are scalar objects.
```

In any case, we always need to make sure that the source and target widths are the same because VHDL does not extend array values automatically.

Element Ordering in Composite Assignments   When an assignment is made between composite objects, the value of each element in the source is assigned to a corresponding element in the target. Because array types and objects can use ascending or descending order, this could cause an ambiguity when the source and target have different orderings. Should the elements be matched by position (the leftmost element in the source assigned to the leftmost element in the target) or index value (the element at position 0 in the source assigned to the element at position 0 in the target)?

To remove any ambiguity, VHDL specifies that assignments involving composite objects are always made in left-to-right order, independent of the object ranges. The following example demonstrates how the leftmost element of the target array is paired with the leftmost element of the source array.

```
variable x: char_vector(3 downto 0);  -- Leftmost element of x is x(3)
variable z: char_vector(0 to 3);       -- Leftmost element of z is z(0)
...

-- Leftmost element of x gets 'a':
x := "abcd";  -- x(3)='a', x(2)='b', x(1)='c', x(0)='d'

-- Leftmost element of z gets the leftmost element of x:
z := x;       -- z(0)=x(3)='a', z(1)=x(2)='b', z(2)=x(1)='c', z(3)=x(0)='d'
```

If this seems hard to grasp or remember, we can always mentally replace the source and target objects by an equivalent aggregate. In our example, because x is declared with the range (3 downto 0), it is equivalent to the aggregate (x(3), x(2), x(1), x(0)). Conversely, because z is declared with the range (0 to 3), its equivalent aggregate is (z(0), z(1), z(2), z(3)). When you put each aggregate on the corresponding side of an assignment, it becomes clear how the individual elements are matched:

```
( z(0), z(1), z(2), z(3) ) := ( x(3), x(2), x(1), x(0) );
```

Assignments to Aggregates   Besides being used on the right side of the assignment symbol, as an expression, aggregates can also be used as the target of an assignment. As expected, each element on the right side gets associated with an element on the left side. In the following example, the single-bit variable carry will get loaded with '1', and the byte-wide variables hi_byte and lo_byte will get loaded with "00000000" and "11111111", respectively:

```
variable carry: std_logic;
variable hi_byte: std_logic_vector(7 downto 0);
variable lo_byte: std_logic_vector(7 downto 0);
constant INITIAL_VALUE: std_logic_vector := "100000000011111111";
...
(carry, hi_byte, lo_byte) := INITIAL_VALUE;
```

This can help make the code more compact by reducing the number of assignments. As another example, the following code breaks an instruction into its component fields using an aggregate as the target:

```
signal instruction: std_logic_vector(31 downto 0);
signal opcode: std_logic_vector(5 downto 0);
signal register_s: std_logic_vector(4 downto 0);
signal register_t: std_logic_vector(4 downto 0);
signal immediate: std_logic_vector(4 downto 0);
...
(opcode, register_s, register_t, immediate) <= instruction;
```

An important point to note about the first example is that the constant INITIAL_ VALUE was declared without an explicit range. In such cases, the leftmost index of the array object will be the leftmost value of its index type. Because the index type of std_logic_vector is natural, the leftmost element of INITIAL_VALUE will be at position 0. Therefore, the declaration is the same as:

```
constant INITIAL_VALUE: std_logic_vector(0 to 17) := "10000000011111111";
```

11.3 Assignments with Timing and Delay

The assignments we have used so far did not include any user-specified time delay; therefore, they were all effective after an infinitesimal delta time unit. However, when writing VHDL models that are not intended for synthesis, it is possible to specify an interval between the moment the assignment is executed and when the target will assume the new value.

Figure 11.7a shows the simplified syntax for a signal assignment with a specific time delay. Figure 11.7b shows how an assignment with a user-specified time delay can model the case where a signal (y) must follow another signal (a) after a 100 ps delay. Figure 11.7c shows how the two signals would appear in a simulation.

Although the value expression in figure 11.7b simply copies the value of signal a, the expression could be as complex as needed to model a block of combinational logic. Moreover, it is not necessary to restrict the assignment to a single transaction. In VHDL, a *waveform element* is the name given to a clause containing a value expression together

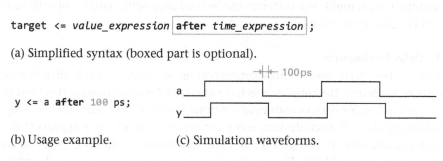

```
target <= value_expression after time_expression ;
```

(a) Simplified syntax (boxed part is optional).

```
y <= a after 100 ps;
```

(b) Usage example.   (c) Simulation waveforms.

Figure 11.7
Signal assignment statement including a user-specified time delay.

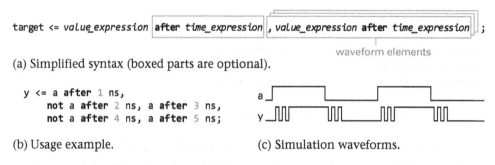

(a) Simplified syntax (boxed parts are optional).

```
y <= a after 1 ns,
    not a after 2 ns, a after 3 ns,
    not a after 4 ns, a after 5 ns;
```

(b) Usage example.

(c) Simulation waveforms.

Figure 11.8
Signal assignment statement with multiple waveform elements.

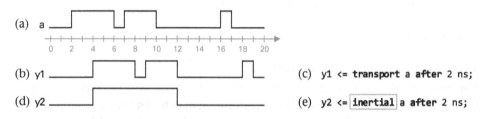

(c) y1 <= **transport** a **after** 2 ns;

(e) y2 <= inertial a **after** 2 ns;

Figure 11.9
Use of the two VHDL delay mechanisms. (a) Input signal. (b) Output using transport delay. (c) Signal assignment using transport delay. (d) Output using inertial delay. (e) Signal assignment using inertial delay (the keyword inertial is optional).

with a time expression in an assignment. A signal assignment statement may include any number of waveform elements, as shown in the simplified syntax of figure 11.8a. The assignment in figure 11.8b schedules several future transactions for signal y, with alternating values based on the value of a, sampled when the statement is executed. As a result, signal y will present a short oscillation every time signal a changes (figure 11.8c).

Assignments with multiple waveforms can be used only with signals (not with variables) and are not synthesizable.

11.3.1 Delay Mechanisms
VHDL offers two delay mechanisms corresponding to different kinds of behavior occurring in hardware. The simplest mechanism is called *transport* delay. This kind of delay is appropriate for devices with nearly infinite frequency response, such as wires or transmission lines. In such devices, any pulse occurring at an input appears at the output, no matter how short. Figure 11.9b illustrates the use of transport delay on an input waveform (figure 11.9a). The corresponding assignment statement is shown in

figure 11.9c. This is the simplest form of delay mechanism because the output is simply a delayed copy of the input signal.

The second mechanism is called *inertial* delay. This kind of delay is appropriate for modeling switching circuits, such as networks of logic gates, whose inputs have an associated capacitance and a threshold level that must be overcome to activate a logic element. This kind of circuit naturally rejects input pulses that are too short. Figure 11.9d illustrates the use of inertial delay on the waveform of figure 11.9a; the corresponding assignment statement is shown in figure 11.9e. Note that the reserved word *inertial* is optional because inertial delay is the default delay model in VHDL.

Inertial delay allows us to specify the minimum width for a pulse not to be rejected, known as the *pulse rejection limit*. If we do not specify a limit, then it will have the same value as the assignment delay. In other words, all pulses shorter than or with the same duration as the specified signal delay will be rejected. If this is the intended behavior, then we can use one of the following two forms:

```
-- The two assignments below use inertial delay with a propagation delay
-- of 'delay_value' and a pulse rejection limit of 'delay_value':
target <= inertial output_value after delay_value;
target <= output_value after delay_value;
```

If we need to specify different values for the pulse rejection limit and the propagation delay, then we can use the following syntax:

```
target <= reject pulse_rejection_limit inertial output_value after delay_value;
```

The pulse rejection limit must be shorter than the delay value of the assignment. We can explain this by reasoning that the delay value is the propagation delay of a circuit element; therefore, if a pulse is wide enough to pass through the element, then it cannot be rejected anymore.

We close this section with a simplified view of how a simulator implements the two delay mechanisms. Recall from section 5.6.2 that a driver is composed of two parts: the current driving value and the projected waveform. The projected waveform is a sequence of time-ordered transactions, where each transaction is a pair consisting of a time value and an output value.

When assignments are made to a driver that already has any scheduled transactions, the new transactions are merged with the existing ones. What happens at the low level depends on the transport mechanism. For both mechanisms, all transactions that are further in time than the earliest transaction being made are deleted. Then all new transactions are added to the signal driver. For the transport mechanism, nothing else must be done. For the inertial mechanism, all transactions that would violate the rejection limit must also be removed from the driver queue.

This is just a simplified view of the process. The LRM specifies the complete algorithm in section 10.5.2.2.

11.4   Force and Release Assignments

We close this chapter with a new feature introduced in VHDL-2008 to help in the creation of testbenches. When we are verifying a model, sometimes it is useful to override the current value of a signal and replace it with a given value. This could be used, for instance, to put the design into a state that would require a complex initialization sequence to be reached. Another possible use is to force the design into an invalid or unreachable state to check its ability to recover itself.

Previously, this had to be done with commands made available by the simulator. Because it knows the values of all signals and is responsible for maintaining them, this has been a traditional simulator role. However, using simulator-specific commands has some disadvantages: it makes the simulation dependent on information from outside the VHDL code, and simulator commands are not portable across vendors.

VHDL-2008 introduced a new kind of signal assignment that is able to override values that would be the natural result of a simulation. This kind of assignment is called a *force assignment*. The corresponding statement that returns control of the signal back to the model is called a *release assignment*.

Let us illustrate the use of these new constructs with an example. Suppose we must write a model for an up-counter that counts from 0 to 9 and displays the current count value in a seven-segment display. We could structure our solution as shown in figure 11.10.

At the outer level, there is the testbench entity, named `counter_display_0_to_9_testbench`. The testbench instantiates our model, which is called `counter_display_0_to_9` and labeled DUV (for *design under verification*). Internally, the DUV is implemented

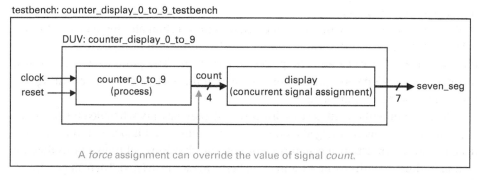

Figure 11.10
Block diagram showing the testbench and model of an up-counter.

as a process and a concurrent signal assignment. The process keeps track of the count value, whereas the concurrent signal assignment converts this value to a signal appropriate for driving a seven-segment display. Listing 11.1 shows the source code for the up-counter model. The process implementing the counter is shown in lines 18–24, and the concurrent signal assignment that drives the display is in line 26.

Listing 11.1 Source code for the up-counter model

```
1    library ieee;
2    use ieee.std_logic_1164.all;
3    use ieee.numeric_std.all;
4    use work.seven_seg_pkg.all;
5
6    entity counter_display_0_to_9 is
7        port (
8            clock: in std_logic;
9            reset: in std_logic;
10           seven_seg: out std_logic_vector(6 downto 0)
11       );
12   end;
13
14   architecture rtl of counter_display_0_to_9 is
15       signal count: unsigned(3 downto 0);
16   begin
17
18       counter: process (clock, reset) begin
19           if reset then
20               count <= (others => '0');
21           elsif rising_edge(clock) then
22               count <= count + 1;
23           end if;
24       end process;
25
26       display: seven_seg <= seven_seg_from_bcd(count);
27
28   end;
```

As shown in listing 11.1, signal count is used internally to communicate between the process and concurrent signal assignment. Because it is internal to the architecture, we cannot control it directly using only normal assignments. To test the outputs for the count value 9, for instance, we would have to reset the counter and wait until it reaches the desired value. In our example, this is not a big problem, but in a large design with thousands of possible states, the simulation could take considerable time. Another problem is that we cannot verify what would appear at the outputs if a corrupted value were provided by the counter. If our specifications required that the display be turned off for values of count beyond 9, then we would not be able to test it.

A force assignment can be used to circumvent these two problems. Listing 11.2 shows the testbench code used to test the counter model. First, the count signal internal to the DUV is given an alias (line 20) to facilitate its use in the testbench. Next, a force assignment is made to simulate the state corruption by forcing the count value to 10 ("1010" in binary) (line 25). Then we verify the expected behavior with an assertion (line 27). After we have finished the test, we must remove the force assignment to give control of the signal back to the model. This is done with the release assignment in line 30. Finally, we wait for a clock pulse and verify that the DUV correctly returns to a valid state (line 34).

Listing 11.2 Testbench code for the up-counter model

```
1    library ieee;
2    use ieee.std_logic_1164.all;
3    use ieee.numeric_std.all;
4
5    entity counter_display_0_to_9_testbench is
6    end;
7
8    architecture testbench of counter_display_0_to_9_testbench is
9        signal clock, reset: std_logic := '0';
10       signal seven_seg: std_logic_vector(6 downto 0);
11   begin
12
13       dut: entity work.counter_display_0_to_9
14           port map(clock => clock, reset => reset, seven_seg => seven_seg);
15
16       reset <= '1', '0' after 20 ns;
17       clock <= not clock after 10 ns;
18
19       process
20           alias duv_count is << signal duv.count: unsigned(3 downto 0) >>;
21       begin
22           -- Put the counter in an invalid state (count = 1010b = 10d)
23           -- and assert that the 7-segment driver is off (all bits are '0')
24           wait until clock;
25           duv_count <= force "1010";
26           wait on seven_seg;
```

```
27          assert seven_seg = "0000000"
28              report "Display is not off for count = 10d"
29              severity failure;
30          duv_count <= release;
31
32          -- Assert that the circuit resets itself after an invalid state
33          wait until clock;
34          assert duv_count = "0000"
35              report "DUV count did not go back to 0 after invalid state"
36              severity failure;
37
38          std.env.finish;
39      end process;
40  end;
```

Force assignments are a useful tool, but we must be careful not to abuse them by writing white-box tests when black-box tests would be possible. The problem with white-box tests is that they verify a specific implementation, not the design specifications. Moreover, they require knowledge about the internal structure of a design and are easier to break; if the implementation changes, then we may have to change the tests as well.

Part IV Types and Objects

12 Categories of Types in VHDL

VHDL is famous for its strongly typed nature. In VHDL, each piece of data must have an unambiguous type, including every signal, variable, parameter, return value, and intermediate result in an expression. The type of a data object is set at its creation and determines which values the object can assume and which operations it allows. On the upside, this strong typing system makes it impossible for an operation to be performed on objects of incompatible types. On the downside, it may require many type conversions, especially if the types used in the design were poorly chosen.

VHDL offers a great set of tools for creating our own types. If we use them to our favor, then we can write code that is less verbose and communicates our intent more clearly. Many designers err on the side of relying almost exclusively on predefined types, missing opportunities to make their code safer, easier to change, and more obviously correct. We can make our job much easier by creating types that match the abstract values processed by our application.

To create our own types and use the ones provided by VHDL effectively, we need to understand the VHDL type hierarchy. This chapter explains how types are organized in VHDL, providing a foundation for the next chapters that will introduce the predefined data types and explain how and when to create our custom data types.

12.1 Overview of Data in VHDL

Before we can discuss the actual types used in VHDL designs, we need to introduce some basic concepts. If we simply reviewed the available types, then it would be hard to understand where they fit in the language type system. Furthermore, some basic terms such as *classes* and *objects* have an unconventional meaning in VHDL. If we draw on our common knowledge of Computer Science, we will often be misled. This section introduces the fundamental concepts that will be needed throughout this chapter.

12.1.1   Types in VHDL

We design circuits to solve real-world problems. Most of the time, a VHDL description is just part of a larger system, and it must interface with other components and the outside world to provide the desired functionality. For this reason, the behavior of a VHDL design is usually specified in terms of the outputs it must produce.

Because VHDL is intended for modeling digital systems, our descriptions communicate with other devices using only the logic levels 0 and 1. However, even if at the physical level all the information is stored and processed in terms of bits, this representation is not always practical for us designers. As humans, we can greatly benefit from a description that presents just the necessary amount of information; it is not practical to be reminded of all the low-level details in every operation we perform. We should be able to specify as much as possible of a design at a high level of abstraction, writing our code in terms of meaningful operations and objects. As seen in section 4.2, abstraction is one of the best tools to manage the complexity of a design.

To create meaningful objects and perform meaningful operations on them, we need types. A *type* is a set of values and a set of operations for working with objects of that type. VHDL provides a number of predefined types for use in our designs. We need to know them well to choose the right type for each situation, and we also need to know when it is better to create our own types. Any modern programming language allows the designer to create new types, and VHDL is no different. In fact, in VHDL, we will be creating a lot of types.

12.1.2   Objects in VHDL

In VHDL, an *object* is a container for a value of a given type. Types are templates for creating objects; objects are the instances of a type. In a design, each variable, signal, constant, and file that we declare are objects. Ports, generics, and subprogram parameters are objects as well.

An object's type determines the set of possible values it can hold, as well as the operations available for manipulating that object. An object also has a *class*, which determines how and where the object can be used in the design (classes are explained in the next topic). Any object that we declare in our code will also have a name (although some objects created implicitly or dynamically do not). While a model is being executed, each object will have a value. For simpler object classes, such as constants and variables, this value is the only important information an object needs to keep. For more complex classes, such as signals and files, the object must keep additional information, such as the previous history of a signal or the current read and write positions of a file. Figure 12.1 shows the information that characterizes a VHDL object, highlighting the roles of its type, class, name, and value.

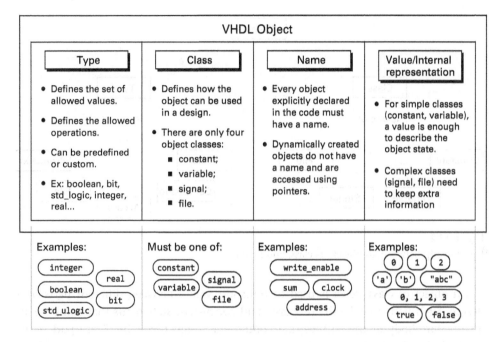

Figure 12.1
Information that defines a VHDL object (type, class, name, and value).

12.1.3 Object Classes

The class of an object determines how and where it can be used in the code. If you think of an object as a box, then the type dictates what kinds of things can be put inside it, while its class dictates how the box can be used in the surrounding code.

There are only four object classes in VHDL: *constant, variable, signal,* and *file.* Sometimes we need to specify an object's class explicitly, whereas at other times the class will be implied from a declaration. For example, every port in an entity is an object of class *signal,* and all generics are *constants* by default.

From the four object classes, *constant* is the simplest. An object of class constant holds the same value for its entire life. This value is set when the object is created and cannot be changed afterward.

Objects of class *variable* are the simplest that can change. The only state associated with a variable is its current value, which can be modified with a *variable assignment* statement. When this statement is executed, the variable immediately assumes the new value. There are no side effects in the model, and no history is kept about the previous values of the variable.

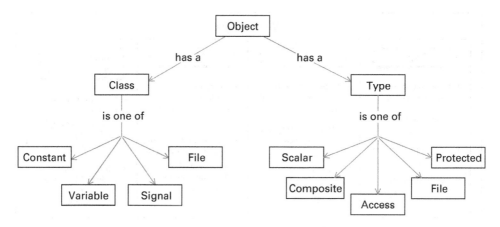

Figure 12.2
Relationship between an object, its class, and its type.

An object of class *signal* is used to transport values between different parts of a design. Such parts may be the processes and concurrent statements that compose an architecture body or the entities that compose a design hierarchy. Every port in an entity declaration is also a signal.

Signals are much more complex than variables. Signal values are never updated immediately after an assignment. Their actual values depend on several other factors, such as other sources that may exist for the signal, its resolution function, and the delay mechanism used in the assignment. Changes in the value of a signal may trigger events in other parts of the code. In a simulation, signals require much more memory and processing resources than variables. However, signals are the basis for the VHDL execution model, which allows deterministic results in a model composed of many parallel units of computation. Signals are indispensable to modeling the behavior of real hardware.

Finally, an object of class *file* is an interface to a file existing in the machine where a simulation is running. File objects are intended for simulation only.

To illustrate the distinction between classes and types, figure 12.2 shows the relationship between an object and its class and type. The figure also shows the four object classes and the five type categories available in VHDL. Object classes will be seen in detail in chapter 14. Type categories are covered in section 12.2.

12.1.4 Use of the Terms *class* and *object* in VHDL and Object-Oriented Languages
Because the terms *class* and *object* mean different things in VHDL and object-oriented programming (OOP) languages, they are a common source of confusion for VHDL learners. In this section, we summarize and contrast the two terms.

In an OOP language, the class is a single construct that specifies an object's under-lying data structure and its behavior. In other words, a class specifies the data fields that comprise an object's state and also the routines that define its operations. In a class, we may choose to make each field or operation public via the class interface or hide it from the client code. The only way to manipulate an instance of the class is via what it chooses to make public. Classes support the principles of information hiding and encapsulation by putting data and behavior together and hiding implementation details from the client code.

In contrast, classes in VHDL are simply one of the four object classes: constant, vari-able, signal, or file. An object's class is not directly related with its type: each object has both a type and a class, which are mostly independent from each other. There are, however, certain restrictions as to which combinations are allowed.

Objects are easier to define. In OOP, an object is simply an instance of a class. In VHDL, an object is an instance of a type. Again, remember that in VHDL, a large part of what can be done with an object depends on its class, in addition to its type. The type determines which values an object can hold and the operations available for objects of that type; the class determines where an object can be declared and used in the code.

For those more familiar with VHDL, a class in OOP is similar to a protected type in VHDL. In a protected type, data and operations are defined in a single construct, and each operation can be made public or not. A class in OOP would also be similar to a VHDL record type if a record could contain subprograms in addition to data fields. Protected types and record types are introduced in section 12.2.

Figure 12.3 summarizes the similarities and main differences between objects and classes in VHDL and OOP languages.

Class, OOP language	*Class, VHDL*
A single construct that defines a type, specifies its data structure, and defines the operations for instances of the class. Similar to a VHDL protected type, or to a record type if it could include routines and hide its data fields.	One of the four object classes: signal, variable, constant, or file. Determines how an object can be used, but does not define its data structure or operations.

Object, OOP language	*Object, VHDL*
An instance of a class.	An instance of a type. Also has a class, a name, and a value.

Figure 12.3
Use of the terms *class* and *object* in VHDL and object-oriented programming (OOP).

12.1.5   What Does Strongly Typed Mean?

VHDL is famous for its strongly typed nature. This feature of the language has both strong supporters and staunch critics; however, few understand what the concept really means. The first thing to understand is that there are two ways to classify a programming language according to its type system. The first classification is between statically and dynamically typed languages. The second, which is a less formal distinction, is between weakly and strongly typed languages. We will see each classification in turn.

Static versus Dynamic Typing   Before applying an operation to an object, a language performs type checks to ensure that the operation supports the object type. The concept of static or dynamic typing is related with when a language performs such type checks.

A language is *statically typed* if it performs type checks during compile time. In such languages, each object must have an explicit type, and the type cannot be changed during the object's lifetime. Without these requisites, the compiler would be unable to perform type checks before actually executing the code. VHDL is statically typed.

A language is *dynamically typed* if object types are checked only at run time. In such languages, it is possible to declare a function without specifying parameter types, and there are no compile-time checks to guarantee that only valid objects will be used with the function. The object type will be checked only when it is used inside the routine. Usually, the check is as lenient as possible: if the routine uses only a small part of the object's data and operations, then only the existence of such features is checked. In this way, the routine can accept and work with objects of different types as long as they have the necessary features in common.

So which one is better? Each approach has advantages and disadvantages. In fact, they require different approaches to programming. In a statically typed language, the compiler and other static analysis tools can perform many checks for us. The compiler can also guarantee that an operation will never be invoked with the wrong type of argument. Also, because we can be certain about certain characteristics of an object, we can reason about a smaller part of the code at a time and guarantee that it is correct. However, static languages force us to be explicit about types, which usually results in numerous type conversions and in more verbose code.

In contrast, dynamic languages are easier to work with at first because they impose fewer restrictions on the structure of our code. They make it easier to write generic programs and can lead to a design with less duplication and boilerplate code. However, we must perform a good amount of testing to ensure that no uncovered type errors will surface when the code is executed.

Weak versus Strong Typing   Unfortunately, there is no consensus on what it means for a language to be strongly typed. Some take it to mean the same as *type safety*, the ability of a language to prevent an operation from being used with values of a wrong type. Others view the classification between weak and strong typing as a scale, combining several factors such as type safety, automatic type conversions, and the need for explicit type casting. We will see each factor in turn.

Type casting means explicitly telling the compiler to use a value of a type as if it were of a different type. There is nothing wrong with type casting per se; the only problem is that in some languages, it can be implemented in an unsafe way. If all that a casting operation does is to call a well-behaved type conversion function, or if it guarantees that objects of two different types can be used interchangeably, the operation is safe. However, if it circumvents compiler checks, such as when a pointer to an integer is used as a pointer to a function, then it can be a dangerous practice. Unsafe type casting is often found in "clever" code that uses knowledge of the underlying implementation of a type in an attempt to make the code shorter or more "optimized" in some other way.

A language is *type safe* if it prevents any operation from being applied to objects of a wrong type. This can be done with checks at compile time, at run time, or both. A language is *type unsafe* if it is possible to work around the type system and to use an object as if it had a different type. This can be done, for instance, by unsafe type casting or direct pointer manipulation.

The subject of implicit type conversions also comes up when discussing strongly and weakly typed languages. An *implicit type conversion*, also called *type coercion*, is allowed in some languages when there is no suitable operation for an operand, but the compiler knows how to convert it to another type, for which there is a version of the operation. For example, during the analysis of a program, a compiler may know two addition operations: one for two integer numbers and another for two floating-point numbers. If the compiler finds an addition between an integer and a floating-point number, then it may be allowed to coerce the integer to a floating-point value and then apply the operation. Although this practice is regarded by some as dangerous, it does not make a language type-unsafe.

Where does VHDL stand in this discussion? VHDL is clearly on the strongly typed side of the spectrum, close to the end of the scale. In VHDL, each object must have a type. The language forces the designer to be explicit about it, and the compiler checks this type every time the object is used in the code. In VHDL, two types are treated as different even if they are declared exactly the same way and have the exact same underlying data structure. Objects of one type cannot be assigned to (or used as) objects of a different type.

In VHDL, type casting is safe because it is allowed only between closely related objects, which are guaranteed to be compatible. Besides, the conversion is not done via raw memory manipulation but rather following strict conversion rules.

Finally, VHDL will not try to guess the type of an object even when it could. Even in a simple assignment between two closely related objects, we must do a type cast for the compiler to accept it. All those features combined mean that we have to be explicit about types at all times; however, it also means that the compiler will make fewer assumptions and throw us fewer surprises.

12.2   VHDL Type Hierarchy

Now that we have reviewed the basic concepts of types, classes, and objects, we can take a closer look at the types that VHDL makes available for use in our models. Figure 12.4 illustrates how types are organized in VHDL. The top row shows the five general categories of types: *scalar*, *composite*, *access*, *file*, and *protected*. Some of the categories include subcategories, shown in the second row.

The rectangular boxes in the top two rows are *categories* of types–they are not actual types that can be used in a design. In contrast, the slanted boxes are actual types and subtypes that can be instantiated in any design. A *subtype* is a type together with a constraint that limits the range of allowed values. The types and subtypes in the figure are defined in package *standard*, which resides in a library called *std* and is included by default in any design.

If we look at the names in the figure, two cases are particular confusing because their names are overloaded. The term *integer* refers to both a category (the category of integer types) and a type (the predefined type *integer*). Also, the term *file* refers to both

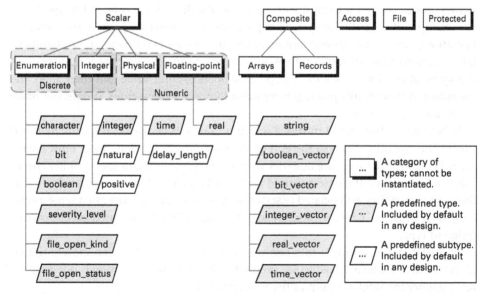

Figure 12.4
VHDL type hierarchy and predefined data types.

a category and an object class. When those terms appear in the text, we need to pay close attention to prevent any confusion.

The focus of this section is to introduce the *categories* of types available in VHDL. The actual types that are instances of such categories are covered in chapter 13. For a list of the most commonly used types, see table 13.1. As each category of type is introduced in the following topics, it is useful to refer back to figure 12.4 to reinforce your understanding of the VHDL type hierarchy.

12.2.1 Scalar Types

Objects of scalar types can hold only a single value at a time. The type definition of a scalar type specifies all the allowed values.

The category of scalar types includes four subcategories: *enumeration, integer, physical,* and *floating-point* types. The subcategory determines the mechanics of how to create a type, the operations that are declared implicitly when we create a type of that category, and how we spell out values of that type in the code. As a memory aid, we can think of scalars as all the numeric and enumeration types (see shaded areas in figure 12.4).

Integer and enumeration types are also *discrete* types; values of such types have an immediate predecessor and a successor. This feature is important because it means that discrete types can be used as loop parameters and array indices.

Integer, physical, and floating-point types are also called *numeric* types. This is important because certain operations (e.g., the basic arithmetic operations) are implicitly created for all numeric types.

We will see each subcategory of scalar types in turn. The focus of this discussion is not on the actual types but rather on understanding each subcategory. The actual types and subtypes will be seen in chapter 13.

Enumeration Types   Enumeration types are defined by giving a name to each of the values that compose the type. The values must be valid identifiers or character literals. The following examples demonstrate the creation of enumeration types:

```
type transmission_mode_type is (park, reverse, neutral, drive, second, low);
type multivalued_logic_type is ('X', 'Z', 'L', '0', 'H', '1');
type mixed_enumeration_type is ('0', zero, '1', one);
```

Enumeration types are used when we know in advance all the values that an object can assume, and we want to refer to those values by name. Good candidates for enumerations include the states of a finite state machine, the operations of an ALU, or configuration modes in a configurable circuit:

```
-- An enumeration type describing the states of an FSM:
type garage_door_state is (opened, closing, closed, opening);

-- An enumeration type for the operations of an ALU:
type alu_operation_type is (
    op_add, op_subtract, op_increment, op_decrement, op_and, op_or, op_nand, op_nor
);

-- An enumeration type for parity configuration in a UART:
type parity_mode_type is (even_parity, odd_parity, no_parity);
```

Enumeration types, like all scalar types, are ordered. This means that all the relational operators come for free with the type when it is declared. Enumerations are also discrete types, so they can be used as ranges in a loop or indices in an array. In the following example, the enumeration color_type is used as the index of the array IS_WARM (declared in lines 3–6) and the range of the loop in line 8.

```
1    type color_type is (red, orange, yellow, green, cyan, blue, violet);
2    type color_to_boolean_type is array (color_type) of boolean;
3    constant IS_WARM: color_to_boolean_type := (
4        red to yellow => true,
5        others => false
6    );
7    ...
8    for i in color_type loop
9        if IS_WARM(i) then
10           write(output, "Color is warm" & LF);
11       end if;
12   end loop;
```

Enumeration types are great for improving readability, especially when they are used to replace enigmatic literal values:

```
if error_condition = 3 then ...      -- Bad example: what is error condition 3?
if error = overflow_error then ...  -- Good example, error is overflow

if alu_mode = 5 then ...             -- Bad example: what is alu_mode 5?
if alu_operation = op_subtract then  -- Good example, operation is subtraction
```

Integer Types The category of integer types includes the predefined type integer and any other integer types that we create. This is an important distinction: we should not mix up integer *types*, which is the category of all types whose values are integer numbers, and the `integer` type, which is a predefined type of VHDL. Assignments and operations between the types we create and the predefined `integer` type are not possible without type conversion.

To create an integer type, we provide a name and a range:

```
type gear_number_type is range 1 to 5;
type day_of_month_type is range 1 to 31;
type countdown_type is range 10 downto 0;
```

All these are integer types, but none of them is a subtype of the predefined type `integer`. They are different types, so none of the following operations is legal:

```
variable gear_number: gear_number_type := 1;
variable day_of_month: day_of_month_type := 5;
...
gear_number := day_of_month; -- Illegal assignment, types are different
(gear_number + day_of_month) -- Illegal expression, no operator "+" for the types
```

If we want to perform operations like the ones in the previous example, then we have a few choices. One is to use subtypes for the objects instead of distinct integer types (subtypes are introduced in section 12.3). Another choice is to use type conversion (covered in section 13.3). Yet another option would be to declare a function accepting objects of different types and name it with the "+" symbol. This feature is called operator overloading, and it was covered in section 8.1.

Physical Types Physical types represent values and measurements written with a number and a unit. Every physical type defines a *primary unit*, which is the smallest quantity that can be expressed with the type. The type can also provide *secondary units*, which must be integral multiples of the primary unit. The following example demonstrates the declaration and use of a physical type for representing frequencies in Hz, kHz, MHz, and GHz:

```
type frequency_type is range 0 to integer'high units Hz;
    kHz = 1000 Hz;
    MHz = 1000 kHz;
    GHz = 1000 MHz;
end units frequency_type;

constant HALF_A_KILOHERTZ: frequency_type := 0.5 kHz;
constant ONE_KILOHERTZ: frequency_type := 1 kHz;
constant ONE_AND_A_HALF_A_GIGAHERTZ: frequency_type := 1.5 GHz;
```

The only predefined physical type in VHDL is `time`, used in places that require a time expression, such as the timeout clause of a *wait* statement or an *after* clause in a signal assignment.

Physical types can be useful when modeling at higher levels of abstraction, but they have one major limitation. Because their values are represented as integer multiples of the primary unit, it is not possible to represent values greater than the largest integer value in primary units. The VHDL standard only requires that a tool support ranges within the inclusive limits −2,147,483,647 and +2,147,483,647. In the `frequency_type` example, this would allow us to express values up to approximately 2.15 GHz. If the primary unit were specified in millihertz, then the maximum value would be a little more than 2.15 MHz. This range is not enough for many practical uses, so many designers simply use floating-point types rather than physical types.

Floating-Point Types   The fourth and last subcategory of scalar types is *floating-point* types. Floating-point types that are not intended for synthesis exist in VHDL since the first version of the language. However, the introduction of synthesizable floating-point packages in VHDL-2008 added another dimension to this subject. To prevent any confusion, we start our discussion by shedding some light on this issue.

Floating-point types that are not intended for synthesis have been around since the first version of VHDL. These types are simply called *floating-point types* in the standard that defines the language; however, to prevent any confusion, we will call them *abstract floating-point types*. The category of abstract floating-point types is built into the language. VHDL predefines only one type of this category: type `real`, declared in package `std.standard`. Whenever we write a literal value in the code using a decimal point (e.g., `-1.0`), this number is an instance of an abstract floating-point type. Such numbers are represented by the compiler using at least 64 bits, which makes them ill-suited for synthesis. The main use for abstract floating-points is in formulas that produce objects of simpler types (such as integers), in literals used as arguments for type conversion functions, or in expressions involving time values.

Now let us change the subject to *synthesizable* floating-point types. These types were introduced in VHDL-2008 as a set of packages for synthesizable floating-point types and operations. Because they are not built into the language (as opposed to type `real`), we need to include those packages explicitly in our designs if we want to use them.

Contrary to abstract floats, synthesizable floats have a strictly defined underlying data structure: they are an array of `std_ulogic` elements (much like a `std_ulogic_vector`) whose meaning is specified in IEEE Std 754-2008. The functions defined in the package operate on such vectors, interpreting them as floating-point values. Because of their underlying data representation, the synthesizable floating-point types are also called *binary-coded* or *vector-based* floating-point types.

An interesting feature of the synthesizable floating-point packages is that the size of each object or type can be tailored to a number of bits that suits the application at hand, allowing the designer to trade off precision for area and speed. Finally, an unpleasant consequence of them being actually vectors of bits is that the compiler cannot convert automatically a literal number with a decimal point to a synthesizable float value. Whenever we need to specify such a value in the code, we must use a conversion function.

To illustrate the differences between abstract and synthesizable floats, we will see three examples of the same calculation: using custom abstract floating-point types, using objects of the predefined type real, and using objects of the synthesizable floating-point type float32.

The first example uses two custom, abstract floating-point types. Because the types of the variables input_sample and analog_gain are different, the multiplication in the last line requires a type cast:

```
type analog_gain_type is range -1.0 to 1.0;
type sample_value_type is range -100.0 to 100.0;
variable analog_gain: analog_gain_type;
variable input_sample, output_sample: sample_value_type;
...
analog_gain := 0.5;
input_sample := 50.0;

-- Note type cast from analog_gain_type to sample_value_type:
output_sample := input_sample * sample_value_type(analog_gain);
```

The second example uses instances of the predefined abstract floating-point type real. Because the two objects have the same base type, a type cast is not necessary:

```
variable analog_gain: real range -1.0 to 1.0;
variable input_sample, output_sample: real range -100.0 to 100.0;
...
analog_gain := 0.5;
input_sample := 50.0;

output_sample := input_sample * analog_gain;
```

Finally, the third example uses instances of the synthesizable floating-point type float32. This type is defined in package float_generic_pkg and uses eight bits for the exponent, 23 bits for the fraction, and one sign bit:

```
variable analog_gain: float32;
variable input_sample, output_sample: float32;
...
analog_gain := to_float(0.5);
input_sample := to_float(50.0);

output_sample := input_sample * analog_gain;
-- Binary representation: 0:10000011:10010000000000000000000
```

Predefined Operations of Scalar Types   When we create a new type, our code behaves as if a large number of operations were declared immediately following the type declaration. Such operations are said to be *implicitly defined* for this new type. The selection of which operations are declared depends on the category of the new type. For instance, all numeric types (integer, physical, and abstract floating-point) get the +, -, *, and / operations. Any array of scalars gets a maximum and a minimum operations, which return the largest and lowest value in the array. Most types get a to_string function, which returns a string representation of an object's value.

Table 12.1 shows some of the operations implicitly declared when we create a scalar type. Some operations are available for all scalar types, whereas others are available only for certain categories.

In general, the numeric operations are predefined for two operands of the same type. However, they are also predefined between a physical type and an integer, and between a physical type and a real. This makes it possible, for example, to multiply a time value by an integer or a real value.

Table 12.1 Selected predefined operations of scalar types

Operations	Predefined for	Meaning
= /= < <= > >=	Any scalar type	Relational operators
num + *num* *num* − *num*	Any numeric type	Addition and subtraction
+ *num* − *num*	Any numeric type	Sign operators
num * *num* *num* / *num*	Any numeric type	Multiplication and division
abs(*num*)	Any numeric type	Absolute value
int **mod** *int* *int* **rem** *int*	Any integer type	Modulo and remainder
int ** *int* *real* ** *int*	Any integer or floating-point type	Exponentiation
maximum(*scalar*, *scalar*)	Any scalar type	Maximum of two values
minimum(*scalar*, *scalar*)	Any scalar type	Minimum of two values
maximum(*array*)	Any array of scalars	Maximum value in the array
minimum(*array*)	Any array of scalars	Minimum value in the array
to_string(*scalar*)	Any scalar type	String representation
to_string(*array*)	Any array of scalars	String representation

Here are some usage examples of the predefined operations of scalar types:

```
type color_type is (red, orange, yellow, green, cyan, blue, violet);
type color_array_type is array (natural range <>) of color_type;

type transmission_mode_type is (park, reverse, neutral, drive, second, low);

type plank_temperature_type is range 0.0 to 1.416833e32;
type plank_temperature_array_type is array (natural range <>) of plank_temperature_type;

variable current_temperature: plank_temperature_type;
variable bool: boolean;
variable str: string;
...
bool := (red > orange);        --> bool := false
bool := (neutral <= drive);    --> bool := true
str := to_string(reverse);     --> str :=  "reverse"
current_temperature := maximum( (1.0, 2.0, 5.0) );   --> current_temperature := 5.0
str := to_string( maximum( (red, green, blue) ) );   --> str := "blue"
```

12.2.2 Composite Types

The previous section introduced the category of scalar types, whose objects can hold a single value at a time. In this section, we introduce *composite* types, whose objects can hold multiple values at a time. Each individual value in a composite object is called an *element*.

Composite types are arrays and records. An *array* is a composite whose elements are all of the same type. The elements in an array are accessed by *indexing*—to access an element, we write the array name followed by an index value in parentheses. When necessary, we can also access a subset of contiguous elements, called an *array slice*. In contrast, a *record* is a composite whose elements may have different types. The elements in a record are accessed by name—to access an element, we write the name of the record object, followed by a dot, and then by the element name.

Composite types are essential to writing clear and concise code because they allow us to treat related elements as a single conceptual unit.

Array Types An *array* is a composite object whose elements are all of the same type. Arrays have many uses in VHDL, but two of the most common cases are to model multibit values, such as buses or registers, and implement memories. Most of the arrays you will find in VHDL will have a single dimension, but the language allows an array to be declared with multiple dimensions. A one-dimensional array is typically called a *vector*.

To declare an array type, we must specify a name, the allowed values for each index (one index per dimension), and the type and any optional constraints for the elements:

```
-- A one-dimensional (1D) array of bits:
type memory_word_type is array (natural range <>) of bit;

-- A one-dimensional (1D) array of memory_word_type:
type memory_type is array (natural range <>) of memory_word_type;

-- A two-dimensional (2D) array of bits:
type bit_matrix_type is array (natural range <>, natural range <>) of bit;
```

The previous array types are said to be *unbounded* or *unconstrained* because the limits for each index were not specified at the type declarations. When we create objects of these types, we need to constrain them:

```
variable memory_word: memory_word_type(127 downto 0);
variable memory_contents: memory_type(0 to 65535)(127 downto 0);
```

Alternatively, we could change our type declarations to constrain their indices:

```
type memory_word_type is array (127 downto 0) of bit;
type memory_type is array (0 to 65535) of memory_word_type;

variable memory_word: memory_word_type;
variable memory_contents: memory_type;
```

The disadvantage of this approach is that it forces all instances of the type to have the same dimensions.

Although it is possible to create 2D arrays of bits, they are rarely used in practice for modeling memories because there is no predefined operation to obtain the value of one row or one column of a 2D array. However, we can create our own function if necessary:

```
function matrix_row(matrix: bit_matrix_type; row_index: integer)
    return bit_vector
is
    variable row: bit_vector(matrix'range(2));
begin
    for j in row'range loop
        row(j) := matrix(row_index, j);
    end loop;
    return row;
end;
```

Table 12.2 Some useful attributes for working with array objects or subtypes

Attribute	Result
array_object_or_subtype'**length**(*N*)	Number of elements in dimension #*N* (an integer)
array_object_or_subtype'**range**(*N*)	Range of dimension #*N* (a range)
array_object_or_subtype'**left**(*N*)	Left bound of index for dimension #*N*
array_object_or_subtype'**right**(*N*)	Right bound of index for dimension #*N*
array_object_or_subtype'**ascending**(*N*)	True if range of dimension #*N* is ascending
array_object_or_subtype'**element**(*N*)	The subtype of the array elements

The 'range attribute is useful when we are working with arrays because it allows us to write subprograms that work with arrays of any size. In the prior example, we used the form 'range(2) to refer to the second dimension of the array or its width. If we wanted the matrix height, then we could have used 'range(1) or simply 'range (because 1 is the default value for this parameter). Other useful attributes for working with array types or array objects are shown in table 12.2.

An interesting point is that the index type need not be numeric. For instance, we can declare an array to define the number of cycles taken by each instruction in a CPU:

```
type operation_type is (nop, add, mul, div);
type operation_cycles_table_type is array (operation_type) of natural;
constant OPERATION_CYCLES_COUNT: operation_cycles_table_type := (
    nop => 1,
    add => 1,
    mul => 2,
    div => 4
);
...
num_cycles := OPERATION_CYCLES_COUNT(div);  -- num_cycles := 4
```

In this example, the array range was defined without writing the box symbol <> as was the case in the previous examples. For this reason, the array object must necessarily have one element for each value of the index type (in this case, four values). We would never use this notation if the index type were integer, or natural, for example.

Besides accessing one element of an array at a time with the index notation, we can extract a contiguous subset of elements using slice notation:

```
signal j_type_instruction: std_ulogic_vector(31 downto 0);
signal opcode: std_ulogic_vector(5 downto 0);
signal target_address: std_ulogic_vector(25 downto 0);
```

Table 12.3 Selected predefined operations of array types

Operations		Predefined for	Meaning
maximum(*array*) minimum(*array*)		1D array of scalar type	Minimum value in the array
and or nand nor xor xnor		Two 1D arrays of logic types	Bitwise logic operators
and or nand nor xor xnor		One scalar and one 1D array of logic types	Extend the scalar to array width, then do bitwise operations
and or nand nor xor xnor		1D array of logic types	Reduction operators
not		1D array of logic types	Bitwise negation
array **sll** *num*	*array* **srl** *num*	1D array of logic types	Logic shift (by *num* bits)
array **sla** *num*	*array* **sra** *num*	1D array of logic types	Arithmetic shift (by *num* bits)
array **rol** *num*	*array* **ror** *num*	1D array of logic types	Rotate (by *num* bits)
array & *array*		Two 1D arrays	Concatenation
array & *scalar*	*scalar* & *array*	One scalar and one 1D array	Concatenation
to_string(*array*)		1D array of character literals	String representation
to_hstring(*array*)		1D array logic types	String representation

```
...
opcode <= j_type_instruction(31 downto 26);
target_address <= j_type_instruction(25 downto 0);
```

Just like for scalar types, when we declare a new array type, the code behaves as if a number of operations were declared following the type declaration. Some operations are declared for any array type, whereas others depend on the element type. The logical operations and, or, nand, nor, xor, and xnor, for instance, are predefined only for arrays of bit or boolean. Although they are not implicitly declared for arrays of std_ulogic, the same operations are explicitly defined in package std_logic_1164. Table 12.3 shows some of the relevant operations that are implicitly or explicitly declared for array types and subtypes.

Here are some examples of how to use the predefined operations:

```
minimum( (0.0, -1.0, 2.0) )    --> -1.0
"10101010" xor "00001111"      --> "10100101"
'1' xnor "00001111"            --> "00001111"
not "11001100"                 --> "00110011"
and "1111"        --> '1'
and "1110"        --> '0'
```

```
"1111" sll 1      --> 1110
"1111" srl 2      --> 0011
"1110" rol 1      --> 1101
```

Record Types A record is a composite object whose elements may have different types. To refer to a record element, we write the name of the record object followed by a dot and the element name.

Records are great for describing groups of objects that are often used, moved, or transformed together. They are also useful for implementing abstract data types (ADTs), which allow us to work with objects at a higher level of abstraction and reduce the complexity of our designs (ADTs are introduced in section 13.4).

To create a record type, we provide a name for the type, plus a name and subtype for each element:

```
-- A record type to represent a point on the screen:
type point_type is record
    x: coordinate_type;
    y: coordinate_type;
end record;
```

```
-- A record type to represent an entry in an ARP
-- (address resolution protocol) table:
type arp_entry_type is record
    ip: ip_address_type;
    mac: mac_address_type;
    is_valid: boolean;
end record;
```

We can access the element values using selected names. We can also assign a value to the entire record at once:

```
if arp_entry.is_valid then
    response_mac := arp_entry.mac;
end if;
```

Record types also help reduce the number of parameters in a routine or ports in an entity. In testbenches, record types are used in bus-functional models (BFMs). A BFM is an interface between a procedural testbench on one side and low-level signals connected to a bus or component on the other side (figure 12.5). A record type is often created to represent a command from the testbench to the BFM or to group a sample of output signals from the component under verification.

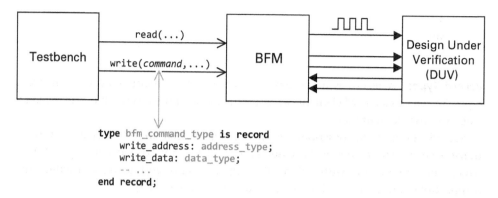

Figure 12.5
A bus-functional model (BFM) controlled by a testbench. A record type is used to model a command to be executed by the BFM.

Like arrays, it is possible to create record types whose elements are unconstrained. The elements may be constrained when declaring a subtype or an object:

```
-- An unconstrained record type:
type color_type is record
    red: unsigned;
    green: unsigned;
    blue: unsigned;
end record;
```

```
-- Elements constrained using a subtype:
subtype color_24bit_type is color_type(
    red(7 downto 0), green(7 downto 0), blue(7 downto 0)
);
variable pixel_24bit: color_24bit_type;
```

```
-- Elements constrained at object declaration:
variable pixel_16bit: color_type(
    red(4 downto 0), green(5 downto 0), blue(4 downto 0)
);
```

12.2.3 Access Types

Objects of access types are similar to pointers in other programming languages. They serve as references to variables created dynamically during a simulation.

The most common way to create an object in the code is with an object declaration (e.g., a signal declaration in an architecture or a variable declaration in a process).

Sometimes, however, it is useful to create new objects while a simulation is running. This is the case, for instance, of data structures whose elements point to each other, such as linked lists or associative arrays. Another use is when we do not know the number of elements in advance, such as when reading data from a file. The total number of lines in the file and the number of characters in a line may be unknown. Finally, another common use for pointers is to implement optimized data structures. We can defer creating an object until we are sure it will be necessary. This technique can be used to model memory pages in a memory model, leaving the memory unallocated until it is really needed in a simulation.

Pointers are notorious sources of trouble in many programming languages. Code that uses pointers is more complicated and error-prone. When we create objects dynamically, we assume the responsibility for managing a region of memory in the machine that is running the simulation—a job that would be much better suited for the simulation environment and the compiler. We need to start worrying about data corruption, memory leaks, and garbage collection. For all those issues, pointers are best avoided unless strictly necessary.

VHDL access types are not as dangerous as, say, pointers in C. For instance, it is not possible to use access types to treat an object as if it had a different type. It is also not possible to create pointers to anything but dynamically created variables. Still it is possible to access memory after it has been deallocated or lose the only reference to an object, creating a memory leak.

One last point to make about access types is that they are not synthesizable because the new objects are created in the memory of the machine running the simulation.

To create a pointer, first we need to declare an access type. For instance, if we want to create pointers to objects of type `integer`, then we can create the following access type:

```
type integer_pointer_type is access integer;
```

The next step is to declare a variable of this new type. This kind of variable is called an *access variable*:

```
variable integer_pointer: integer_pointer_type;
```

When an access variable is created without an explicit initial value, it gets the special value `null`. To make this new variable point to an object, we must create a new object using a special operation called an *allocator*. An allocator consists of the keyword `new` followed by a type name or qualified expression:

```
-- Allocator using a type name; the new object gets
-- the default initial value defined by its type:
```

```
integer_pointer := new integer;

-- Allocator using a qualified expression and a literal value:
integer_pointer := new integer'(5);
```

Finally, we can access the value of the referenced variable using the suffix .all:

```
print( integer_pointer.all );    -- output: 5
```

Besides pointers to scalar objects, it is possible to create pointers to arrays. In this way, the number of elements can be defined at the last moment, when the object is created with an allocator:

```
type integer_vector_pointer_type is access integer_vector;
variable integer_vector_pointer: integer_vector_pointer_type;
...
-- Create a vector with three integer objects:
integer_vector_pointer := new integer_vector'(1, 2, 3);
...
-- Create a vector with room for 10 integer objects:
integer_vector_pointer := new integer_vector(1 to 10);
```

When the access variable refers to an array or record, we can read and write to its elements without using the .all suffix:

```
-- Same as integer_vector_pointer.all(1) := 5;
integer_vector_pointer(1) := 5;

-- Same as rectangle_pointer.all.top_left := (10, 10);
rectangle_pointer.top_left := (10, 10);
```

Example: A Dynamically Allocated Memory Model   Listing 12.1 shows an example of how access types and dynamic memory allocation can optimize memory usage during a simulation. The entity describes a memory whose basic storage unit is a "block" and whose size is defined via the generic BLOCK_SIZE_IN_BITS. The number of blocks is defined with the generic NUM_BLOCKS. The total memory capacity in bits is then the product NUM_BLOCKS times BLOCK_SIZE_IN_BITS, which can get quite large if real-world memory sizes are used.

Listing 12.1 A memory model using access types and dynamic memory allocation

```
 1 library ieee;
 2 use ieee.std_logic_1164.all;
 3
 4 entity dynamically_allocated_memory is
 5     generic (
 6         NUM_BLOCKS: integer := 1024;
 7         BLOCK_SIZE_IN_BITS: integer := 1024
 8     );
 9     port (
10         block_address: in integer range 0 to NUM_BLOCKS-1;
11         write_enable: in std_logic;
12         data_in: in std_ulogic_vector(BLOCK_SIZE_IN_BITS-1 downto 0);
13         data_out: out std_ulogic_vector(BLOCK_SIZE_IN_BITS-1 downto 0)
14     );
15 end;
16
17 architecture behavioral of dynamically_allocated_memory is
18 begin
19     process (all)
20         subtype block_type is std_ulogic_vector(BLOCK_SIZE_IN_BITS-1 downto 0);
21         type block_pointer_type is access block_type;
22         type block_pointers_type is array (natural range <>) of block_pointer_type;
23         variable memory_contents: block_pointers_type(0 to NUM_BLOCKS-1);
24     begin
25         if memory_contents(block_address) = null then
26             memory_contents(block_address) := new block_type;
27         end if;
28
29         if write_enable then
30             memory_contents(block_address).all := data_in;
31         end if;
32
33         data_out <= memory_contents(block_address).all;
34     end process;
35 end;
```

To optimize memory usage during the simulation, we can use access variables for the memory blocks. If we simply declared an array of memory blocks, then all memory would have to be allocated at once. Using an array of access variables, the memory for each block is allocated only when it is really needed. In any case, the functional behavior is exactly the same from an external point of view.

We start by creating an access type that points to a memory block (line 21) and an array of such pointers (lines 22–23). Whenever the memory is accessed, we check whether the given block has already been allocated (line 25). If not, a new block is allocated dynamically (line 26). The client code never gets to know that memory is being allocated dynamically; the only difference is that the simulation uses less memory from the host machine.

In the previous example, we never deallocate any memory, so the used memory is under our responsibility until the simulation is finished (more precisely, while the name of the access variable can be reached). If we wanted to explicitly release the memory used by a block, then we could use the `deallocate` procedure:

```
-- Release the used memory and set pointer value to null:
deallocate( memory_contents(0) );
```

12.2.4 File Types

File types describe the information layout of files that exist in the machine running a simulation. To manipulate those files, we create instances of file types, called *file objects*. Each file object is an interface to a specific file in the host machine.

There are two ways to use files in VHDL. The first is to read and write VHDL objects directly to and from a file in a format that is chosen by the used simulator and is therefore nonportable. This approach uses only operations that are built into the language and does not require any package. The second method is to read and write data in a textual, human-readable format. This approach uses the `text` file type declared in package `std.textio`. Section 14.6 shows examples of the two approaches.

Because file types, file objects, and file operations are intended for accessing files in the machine that is running a simulation, they cannot be used in synthesizable models.

12.2.5 Protected Types

The last category in the VHDL type hierarchy is that of protected types. VHDL-93 introduced a new category of variables, called *shared variables*, which can be accessed by multiple processes (shared variables are covered in section 14.5). However, that version of the standard did not specify what would happen if two processes tried to access a shared variable at the same time. Depending on the model and the simulator, this could cause nondeterministic behavior or data corruption.

To fix this problem, VHDL-2002 introduced *protected types*, which provide a safe way for multiple processes to access the same variable. If a variable is of a protected type, only one process at a time is allowed to access its data. This protection is achieved by preventing direct access to the variable data. Unlike records, the underlying data of an object of a protected type is accessible only via subprograms provided in the type declaration. Such subprograms are called *methods*. When a method is invoked, the calling process is granted exclusive access to the variable, and all other processes requiring access to the variable must wait.

The 2002 version of the language solved the problem of potential corruption of shared variables by requiring that all shared variables be of a protected type. However, the way protected types were engineered provides two other desirable characteristics: encapsulation of data and behavior and information hiding. Those two characteristics make protected types similar to classes in object-oriented programs. For those reasons, protected types are popular for implementing complex and dynamic data structures used in simulation, such as linked lists, queues, and data dictionaries. Although shared variables must be of a protected type, the opposite is not true: a normal (nonshared) variable can also be of a protected type. When used with nonshared variables, the benefits of exclusive access to the variable are not relevant, but the benefits provided by information hiding and encapsulation still apply.

Listing 12.2 shows an example of a protected type for a behavioral model of a digital clock. This type is intended to make the modeling job easier for the client code by providing a simple and narrow interface: an `increment` method that updates the clock every second, and a `digits` method that returns an array of integers representing the individual clock digits.

Listing 12.2 A protected type encapsulating data and operations for a behavioral model of a digital clock

```
1    package clock_time_pkg is
2        -- State and operations for a behavioral model of a digital clock
3        type clock_time_type is protected
4            procedure reset;
5            procedure increment;
6            impure function digits return integer_vector;
7            impure function is_pm return boolean;
8            impure function to_string return string;
9        end protected;
10   end;
11
12   package body clock_time_pkg is
13
14       type clock_time_type is protected body
15           variable current_time: time;
16
17           procedure reset is begin
18               current_time := 0 sec;
19           end;
20
```

```
21          procedure increment is begin
22              current_time := current_time + 1 sec;
23          end;
24
25          impure function is_pm return boolean is begin
26              return current_time >= 12 hr;
27          end;
28
29          impure function digits return integer_vector is ...
30              return digits_vect;
31          end;
32
33          impure function to_string return string is ...
34              return time_string;
35          end;
36      end protected body;
37
38   end;
```

Lines 3–9 declare the protected type interface, which is the only way for the client code to interact with an object of the type. In this example, the object state is kept in a single variable (current_time, in line 15), but a protected type can have its state defined by any number of variables of arbitrarily complex types. In any case, the underlying representation (and its complexity) are hidden from the client code. The only way to get the time value is via the digits function (line 6), or with the to_string method, which provides a string representation for debug purposes. The implementation of each method is given in the protected type body (lines 14–36).

Listing 12.3 shows a usage example for the protected clock_time_type. A variable (nonshared) is declared local to a process (line 2). To modify the variable, the client code uses the only two methods that can change its state. The reset method is called when the reset signal is asserted (line 5), and the increment method is called every second (line 5). After incrementing, the client code copies the value returned by the digits method to a signal called clock_digits.

Listing 12.3 Usage example of the clock_time_type from listing 12.2

```
1   process (clock_1_sec, reset)
2       variable clock_time: clock_time_type;
3   begin
4       if reset then
5           clock_time.reset;
6       elsif rising_edge(clock_1_sec) then
```

```
 7              clock_time.increment;
 8              clock_digits <= clock_time.digits;
 9              print(clock_time.to_string);
10         end if;
11     end process;
```

According to the standard IEEE Std 1076.6-2004—*IEEE Standard for VHDL Register Transfer Level (RTL) Synthesis*, shared variables and protected types are not officially supported for synthesis; however, several synthesis tools already offer some level of support. For a definite answer, check your tool documentation.

12.3 Subtypes

Up to this point, we have been using the concept of subtype only informally, meaning a type together with some sort of constraint. Here we properly introduce the term and take a closer look at this subject.

So far we have seen how to create types of the five categories available in VHDL: scalar, composite, access, file, and protected. Types are useful in our designs because they help to organize our code, document our intent, and prevent unwanted operations, among other reasons. However, if we created one type for each kind of data in our code, then our objects would be harder to use. Remember that in VHDL, two types are treated as different even if they are declared exactly the same way and with the exact same structure. Therefore, it would be harder to make assignments between objects, and there would be no predefined operations that could mix them. If we want to make our objects more compatible with one another, then we need a different approach.

A *subtype* is a type together with a constraint that limits the values allowed for objects of that subtype. Subtypes are important for several reasons. The first is that a subtype is compatible with the type that originated it (called its *base type*). This means that objects of the type and subtype can be used interchangeably in expressions, function calls, and predefined operations. Second, just like a type, a subtype can be used to limit the range of values an object can assume. If we never expect a count value to be larger than 99, then we can create an `integer` subtype to enforce this restriction and detect errors in the model. Third, synthesis tools can use the subtype constraints to generate hardware that uses only the resources necessary to accommodate the allowed values.

We can create a subtype with a *subtype declaration*, in which we specify a base type and one or more constraints to limit the allowed values. The kind of constraint depends on the category of the original type. If the original type is a scalar, then the constraint is a range of values. If the original type is an array, then the constraint is a range of indices. If the original type is a record, then the constraint is a combination of individual constraints for the record elements, which can be scalars or arrays. Subtypes of access, protected, and file types cannot have constraints.

Here are some examples of subtypes for scalar types using range constraints:

```
-- A subtype of the predefined type integer:
subtype register_address_type is integer range 0 to 31;

-- A subtype of the predefined type real:
subtype analog_gain_type is real range -1.0 to 1.0;

-- A custom enumeration type and a corresponding subtype:
type color_type is (red, orange, yellow, green, cyan, blue, violet);
subtype warm_color_type is color_type range red to yellow;
```

The following are some examples of array subtypes using array constraints:

```
-- A subtype of the standard logic type std_ulogic_vector:
subtype opcode_type is std_ulogic_vector(5 downto 0);

-- A custom type for a memory and a subtype constraining its size:
type memory_type is array (natural range <>) of std_logic_vector;
subtype memory_1k_words_type is memory_type(0 to 1023);
```

All operations that work with objects of a base type will accept objects of a subtype as well. However, for predefined operations, the return value will be of the base type. Any constraints will be checked when the object is the target of any assignment. In the following example, the value of a + b is 10; this is within the constraints of c, but it is too large to be assigned back to a or b. In a simulation, this assignment would cause an out of range error. In a synthesized circuit, the most likely result would be an overflow, and a would get the value 2 instead of 10.

```
subtype small_int is integer range 0 to 7;
subtype large_int is integer range 0 to 1023;
variable a, b: small_int;
variable c: large_int;
...
a := 5;
b := 5;
c := a + b;     -- Ok: c is large enough to hold 5+5
a := a + b;     -- Error: 5+5 is out of range 0 to 7
```

By definition, a type is a subtype of itself. This is important because there are many places in VHDL where a subtype is required (e.g., in every signal, variable, constant, or file declaration). In all those cases, we can use a type as well as a subtype. This also explains why the two terms are used interchangeably in many occasions.

Unconstrained and Constrained Array Types An array type is unconstrained if no constraint is given for its indices and no constraints are given for its elements (if the elements are also of a composite type). Such arrays can be constrained when we declare an object or by creating a subtype. To create an unconstrained array type, instead of specifying a fixed range, we use the name of the index type followed by the range keyword and the "box" symbol <>:

```
type coefficients_type is array (natural range <>) of real;
```

We cannot declare an unconstrained or a partially constrained object. To create an object of an unconstrained array type, we could create a constrained subtype first and then declare an object of that subtype:

```
-- Declare a constrained subtype of coefficients_type:
subtype filter_coefficients_type is coefficients_type(0 to 4);
```

```
-- Declare an object of the constrained subtype:
variable notch_filter_coefficients: filter_coefficients_type;
```

Alternatively, we could provide the constraints when the object is declared:

```
-- Declare a constrained object of the unconstrained subtype
variable notch_filter_coefficients: coefficients_type(0 to 4);
```

Finally, if the object is a constant, then we can let the compiler figure out the range for itself:

```
constant NOTCH_FILTER_COEFFICIENTS: coefficients_type := (
    0.4619, 1.8104, 3.5447, 1.8104, 0.4619
);
```

In this last example, the range will be 0 to 4 because 0 is the leftmost value of the index type (natural), and the order associated with type natural is ascending. If we wanted the constant to have a different range, then we could specify the indices nominally using an array aggregate or specify a range constraint as shown in the previous examples.

If the array has more than one dimension or the array elements can also be constrained, then the array type may be unconstrained, partially constrained, or fully constrained. In the following example, memory_type is unconstrained because both the array and element types are unconstrained:

```
type memory_type is array (natural range <>) of std_logic_vector;
```

One thing to note in passing is that in pre-2008 versions of VHDL, this declaration was illegal because it was not possible for an element type to be unconstrained.

When we declare subtypes of the previous type, we can constrain one or both of the indices. If we specify only one of the ranges, then the subtype will be partially constrained. To leave one of the constraints unspecified, we use the reserved word open instead of providing a range. In the following example, memory_32bit_type constrains only the array elements, which will have to be 32-bit std_logic_vectors. In contrast, memory_1k_words_type constrains only the size of the array but leaves the word size unspecified.

```
subtype memory_32bit_type is memory_type(open)(31 downto 0);
subtype memory_1k_words_type is memory_type(0 to 1023)(open);
subtype memory_1k_words_type is memory_type(0 to 1023);  -- (open) is optional
```

If the unspecified dimension comes last, the keyword open is optional.

Finally, we can declare a fully constrained subtype by providing both ranges:

```
subtype memory_1k_32bit_type is memory_type(0 to 1023)(31 downto 0);
```

To declare objects of the previous types, we need to provide all the missing information to make the object fully constrained. If the subtype is already fully constrained, then we do not need to provide any more information:

```
variable memory_sector_1: memory_1k_32bit_type;
```

If the subtypes are partially constrained, then we need to provide only the missing parts:

```
variable memory_sector_2: memory_32bit_type(0 to 1023)(open);
variable memory_sector_3: memory_32bit_type(0 to 1023); -- (open) is optional
variable memory_sector_4: memory_1k_words_type(open)(31 downto 0);
```

If the type is unconstrained, then the object declaration must provide all the constraints:

```
variable memory_sector_5: memory_type(0 to 1023)(31 downto 0);
```

A particularly interesting use for unconstrained arrays is in the ports of an entity. If we leave the dimensions unspecified, the array port will get its size from the actual object associated with it when the entity is instantiated. This way we can write entities that are flexible without needing to declare a generic in the entity interface.

Finally, if we want the array to have an element for each value of the index type, it is not necessary to use the range keyword in the type declaration. This is sometimes useful with arrays of enumeration types. However, we would not want to use this with types that have a large number of values, such as integer or natural:

```
-- Objects of type logic_values_matrix will have 4x4 elements:
type logic_values is ('X', '0', '1', 'Z');
type logic_values_matrix is array (mvl4, mvl4) of mvl4;

-- Warning! This could easily exceed your simulator memory limits...
type huge_array_type is array (integer, integer) of integer;
```

12.4 Resolved Types

In most of our designs, especially those intended for synthesis, we avoid driving a signal from different sources. This can happen, for instance, if we connect two output ports to the same signal or if we make assignments to the same signal from two different processes.

If we want to model a circuit where multiple output ports are connected to the same signal, such as high-impedance, wired-and, or wired-or buses, then VHDL allows us to specify what should happen when multiple values are driven on the same line. This algorithm is specified in a *resolution function*—a function that defines how the values of multiple sources are combined to provide a single value for a signal. A resolution function has a single input parameter, which is an unconstrained array of a given type, and returns a single value of that type.

A signal that has an associated resolution function is called a *resolved signal*. We can specify a resolution function when a signal is declared, or we can associate the function with a subtype and then create signals of that subtype. In the latter case, all signals of that subtype will use the same resolution function. This is the recommended approach because the resolution function becomes part of the abstraction provided by the type. For instance, IEEE standard 1164 offers two types for modeling multivalued logic: std_ulogic (which is unresolved) and std_logic (which is resolved). Users of the first type should not drive signals from multiple sources, whereas this is allowed for the second type. The multiple-valued logic system defined in this standard is an instance of an *MVLN system*. MVLN is the generic name used for multivalued logic systems whose objects can assume one of N possible values. std_logic_1164 is an MVL9 system because it has nine possible values.

The mechanism behind resolved types, resolution functions, and resolved signals is best illustrated with an example. In this example, we will create our own multivalued logic type, which can be used for modeling tri-stated buses similarly to std_logic. Because our only goal is to model tri-state buses, we can make do with a simpler type than std_logic. In our case, an MVL4 type will be enough. The mvl4_pkg in listing 12.4 shows the implementation of this type.

We start the package declaring an unresolved type with four values: an unknown value 'X' used for indicating errors and uninitialized signals, the two logic levels '0' and '1', and a high-impedance value 'Z'. The way our resolution algorithm handles the 'Z' value will provide the look and feel of a tri-stated bus. This type is called mvl4 and is declared in line 2.

Next, we declare a 1D array of mvl4 values, called mvl4_vector (line 3). This type is used for two purposes: to declare nonresolved vectors of mvl4 values in the client code, and as the input type to the resolution function. During a simulation, the resolution function is used behind the scenes: the simulator assembles an array with the values of all drivers for a signal and then passes it to the resolution function. The return value is then used as the effective value for the resolved signal.

Our resolution function (named resolved) is defined in lines 19–26. It works by iterating over all the values of an input parameter called sources, which is an array provided by the simulation environment. Instead of implementing the algorithm with conditional statements such as *if* and *case*, we can implement it efficiently with a lookup table (lines 11–17). Each row in the table corresponds to one value of the provisional resolved value, which is calculated iteratively. Each column corresponds to the value of the current source being analyzed in the loop.

To calculate the resolved value, the code iterates over the array of sources (lines 22–24). The provisional value of the resolved signal is initialized with 'Z' (line 20), and it is updated at each iteration with a value looked up from the table. For example, if resolved_value is currently 'X', then we read the first table row and see that the only possible next value is also 'X'. After all elements have been processed, the current value of resolved_value is returned (line 25).

Listing 12.4 Package for a multivalued logic type with a resolution function

```
1    package mvl4_pkg is
2        type mvl4 is ('X', '0', '1', 'Z');
3        type mvl4_vector is array (natural range <>) of mvl4;
4        function resolved(values: mvl4_vector) return mvl4;
5        subtype mvl4_resolved is resolved mvl4;
6    end;
7
8    package body mvl4_pkg is
9
10       type mvl4_matrix is array (mvl4, mvl4) of mvl4;
```

```
11        constant RESOLUTION_MATRIX: mvl4_matrix := (
12        --   'X'  '0'  '1'  'Z'
13             ('X', 'X', 'X', 'X'), -- 'X'
14             ('X', '0', 'X', '0'), -- '0'
15             ('X', 'X', '1', '1'), -- '1'
16             ('X', '0', '1', 'Z')  -- 'Z'
17        );
18
19        function resolved(sources: mvl4_vector) return mvl4 is
20            variable resolved_value: mvl4 := 'Z';
21        begin
22            for i in sources'range loop
23                resolved_value := RESOLUTION_MATRIX(resolved_value, sources(i));
24            end loop;
25            return resolved_value;
26        end;
27
28    end
```

We could call the resolution function explicitly in our code if we wanted, but this should never be necessary—the simulation engine does it automatically when there are multiple active drivers for a signal. If we did call it manually, then we would see the following return values:

```
resolved( ('Z', 'Z', '1') )  -- returns '1'
resolved( ('Z', '1', '0') )  -- returns 'X'
```

To see the resolution mechanism at work, we can declare a signal of type mvl4_resolved and assign different values to it from two concurrent signal assignments:

```
architecture testbench of mvl_4_tb is
    signal target: mvl4_resolved;
begin
    target <= '0';
    target <= '1';
end;
```

If we check the value of signal target with a simulator, then it will show an 'X'.

As a reference, listing 12.5 shows the resolution table and resolution function for type std_logic from package std_logic_1164. The scalar and array versions of the type are declared as:

```
subtype std_logic is resolved std_ulogic;
subtype std_logic_vector is (resolved) std_ulogic_vector;
```

For the array type, the parentheses around the name of the resolution function mean that the function should be applied to each element of an array that must be resolved. This allows the same resolution function to be used for vectors and single-element values.

Listing 12.5 Resolution table and resolution function for type std_logic (from package std_logic_1164)

```
1   ...
2   type stdlogic_table is array(std_ulogic, std_ulogic) of std_ulogic;
3
4   ------------------------------------------------------------------------
5   -- resolution function
6   ------------------------------------------------------------------------
7   constant resolution_table : stdlogic_table := (
8       --     ------------------------------------------------------------
9       --      |  U    X    0    1    Z    W    L    H    -     |  |
10      --     ------------------------------------------------------------
11              ('U', 'U', 'U', 'U', 'U', 'U', 'U', 'U', 'U'),  -- | U |
12              ('U', 'X', 'X', 'X', 'X', 'X', 'X', 'X', 'X'),  -- | X |
13              ('U', 'X', '0', 'X', '0', '0', '0', '0', 'X'),  -- | 0 |
14              ('U', 'X', 'X', '1', '1', '1', '1', '1', 'X'),  -- | 1 |
15              ('U', 'X', '0', '1', 'Z', 'W', 'L', 'H', 'X'),  -- | Z |
16              ('U', 'X', '0', '1', 'W', 'W', 'W', 'W', 'X'),  -- | W |
17              ('U', 'X', '0', '1', 'L', 'W', 'L', 'W', 'X'),  -- | L |
18              ('U', 'X', '0', '1', 'H', 'W', 'W', 'H', 'X'),  -- | H |
19              ('U', 'X', 'X', 'X', 'X', 'X', 'X', 'X', 'X')   -- | - |
20              );
21
22  function resolved (s : std_ulogic_vector) return std_ulogic is
23    Variable result : std_ulogic := 'Z';   -- weakest state default
24  begin
25    -- the test for a single driver is essential otherwise the
26    -- loop would return 'X' for a single driver of '-' and that
27    -- would conflict with the value of a single driver unresolved
28    -- signal.
29    if (s'length = 1) then return s(s'low);
30    else
31      for i in s'range loop
32        result := resolution_table(result, s(i));
33      end loop;
34    end if;
35    return result;
36  end function resolved;
```

13 Predefined and User-Defined Types

After reviewing the VHDL type system and its type categories in the previous chapter, we are ready to study the actual types that we can instantiate and use in our designs. This chapter introduces the predefined and standard data types available in VHDL and presents recommendations for their use. It also shows how the predefined types can be used to create user-defined types that simplify our designs and improve the quality of our source code. Finally, it presents the mechanisms for converting between types in VHDL.

13.1 Predefined and Standard Types

In a sense, all types that come with a tool or language could be called "predefined" because they are ready-made and available out of the box for our use. In the case of VHDL, they could also be called "standard" because they are included in the LRM, either textually or via a download link. However, the two terms are used in the LRM in a stricter sense, so we need to define them more precisely.

Predefined types are those declared in a package called standard.[1] Because this package is implicitly included in any design, such types are always available for use in any model. The list of predefined types includes boolean, bit, character, integer, and real, among others. Figure 12.4 shows all the predefined data types and their place in the VHDL type hierarchy.

Standard types are also part of the manual. The LRM provides a link for a series of packages that are considered part of the standard.[2] Contrary to predefined types, however, such types are not included by default; they are usually focused on specific design needs and should be included as required by the designer. Examples of standard types include std_logic, signed, unsigned, and float. They are declared in the following packages:

- Standard mathematical packages: math_real and math_complex;
- Standard multivalue logic package: std_logic_1164 and std_logic_textio;

- Standard synthesis packages: `numeric_bit`, `numeric_std`, `numeric_bit_unsigned`, and `numeric_std_unsigned`;
- Fixed-point and floating-point packages: `fixed_float_types`, `fixed_generic_pkg`, `fixed_pkg`, `float_generic_pkg`, and `float_pkg`.

A good knowledge of the types available in VHDL is essential for the creation of high-quality designs. Because there is always more than one way to model a system, it is important to know the features and shortcomings of each type. We can often prevent unnecessary complexity by choosing the most appropriate type for each situation. Another reason to know the predefined and standard types well is because they are the basic building blocks of our user-defined types.

Table 13.1 shows some of the most used predefined and standard data types. We will see each one in turn.

Table 13.1 Common predefined and standard data types

Types	Values	Defined in
`boolean`	`false`, `true`	package *standard* (library *std*)
`bit`	`'0'`, `'1'`	package *standard* (library *std*)
`std_ulogic`	`'U'`, `'X'`, `'0'`, `'1'`, `'Z'`, `'W'`, `'L'`, `'H'`, `'-'`	package *std_logic_1164* (library *ieee*)
`std_ulogic_vector`	Array of `std_ulogic`	package *std_logic_1164* (library *ieee*)
`std_logic`	Same as `std_ulogic`	package *std_logic_1164* (library *ieee*)
`std_logic_vector`	Same as `std_ulogic_vector`	package *std_logic_1164* (library *ieee*)
`integer`	At least -2,147,483,647 to +2,147,483,647	package *standard* (library *std*)
`unsigned`, `signed`	Array of `std_ulogic`, or array of bit	packages *numeric_std* or *numeric_bit* (library *ieee*)
`ufixed`, `sfixed`	Array of `std_ulogic`, with decimal point between the indices "0" and "-1"	packages *fixed_pkg* and *fixed_generic_pkg* (library *ieee*)
`float`, `float32`, `float64`, `float128`	Array of `std_ulogic`, with meaning given by IEEE Std 754 and IEEE Std 854	packages *float_pkg* and *float_generic_pkg* (library *ieee*)
`character`	NUL, ..., `'0'`, `'1'`, `'2'`, ..., `'A'`, `'B'`, `'C'`, ..., `'a'`, `'b'`, `'c'`, ..., `'ÿ'`	package *standard* (library *std*)
`string`	Array of `character`	package *standard* (library *std*)
`real` (non-synthesizable)	At least -1.80×10^{308} to $+1.80 \times 10^{308}$ with approximately 15 decimal digits	package *standard* (library *std*)
`time` (non-synthesizable)	At least -2,147,483,647 fs to +2,147,483,647 fs	package *standard* (library *std*)

Type *boolean* Type boolean is the simplest type available in VHDL. This is a positive thing: few values and few operations make for a type that is easy to understand and hard to misuse. If you do not use booleans routinely in your designs, then you may be missing a good opportunity to simplify your code. Constants, variables, signals, and generics can all be of type boolean. For synthesis, the value false is synthesized as a logic zero and true as a logic one.[3]

An interesting use for a boolean variable is to make a conditional statement more self-evident. Boolean objects can be used directly in the condition of statements such as *if*, when, or *while*. Instead of writing a complex test in the condition of an *if* statement, as in:

```
if (data_in_sync(7 downto 0) = "01000111" or
    data_in_sync(7 downto 0) = "00100111"))
then
    locked <= '1';
    ...
end if;
```

we could store the test result in a boolean variable with a meaningful name:

```
preamble_detected := (
    data_in_sync(7 downto 0) = "01000111" or
    data_in_sync(7 downto 0) = "00100111") );

if preamble_detected then
    locked <= '1';
    ...
end if;
```

Boolean objects are also useful for breaking down a complex boolean expression into smaller chunks that are easier to understand. Each chunk helps document our intent in the code, besides making it more readable. Additionally, expressions using booleans are easier on the eye than expressions requiring multiple comparisons with ones and zeros. Instead of writing:

```
if (sync_counter = 63 and (data_in_sync(7 downto 0) = "01000111" or
    data_in_sync(7 downto 0) = "00100111") and (current_state = acquiring_lock or
    current_state = recovering_lock)
) then ...
```

we can break down the test into three meaningful boolean variables:

```
if waiting_lock and last_sync_bit and preamble_detected then ...
```

Boolean constants are also useful as configuration switches in packages or generics. They read more naturally as the condition of an *if* or *if-generate* statement, and they are harder to misinterpret because the only possible values are *true* or *false*. If we used an integer for the same purpose, then how could a reader be sure that 0 and 1 are the only available choices?

```
generic (USE_MULTIPLIER: integer := 1);     -- What are the possible values?
generic (USE_MULTIPLIER: boolean := true);  -- Ok, only true/false allowed.
```

However, the simplicity that makes booleans so attractive is also a shortcoming in certain cases. With a boolean, it is not possible to distinguish between an uninitialized object and an object with a legitimate value of `false`. Additionally, there are no values to indicate invalid operations or propagate invalid results. We will get back to this and other related problems when we discuss multivalued logic types later in this chapter.

Type *bit* Type `bit` is the simplest type intended for modeling digital circuits. Being an idealized type, its only values are `'0'` and `'1'`. Precisely for its simplicity, it may provide less detail than we would like for modeling a digital circuit. Just as for type `boolean`, it is not possible to differentiate between a `'0'` and an uninitialized value, and it is not possible to indicate invalid operations or bad results. Also, there is no way to model high-impedance values or "don't care" inputs and outputs. Moreover, it is not possible to convert between bits and multivalued logic types automatically or with type casting; such conversions require a type conversion function.

Because of these inconveniences, and because the standard logic types are often required anyway in many designs, most designers shun away from type `bit` in favor of multivalue logic types. However, `bit` and the corresponding array type `bit_vector` may still find a place in higher level and behavioral modeling, where they provide better simulation speed.

Types *std_logic* and *std_ulogic* To overcome the limitations of type `bit` and provide for better interconnectivity between models, a new multivalued logic system was added to VHDL, first as a separate standard (IEEE 1164) and later incorporated into the VHDL-2008 standard. The main types in the VHDL-2008 version of the `std_logic_1164` multivalue logic system are shown in figure 13.1.

Type `std_ulogic` is the base type of the `std_logic_1164` logic system. It is an enumeration composed of nine values. The main improvements over type `bit` include:

- A *don't care* value (`'-'`). For synthesis, this value can represent *don't care* inputs and outputs aiming at logic optimization. For the source code in general, it allows more compact comparisons because a single bit-string literal can represent multiple values at once.

```
1    -- Multivalued logic type (9 values), unresolved
2    type std_ulogic is (
3        'U',      -- Uninitialized      -- Used as a default value
4        'X',      -- Forcing Unknown    -- Bus contentions, error conditions, etc.
5        '0',      -- Forcing 0          -- Transistor driven to GND
6        '1',      -- Forcing 1          -- Transistor driven to VCC
7        'Z',      -- High Impedance     -- 3-state buffer outputs
8        'W',      -- Weak Unknown       -- Bus terminators
9        'L',      -- Weak 0             -- Pull down resistors
10       'H',      -- Weak 1             -- Pull up resistors
11       '-'       -- Don't care         -- Used for synthesis and advanced modeling
12   );
13
14   -- Unconstrained array of unresolved std_ulogic elements
15   type std_ulogic_vector is array (natural range <>) of std_ulogic;
16
17   -- Multivalued logic type (9 values), resolved
18   subtype std_logic is resolved std_ulogic;
19
20   -- Unconstrained array of resolved std_ulogic elements
21   subtype std_logic_vector is (resolved) std_ulogic_vector;
```

Figure 13.1
Type declarations from the std_logic_1164 package (VHDL-2008 version). Comments were edited to make the code more self-explanatory.

- A high-impedance value ('Z') to represent sources that are logically disconnected and do not contribute an effective value to the resolution function.
- Unknown or indeterminate values for modeling bus contention and other error conditions (the value 'X' means an unknown with forcing strength, and 'W' an unknown with weak strength).
- A value for representing uninitialized objects ('U'). Because this is the first value in the enumeration, it is assigned by default to any object that does not provide an initial or a default value.
- Weak logic levels for modeling pull-up ('H') and pull-down ('L') resistors.

The values 'U', 'X', 'W', and '-' are called *metalogical values* because they convey information beyond logical levels. Such values describe the behavior of the model and provide additional information during a simulation or to the synthesis tool. Therefore, type std_ulogic is composed of four *logical values* ('0', '1', 'L', and 'H'), four *metalogical values* ('U', 'X', 'W', and '-'), and the *high-impedance value* 'Z'.

Type std_ulogic is an unresolved type, meaning that it is intended for modeling signals that have a single source. The standard also defines a resolved subtype called std_logic. This subtype is intended for modeling circuits where multiple signals are wired together, such as the signals that compose a high-impedance bus. Because std_logic is a subtype of std_ulogic, signals of one type can be freely assigned to

```
1  type stdlogic_table is array(std_ulogic, std_ulogic) of std_ulogic;
2
3  constant resolution_table : stdlogic_table := (
4     --    ---------------------------------------------------------------
5     --    |  U   X   0   1   Z   W   L   H   -           |  |
6     --    ---------------------------------------------------------------
7           ('U', 'U', 'U', 'U', 'U', 'U', 'U', 'U', 'U'),  --  |  U  |
8           ('U', 'X', 'X', 'X', 'X', 'X', 'X', 'X', 'X'),  --  |  X  |
9           ('U', 'X', '0', 'X', '0', '0', '0', '0', 'X'),  --  |  0  |
10          ('U', 'X', 'X', '1', '1', '1', '1', '1', 'X'),  --  |  1  |
11          ('U', 'X', '0', '1', 'Z', 'W', 'L', 'H', 'X'),  --  |  Z  |
12          ('U', 'X', '0', '1', 'W', 'W', 'W', 'W', 'X'),  --  |  W  |
13          ('U', 'X', '0', '1', 'L', 'W', 'L', 'W', 'X'),  --  |  L  |
14          ('U', 'X', '0', '1', 'H', 'W', 'W', 'H', 'X'),  --  |  H  |
15          ('U', 'X', 'X', 'X', 'X', 'X', 'X', 'X', 'X')   --  |  -  |
16       );
```

Figure 13.2
Resolution table used with the resolution function of type std_logic (package std_logic_1164).

signals of the other type, without any type conversion (although this was not possible before VHDL-2008). Both types have exactly the same nine values—the only difference between them is that std_logic is associated with a resolution function.

Figure 13.2 shows the resolution function used with values of type std_logic. The table specifies what happens when two values are sourced to the same signal. One value is represented by a row and the other by a column. The table is symmetric around its diagonal, so the order of the arguments is irrelevant. For an explanation of how a resolution function works, see section 12.4. In any case, it is easy to draw some conclusions by direct inspection of the table. For example, a 'U' with any other value yields a 'U', thus propagating uninitialized values through a circuit. An 'X' with any other symbol (except a 'U') yields an 'X', so unknown values dominate any logical values.

The recommendation for choosing between std_ulogic and std_logic is to use std_ulogic for single-bit signals that have a single source and std_logic for single-bit signals that have multiple sources. The main motivation for choosing std_ulogic over std_logic for single-sourced signals is that if we unintentionally connect multiple sources together, then the compiler will detect this error soon and prevent the design from being analyzed and elaborated. If we had used std_logic, then the design would compile, but its behavior would be wrong. We would need to simulate and debug the design to find this kind of error.

Types *std_logic_vector* and *std_ulogic_vector* Each of the two scalar types defined in std_logic_1164 has a corresponding one-dimensional array type. The two declarations (for VHDL-2008) are reproduced here:

```
type std_ulogic_vector is array (natural range <>) of std_ulogic;
subtype std_logic_vector is (resolved) std_ulogic_vector;
```

Type std_ulogic_vector is an *unresolved* array of std_ulogic elements. This means that it is intended for modeling array signals that do not have multiple sources (which is the usual case in synthesizable models). Type std_logic_vector is a *resolved* array of std_ulogic elements, meaning that it is intended for array signals driven from multiple sources, such as output ports that are wired together, or signals that are intentionally driven from multiple processes.

Note that both array types have the same element type (std_ulogic); the only difference between them is the resolution function, used in the resolved version. Because one is a subtype of the other, signals of one type can be freely assigned to signals of the other type without any type conversion.

There is some confusion about the choice of std_logic_vector or std_ulogic_vector for modeling multibit signals and ports because the recommendation in the standards changed between the 1993 and 2008 versions. The former guideline was to use std_logic_vector for vector signals and ports. Now the current guideline is to use std_ulogic_vector for vector signals and ports driven by a single source (which is the usual case) and std_logic_vector only for vector signals and port driven by multiple sources (e.g., when modeling high-impedance buses). Table 13.2 summarizes these guidelines. The text of the current recommendations can be found in Clause G.2.11 of IEEE Std 1076-2008–*IEEE Standard VHDL Language Reference Manual*.

In practice, if your tools do not fully support VHDL-2008 yet, then using both forms of the vector type (std_ulogic_vector and std_logic_vector) may lead to many type conversions because they are not directly assignable to each other in pre-2008 versions of the language. In this case, you may want to wait a little before using the types as intended and stick with std_logic and std_logic_vector in your designs. This practice is still adopted by many designers.

Another common source of confusion is whether it is necessary or recommended to use standard logic types for all ports in every design entity. No, it is not. The focus of std_logic_1164 is interoperability and interconnection *between* VHDL models. If your design is composed of several entities, and some of the ports are only used internally, then there is no need for those ports to be of a standard logic type. Instead, use the types that make the most sense for your design, including constrained integers, arrays, and record types.

Table 13.2 Choosing between std_ulogic, std_logic, std_ulogic_vector, and std_logic_vector

	The signal has a single source (Note: this is the usual case)	The signal is short-circuited with other signals to model a high-impedance bus
The signal is scalar (single-bit)	Use std_ulogic	Use std_logic
The signal is a vector	Use std_ulogic_vector	Use std_logic_vector

Type *integer* Type *integer* is the predefined data type that VHDL provides for representing integer values. It should not be confused with the *category* of integer types, seen in the previous chapter. The predefined type `integer` and all the integer types that we create are instances of that category.

When creating objects of type `integer`, it is important to remember that its unconstrained range is at least just under 32 bits (tools are required to support the range $-2^{31}+1$ to $+2^{31} - 1$ but are allowed to support wider ranges if they so choose). In practice, this means that unconstrained integers will use at least 32 bits. Unused bits may or may not be optimized away by synthesis tools, so it is good practice to always constrain the range of integers in synthesizable code.

To constrain an `integer`, we have two choices. The first is to instantiate the `integer` type directly and limit the object with a range constraint at its declaration:

```
variable count_value: integer range 0 to 9;
```

The second choice is to create a constrained subtype of `integer` and then instantiate that subtype:

```
subtype count_value_subtype is integer range 0 to 9;
variable count_value: count_value_subtype;
```

This second approach is a good idea when we need to create a number of objects that share the same range. Often this is a sign that the objects have something in common that deserves to be made explicit in the code. This also makes the code easier to change.

Another point to notice is that type `integer` has two predefined subtypes, `natural` and `positive`:

```
subtype natural is integer range 0 to integer'high;
subtype positive is integer range 1 to integer'high;
```

When we declare objects that cannot assume negative values, it is good practice to use one of the above subtypes to document our intent in the code.

Finally, if we need to represent values that are 32 bit or larger, then we should use the types `signed` or `unsigned` described in the next topic.

Types *signed* and *unsigned* Types signed and unsigned represent integer numbers using arrays of bit or std_ulogic elements (the element type depends on the used package, numeric_bit or numeric_std). The most significant bit is always at the leftmost position. For type signed, the values are represented in two's complement notation.

A nice feature of types `signed` and `unsigned` is that they overload most of the arithmetic functions and operators that can be performed on integers. They also overload most of the functions and operators that can be performed on vectors of logic values, such as shifts, rotations, and logical operations.

Types `signed` and `unsigned` are recommended when we need to be aware of the bit representation of an integer value or when we need to represent values above the guaranteed range of integers (just under 32 bits). They are also recommended for arithmetic circuits and for implementing datapaths. See "Types and Packages for Integer Values and Operations" in section 12.3 for a discussion on the merits and limitations of different integer representations.

Just like `std_ulogic_vector` and `std_logic_vector`, there are resolved and unresolved versions of the vector-based numeric types. The unresolved versions, `u_signed` and `u_unsigned`, are intended for signals that have a single source. The resolved versions, `signed` and `unsigned`, should be used when the signal has multiple sources. In practice, the resolved and unresolved versions are compatible (one is a subtype of the other), so the only disadvantage of always using the resolved versions (`signed` and `unsigned`) is that we miss some checks that could be performed by the compiler regarding the presence of multiple drivers.

Finally, an important remark about the `signed` and `unsigned` types is that they are defined in more than one package, and these packages are mutually incompatible. The guidelines for choosing the right package are presented in "Choosing the appropriate numeric packages" in section 13.2.

Type *real* (Abstract Floating-Point)   Type `real` is the only predefined type in the category of abstract floating-point numbers. Unlike type `integer`, which is synthesizable, type `real` is not intended for synthesis. It can, however, be used to calculate other values that will be converted to a different representation and then used in a synthesized circuit. For example, it is possible to create a sine table using trigonometric functions for type `real` and then convert the results to a synthesizable fixed-point type. In the example of listing 13.1, the values of the sine table are calculated using the `sin` function from the `math_real` package, as well as multiplications and divisions from type `real` (line 24). The results are converted to values of type `sfixed`, which are then stored in a constant table that can be synthesized.

Listing 13.1 Use of type *real* and corresponding operations in synthesizable code—a sine LUT

```
1 library ieee;
2 use ieee.std_logic_1164.all;
3 use ieee.fixed_pkg.all;
4 use ieee.math_real.all;
```

```
 5
 6 -- Calculate sine(phase) using precalculated values.
 7 entity sine_lut is
 8     port (
 9         -- phase = 0 -> 0 rad; phase = 127 -> (pi / 2) rad
10         phase: in integer range 0 to 127;
11         sine: out sfixed(0 downto -15)
12     );
13 end;
14
15 architecture rtl of sine_lut is
16     type sine_table_type is array (natural range <>) of sfixed;
17
18     function make_sine_table return sine_table_type is
19         variable table: sine_table_type(0 to 127)(0 downto -15);
20     begin
21         for i in table'range loop
22             -- Calculate sine table using operations from type real
23             -- (multiplication, division, and sine)
24             table(i) := to_sfixed( sin(MATH_PI_OVER_2 * real(i)/127.0), table(i) );
25         end loop;
26         return table;
27     end;
28
29     constant SINE_TABLE: sine_table_type := make_sine_table;
30 begin
31     sine <= SINE_TABLE(phase);
32 end;
```

Type *float* (Synthesizable Floating-Point)   Package float_generic_pkg defines the types float (resolved) and u_float (unresolved), which are suitable for synthesis. This float representation follows the same principles from IEEE Std 754[4]: each number is represented with a sign bit, an exponent field, and a fraction field. The main difference is that the types allow us to choose, for each field, a number of bits that suits our application. Under the hood, float and u_float objects are arrays of std_ulogic elements.

Besides the size of the exponent and fraction fields, which can be set individually for each float object or type, we can configure other package-wide parameters that will be the same for all floats. Among other configurations, we can choose whether to use denormalization (via the generic parameter float_denormalize) or whether the operations should detect invalid numbers (generic parameter float_check_error). Because float_generic_pkg is a generic package, the client code is not allowed to instantiate it directly. To use the synthesizable floating-point types, we have two choices: we can use a preconfigured version of the package called `float_pkg` or create our own version by instantiating and configuring the generic package. The latter should be done if we need to change one of the default configuration options.

The size of each float is specified with a descending range, in which the leftmost index defines the width of the exponent field and the rightmost index defines the width of the fraction field. This is the definition of subtype `float32` from `float_generic_pkg`, which corresponds to IEEE 754 *binary32* representation:

```
subtype float32 is float (8 downto -23);
```

In this representation, bit 8 is the sign bit, bits 7 to 0 are the exponent field, and bits -1 to -23 are the fraction field. Therefore, the type uses 8 exponent bits and 23 fraction bits.

Types *ufixed* and *sfixed* (Synthesizable Fixed-Point) Fixed-point arithmetic is a way to perform operations on fractional numbers without the overhead of a floating-point representation. In a fixed-point representation, we can specify the number of (binary) digits before and after the decimal point. Because a fixed-point value is essentially an integer with a decimal point at a given position, some operations are as fast as with integer numbers. Addition and subtraction are performed as if the numbers were integers. Multiplication and division can also be performed as for integers, but the result must be interpreted with a shifted decimal point. In VHDL, the operations are performed with full precision (without loss of information) by default, so the client code must account for the different input and output sizes that may result from each operation.

VHDL offers the types `u_ufixed` (unresolved) and `ufixed` (resolved) for fixed-point operations on unsigned values and `u_sfixed` and `sfixed` for fixed-point operations on signed values. The types are defined in `fixed_generic_pkg`, a generic package that can be configured by the user. The configurable parameters include the rounding mode, the overflow behavior, and the number of guard bits. If we accept the default configuration, then we can use package `fixed_pkg,` which already comes preconfigured.

Type *time* Type `time` is the only predefined physical type in VHDL. The main uses for type `time` are to control the passage of time in a simulation, generate waveforms, and simulate the timing behavior of a circuit. This includes, for instance, modeling propagation delays or verifying timing violations, such as setup and hold times.

Type `time` is nonsynthesizable. A synthesis tool should ignore its occurrence in any *after* clause used in signal assignments. This means that models which specify timing delays can be synthesized, but the resulting circuit will ignore those values. The timing characteristic of the synthesized circuit will be given by the target technology and the placement and routing of the circuit elements.

Note, however, that type time can still be used in expressions that produce objects of other types, and then the overall expression may be synthesizable. For example, we could create a timer component that counts a number of pulses; this number could be calculated from a generic constant specified using type time.

Type *string*   Type string is defined in package standard and is simply a one-dimensional array of characters. Its main use is for file IO and text output in simulations. The to_string operation is predefined for most of the types we use and also for every custom scalar type (numeric types or enumerations) or array of characters that we declare. This operation is useful for printing information during a simulation.

Unlike strings in other programming languages, string objects in VHDL must have a fixed length, defined when the object is created. If we need to work with strings of variable length, then the solution is to use the type line, which is a pointer to a string, and then allocate and deallocate new string objects whenever we need to change its size. Because the index of type string is of type positive, the first element in a string has an index position of 1.

Types *line* and *text*   Types line and text are used to perform file input and output with text-mode files, using package textio. Type line is actually a pointer to a string. Type text is a file of strings. The two types are defined in textio as:

```
-- A line is a pointer to a string value:
type line is access string;
```

```
-- A text file is a file of variable-length ASCII records:
type text is file of string;
```

A detailed example of how to perform file operations with text files will be shown in section 14.6 (listing 14.4). To write to a text file, we must first assemble a line by executing write operations that take the line and an object of a predefined type as parameters. When the line is ready, we write it to the file. To read information, we read a line from the file and then parse it using individual read operations for each of the objects that compose the line. For an example, see section 14.6.

13.2   Types and Packages for Integer Values and Operations

Because VHDL offers so many standard types and operations, there is always more than one way to solve any problem. Doing integer math is one of such cases. The first step is to choose a type to represent the numbers. There are three immediate choices: integer types built into the language, the binary-coded signed and unsigned types, or plain logic vectors (such as bit_vector or std_logic_vector). If the chosen representation is not predefined, then we also need to choose a package that provides the desired types and operations.

There are several factors to this decision. We should consider, for instance, the maximum value that can be represented, the need to access individual bits, and the simplicity and readability of the resulting code, including the amount of type conversions that could be avoided by using the right type. The hardware will be the same as long as we do our job correctly, but the difference in readability and maintainability may be significant. Here we discuss the advantages and disadvantages of each alternative.

Using Built-In Integer Types   In many cases, we declare numeric objects to hold integer values and perform basic arithmetic operations on them, such as counting or incrementing. Although we are concerned about the object being implemented as efficiently as possible, with the minimum number of bits, the actual number of bits is not directly related with our problem. In those cases, using an integer type offers several advantages.

Integers allow us to keep the code closer to the problem domain. For example, if we need an index for the current line in a video frame, then we can use an integer with a range constraint from, say, 0 to 525, two values that come directly from the problem domain. The alternative, calculating the number of bits needed to hold this count and then implementing it as a vector of bits, makes us work closer to the solution domain. This tends to make a design more complex than it needs to be.

Using integers also improves readability in certain places. For instance, integers can be used directly as indices in an array; any other representation requires a conversion function. Moreover, integer literals can be used directly in assignments to integer objects. If we want to use a literal with an object of type unsigned, for instance, then it must be a string or bit-string literal. This kind of literal has a fixed length and does not adapt automatically to changes in the object size. The following examples illustrate these two situations:

```
-- Using integers:
ball_x_pos := 400;
ball_y_pos := 300;

-- Using unsigned:
ball_x_pos := to_unsigned(400, ball_x_pos'length);
ball_y_pos := to_unsigned(300, ball_y_pos'length);

-- Using integers:
read_data := memory(row_address, col_address);

-- Using unsigned:
read_data := memory(to_integer(row_address), to_integer(col_address));
```

The fact that the limits are part of an integer subtype can help us write code that is more generic and adaptable. When we write a loop, we can use the `'subtype` attribute to define the iteration limits (`for i in count'subtype loop ...`). When we need to find the largest value allowed for the subtype, we can use the `'high` attribute. Using attributes, we can write code that adapts itself without resorting to constants or generics.

Finally, simulations using integer types run faster and use less memory, and the values may be easier to inspect during a debugging session because they show as numbers instead of sequences of bits.

For all their virtues, integers are not always the best choice. One of their main shortcomings is that the guaranteed maximum value for an integer is only $2^{31} - 1$. Tools are allowed to support a larger range, but your code will not be portable if you count on it. Another disadvantage is that there is no way to specify uninitialized or unknown values in simulations. Uninitialized values make it easier to find errors such as the lack of a reset operation, and unknown values can be used to propagate errors through a datapath making them easier to spot. Finally, we must be careful to always constrain the range of integers that are intended for synthesis. In the absence of any constraints, synthesis tools are likely to use 32-bit integers.

To summarize: if we do not need direct access to the bits, and if we do not need values larger than $2^{31} - 1$, then an integer may be a good choice. If you are not in the habit of using integers, then try experimenting with them in some parts of your code; you will soon get a feel for when they are beneficial. If you feel that they are getting in your way, then it is easy to change them to `signed` or `unsigned`. Finally, remember to always constrain the size of integers in synthesizable models.

Using Vector-Based Integer Types (signed and unsigned)   Sometimes we need to access the underlying representation of a number. This is common in designs that are intimately connected to a binary representation, such as a datapath or an ALU. In other cases, we may need to represent values larger than the maximum integer value guaranteed by the standard ($-2^{31}+1$ to $+2^{31} - 1$). In those cases, the vector-based numeric types `signed` and `unsigned` are good choices.

Types `signed` and `unsigned` have several nice features. They are standardized and part of the language. They are proper types, so the objects carry meaningful type information with them (as opposed to an `std_logic_vector`, which is just a vector of bits). Their underlying binary vectors can be of type `bit`, `std_logic`, or `std_ulogic`, meaning that they can be type cast to vectors of the same base element. They have both the arithmetic and logical operators overloaded, so they are great for implementing datapaths and arithmetic circuits.

In any event, before concluding that we need a binary-coded type, we should consider a few things. Does our design really ask for it, or are we only interested in performing clever bit-twiddling hacks? It may look clever to test the sign of a number by looking at the most significant bit. However, the test int > 0 does the same job in a more natural way. It is clever to rely on overflow to implement a counter that rolls over automatically. However, it makes the code a bit more mysterious because this is a side effect that does not show in any statement in the code. The same could be said about checking for even or odd values by looking at the least significant bit (the operation mod 2 does that for integers) or performing a multiplication by a constant via shifts and additions. Synthesis tools are generally bright enough to perform those tricks automatically.

As for readability and maintainability, another mild problem is that we cannot directly assign integer literals or objects to signed or unsigned objects. Such assignments require a conversion function or a string literal instead of an integer literal. The problem with using string or bit-string literals is that they must be specified with an exact size and cannot adjust to the assignment target automatically (which would work fine with integer objects and integer literals). This turns signed and unsigned objects into a leaky abstraction: sometimes you can treat them just like integers, whereas at other times you cannot. The following examples illustrate these situations.

```
variable x: unsigned(9 downto 0);
...
if x > 15 then ...       -- Works OK, operator">" is overloaded for signed/unsigned
x := 15;                 -- Does not work; assignment symbol":=" cannot be overloaded
x := 10d"15";                    -- Works OK but size is fixed
x := to_unsigned(15, x'length);    -- Works OK but is more verbose
```

To summarize: use the numeric types signed and unsigned when you need access to individual bits, such as when your design is intimately connected with the underlying binary representation or when you cannot use integers because of their size limitations (when you need numbers outside the range $-2^{31} - 1$ to $+2^{31} - 1$). They are also a good choice for implementing datapaths and arithmetic circuits.

Using Standard Logic or Bit Vectors (std_ulogic_vector, std_logic_vector, or bit_vector)
Doing arithmetic operations directly with objects of types std_logic_vector, std_ulogic_vector, or bit_vector should be avoided. The problem with using a logic vector in arithmetic operations is that the object does not contain enough information to represent a number. We need to interpret the element values according to some predefined format, which is not enforced in the code. For instance, the bit vector "1010" could mean the value 10 if interpreted as an unsigned, minus 6 in two's complement notation, or minus 2 in sign and magnitude notation. In

contrast, if we use types intended for numeric operations, we can rely on VHDL's strong typing to guarantee that operations are never called with the wrong kinds of arguments.

If you are concerned about having to perform many type conversions because your input ports are std_logic_vectors, then the best approach is to convert the types in a place as high as possible in the design hierarchy. If you choose a more meaningful type for every signal as soon as an input enters your design, then you will be minimizing the amount of unnecessary type conversions in the rest of the code.

Choosing the Appropriate Numeric Packages If you choose to use integers, often there is no need for extra packages because the types and operations are predefined and built into the language. If you choose to use vector-based numbers, there are four standard packages to choose from: numeric_bit, numeric_bit_unsigned, numeric_std, and numeric_std_unsigned. The four packages are mutually incompatible, so we should never use more than one of them. Also, we should never mix them with nonstandard packages such as std_logic_arith, std_logic_signed, or std_logic_unsigned.

The choice between the bit and the std version of the packages is based on the basic logic type used in your design. If you use std_logic or std_ulogic, then choose the _std version. If your design uses bit, choose the _bit version.

The choice between the packages with and without the suffix _unsigned is based on whether you intend to use the types signed and unsigned or you want to work directly on objects of logic vector types. Packages with the _unsigned suffix do not define the types signed and unsigned. Rather, they perform arithmetic operations on bit vectors or standard logic vectors directly, treating them as unsigned numbers in vector form. Note that this approach is not recommended; a better choice is to use the vector-based numeric types signed and unsigned, in which case we would use the package versions without the _unsigned suffix.

Table 13.3 summarizes the choices involved in selecting a package for numeric operations. The most common choice is to use numeric_std because it is intended for designs that use std_logic or std_ulogic as the basic logic type, and it allows operations with both signed and unsigned numbers.

In any case, do not use nonstandard packages such as std_logic_arith, std_logic_signed, or std_logic_unsigned. std_logic_arith has its own definitions of signed and unsigned, and it is incompatible with numeric_std. If you are using one of these packages, then switch to numeric_std instead. Note that std_logic_unsigned can be directly replaced by numeric_std_unsigned in many cases; however, a better approach is to migrate the code to use numeric_std if possible.

Table 13.3 Choosing a package for numeric types and operations

	Logic type used in the design is std_ulogic	Logic type used in the design is bit
Operations use types signed and unsigned (recommended)	Use numeric_std (this is the usual case)	Use numeric_bit
Operations work directly with vectors of bits	Use numeric_ std_unsigned	Use numeric_bit_ unsigned

13.3   Converting between Types

Often the same value needs to be used in more than one form. For instance, a numeric value may be presented as a vector of bits in an input, but we need it in a numeric form to do arithmetic operations inside a component or to use it as a memory address. In such cases, we need to convert between the two types.

Strictly speaking, there are two ways to convert objects to a different type in VHDL. If the types are closely related, then we can perform a *type cast*. Otherwise the conversion is done via a *conversion function*, which may be predefined, provided with a standard type, or written by the designer. Also, remember that no conversion is necessary between subtypes that have a common base type. These three cases, plus other special cases, are shown in figure 13.3. We will discuss each rule and special case in turn.

No Conversion Is Necessary When Object Subtypes Have a Common Base Type

If two objects belong to subtypes of the same base type, then we can freely assign between them. We can also use either of them in an operation that requires the subtype or base type. However, an error will happen if we try to assign a value that is not within the subtype range.

In the following example, short_int and long_int belong to subtypes of the base type integer, so we can freely assign between them:

```
subtype short_int_type is integer range 0 to 10;
subtype long_int_type is integer range 0 to 10000;
variable short_int: short_int_type;
variable long_int: long_int_type;
...
short_int := 10;
-- OK, short_int and long_int have the same base type and
-- the value (10) is within the range of long_int_type:
long_int := short_int;
```

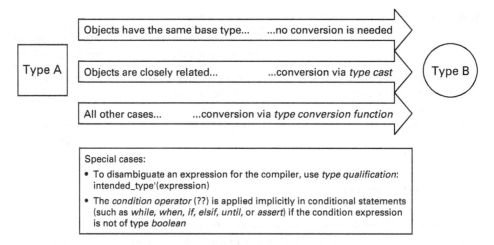

Figure 13.3
Overview of type conversions in VHDL.

```
long_int := 10000;
-- Error: value (10000) is outside the range of short_int_type:
short_int := long_int;
```

If the Object Types Are Closely Related, Then Use *Type Casting*

Some types are similar enough that conversion between them is trivial. Such types are said to be *closely related*. By definition, a type is closely related to itself. There are exactly two other cases of closely related types in VHDL:

- Any abstract numeric types (i.e., integer and abstract floating-point types) are closely related.
- Array types with the same number of dimensions whose element types are also closely related.

An example of the first case is the predefined types `integer` and `real`. The second case includes the standard types `std_ulogic_vector`, `signed`, and `unsigned`, which are all 1D arrays of `std_ulogic` elements.

To convert between closely related types, we use a special operation called *type casting*. To perform a type cast, we write a *type mark* (the name of a type) and an expression of a closely related type in parentheses:

```
intended_type(value_of_original_type)
```

In the following example, we declare two arrays with the same dimensions. Because their element types are closely related (`integer` and `real`), we can assign between them using type casting.

```
variable int_vector: integer_vector(1 to 3);
variable real_vect: real_vector(1 to 3);
...
real_vect := ( 1.0, 2.0, 3.0 );
int_vector := integer_vector( real_vect );
real_vect := real_vector( int_vect );
```

Before VHDL-2008, the definition of closely related array types was stricter: the elements had to be of the same type, and the index types had to be the same or closely related.

A note about terminology: the term *type cast* is borrowed from other programming languages. In VHDL, the official name for this operation is called *type conversion*. However, because "type conversion" is a too general term and could be confused with other type operations, we will call this operation *type cast* in our discussions, as do many developers and instructors in the real world.

If the Object Types Are Not Closely Related, Then Use a *Type Conversion Function*
When two types are not closely related, we need a *conversion function* to convert between them. Type conversion functions may be predefined by the language, provided with a type or package, or written by the designer.

Usually, a conversion function has a single parameter, but sometimes more information is needed. For example, to convert from an unsigned to an integer, the only parameter needed is the value to be converted. However, to convert an integer to an unsigned, we also need to know the size of the target object, which is specified as a second parameter.

```
-- The conversion function "to_integer" from unsigned to natural is provided in numeric_std:
count_int := to_integer( count_uns );

-- The conversion function "to_unsigned" from natural to unsigned is defined in numeric_std.
-- This function takes an extra argument indicating the width of the return value:
count_uns := to_unsigned( count_int, 32 );

-- The conversion function "to_bit" from std_logic to bit is provided in std_logic_1164:
enable_bit := to_bit( enable_sl );

-- The conversion function "to_float" from real to u_float is provided in float_generic_pkg.
-- This function takes two extra arguments indicating the exponent and fraction widths:
gain_float32 := to_float( gain_real, exponent_width => 8, fraction_width => 23 );
```

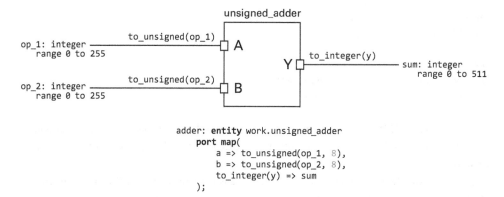

Figure 13.4
Conversion functions used in the association list of an entity instantiation.

An interesting use for conversion functions is to convert between types in association lists, such as in the port mapping of an entity instantiation. The function can be applied directly to the input and output signals, eliminating the need for auxiliary signals in the instantiating code. Figure 13.4 shows an example where integer signals are converted to work with an entity whose ports are of type unsigned. Note that the conversion functions are applied to local signals for input ports (signals op_1 and op_2 in the example) and to ports of the instantiated entity for the outputs (port y in the example). That is to say, conversion functions are always applied on the side that produces the value in an association.

If you are afraid that conversion functions could generate any extra hardware, this usually does not happen in practice. If all a conversion function does is change how a value is interpreted, it does not incur any extra hardware. In other words, if the underlying bits are still the same, then the synthesis tools are bright enough to do the conversion without generating any hardware. Type conversions are a device for programmers and compilers, not an actual hardware construct.

There are many ways to write a type conversion function. For predefined and standard types, nearly all the conversion functions you will need already exist, so look them up before you decide to write your own.

Type Qualification   In some cases, the compiler will not be able to identify the type of an expression, a literal value, or an aggregate. A common cause is the existence of multiple subprograms that have the same name but accept different types of arguments. In such cases, we can provide a hint for the compiler by telling what should be the resulting type of an expression. This is done with *type qualification*: we write the intended type of an expression, followed by an apostrophe (tick) symbol and then the expression in parentheses:

```
type_name'(ambiguous_expression)
```

Consider the following example, where we declare two versions of a function called distance. One of them works with unsigned arguments and the other with std_logic_vectors.

```
function distance(left_op, right_op: unsigned) return integer is ...
function distance(left_op, right_op: std_logic_vector) return integer is ...
```

If we call distance with two literals as arguments, then the compiler will not know which one to use. Instead of guessing, it throws an error and aborts the compilation.

```
distance("1000","0001");    -- Error: subprogram "distance" is ambiguous
```

To disambiguate the function call, we can use type qualification to help the compiler decide between the two functions:

```
distance(std_logic_vector'("0001"),"1000");  -- Ok, use 1st version
distance(unsigned'("0001"),"1000");          -- Ok, use 2nd version
```

Of course type qualification can never be used when the type name is different from the result type of the expression. A type qualifier does not convert between types; it only helps the compiler to decide between multiple possibilities.

Condition Operator (operator ??) A condition is a boolean expression that is part of a conditional statement, such as *if*, *when*, or *while*. The VHDL syntax requires that these expressions return a value of type boolean. In the following expression, the signals are of type std_ulogic, and each equals sign produces a boolean value. Then the booleans are combined with the *and* operators, resulting in a value of type boolean.

```
if en = '1' and cs = '1' and req = '1' and busy = '0' then ...
```

If we wanted to clean up the expression a bit, we could perform the logic operations directly on std_ulogic values:

```
(en and cs and req and not busy)  --> result is an std_ulogic
```

But then the result would be an std_ulogic value, which cannot be used directly as the condition of an *if* statement. In VHDL-2008, the condition operator ?? (introduced in section 8.1.1) is applied implicitly to conditional statements in certain cases. If the compiler cannot interpret the expression as producing a boolean, then it looks for condition operators that can convert the possible expression types to boolean. If only one is found, then this operator is applied implicitly. Using this feature, the original expression can be rewritten as:

```
if en and cs and req and not busy then ...
```

As you write conditional statements, keep an eye out for opportunities to clean up your code using the condition operator and its implicit application.

Converting between Numeric Representations We can illustrate the difference between type casting and type conversion functions by looking at the operations available for converting between the different integer representations. Figure 13.5 shows how to convert between integer, signed, unsigned, and std_logic_vector using the types and operations provided in numeric_std.

Note that using only the operations in numeric_std, it is not possible to go straight from an std_logic_vector to an integer or vice versa because the conversion depends

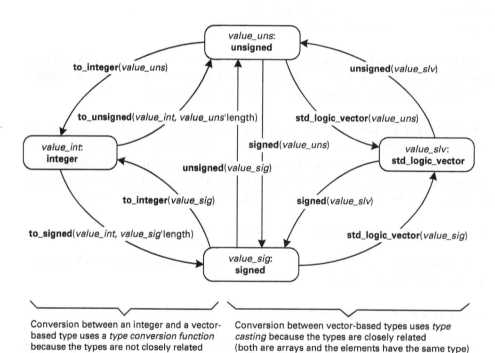

Conversion between an integer and a vector-based type uses a *type conversion function* because the types are not closely related (one is a scalar and the other is an array)

Conversion between vector-based types uses *type casting* because the types are closely related (both are arrays and the elements have the same type)

Figure 13.5
Type conversion between integer, signed, unsigned, and std_logic_vector.

on whether the value represented by the std_logic_vector is signed or unsigned. If we need to convert between those two types, then we can chain the two conversions. If the std_logic_vector is in unsigned form, then we could write:

```
value_int := to_integer( unsigned( value_slv ) );
```

Conversely, if the std_logic_vector is in signed two's complement notation, then we could write:

```
value_int := to_integer( signed( value_slv ) );
```

13.4 Abstract Data Types

An abstract data type (ADT) is a type together with a set of operations for working with objects of that type. ADTs are great for managing complexity because they allow us to work with meaningful entities instead of low-level pieces of data. In a graphics pipeline, we can write our code in terms of vertices and triangles, instead of arrays of floating-point values. In a CPU design, we can manipulate instructions and cache entries instead of bit vectors. The client code becomes cleaner because low-level details get out of the way. The ADT code is also simpler because it is well focused on solving specific tasks and can safely ignore the rest of the application. Overall, the code becomes clearer and more obviously correct.

In VHDL, an ADT is often implemented using a package. The package declaration defines one or more types that represent the underlying data structure of the ADT, as well as functions and procedures to manipulate instances of the ADT. The package body contains the implementation of these subprograms, plus any other subprograms that are used inside the package body but do not need to be publicly visible.

Ideally, the only way to manipulate an instance of the ADT should be via the operations it offers. Unfortunately, in VHDL it is not possible to hide the underlying data structure from the client code because the data structure is a record or an array that the client code instantiates. Because the client code has access to the object, there is nothing to stop the client from manipulating the data directly. In other words, packages do not provide the encapsulation and information hiding that is necessary to implement true ADTs. The only solution is to rely on the user's discipline, who should not manipulate the data structure directly and, while writing the code, should not use any knowledge of how the operations are implemented.

Another alternative for implementing ADTs in VHDL is to use protected types (introduced in section 12.2.5). Protected types provide true encapsulation and information hiding, but they can be used only with variables. When this is not an issue, they are the best choice for implementing an ADT. When we need a more general approach, a package is still the best solution.

Finally, ADTs are more of a way to organize our code rather than a language feature. There is nothing to prevent their use in synthesizable models.

Why Use an ADT?   ADTs help improve the quality of the source code in many different ways. Here we summarize the main advantages of using ADTs:

- **ADTs can help manage and reduce the code complexity.** An ADT reduces the complexity of the client code by hiding the details of how operations are performed. It reduces the complexity of the operations because they can be focused on specific tasks and are separated from the application code. The upshot is that an ADT helps exchange one complex piece of code for two simpler pieces, which is generally a good deal.
- **ADTs help divide the implementation effort.** First, we can split the functionality of a design across several ADTs. Then we can split the development effort between application code and ADT code.
- **ADTs are great for testing.** First, an ADT is composed of smaller routines that can be tested individually. Second, testing ADT operations is easier than testing logic that is buried inside a module. For instance, while we are testing an operation, there is no need to worry about synchronism and reset signals; we only have to provide an input and check that the outputs are correct. After we have tested the ADT and confirmed that it works correctly, we can integrate it in the design with much more confidence and perform additional tests as necessary.
- **ADTs are great for incremental development.** Incremental development is a strategy in which the system is developed one small part at a time. Each piece is then integrated into a partially working system, which gets tested at every step to confirm that it works with the parts already in place. The alternative to incremental development is big-bang integration—to put everything together at once and cross our fingers hoping that things will work out. ADTs help in incremental development because we can implement and test the system one operation or one ADT at a time. In this way, when a bug comes up, we have a much smaller surface to search for it.
- **ADTs are great for readability.** With an ADT, the application code can be written in terms of meaningful objects and operations. The application code is also cleaner because it hides away the low-level implementation details.
- **ADTs are great for reuse.** ADTs usually help draw a line between application code, which is less reusable, and generic operations, which are implemented with ADTs. Many ADTs implement generic data structures that can be reused across different projects with few or no modifications.
- **ADTs promote good design principles.** ADTs promote separation of concerns by splitting the task of using the right operations, which is a responsibility of the application code, from the task of performing the operations, a duty of the ADT. An ADT promotes information hiding by tucking away the details of how low-level

operations are performed. It promotes abstraction by allowing us to work with meaningful entities and operations that are closer to the problem domain.

An Example Initially, it may be hard to find out which parts of an application are good candidates for being implemented as an ADT. Here we provide a series of examples to shed some light on this subject. Table 13.4 lists some application-specific entities that are good candidates for ADTs. The table provides a brief description of what the underlying data structure could be in each case—usually a record or an array type. The table also shows typical operations that could be provided by the ADT. Operations named with a question mark are predicate functions; such operations are often named as a question and return a boolean value. For example, `is_pm?(clock_time)` is a possible operation for a clock time ADT. However, in VHDL, the question mark would have to be left out from the function name because it is not allowed in an identifier.

An important point is that an ADT does not need to be reusable across applications to be worthwhile. Even if it is used only once in a single application, it may still provide enough benefits to be worth implementing.

Table 13.4 Examples of candidates for implementation as an ADT

ADT	Underlying data structure	Operations
Complex number	Record: real part, imaginary part	magnitude, argument, +, *
Floating point number	Record: sign, exponent, fraction	+, *, >, sqrt, to_string
Numeric matrix	2D array of numeric type	+, *, transpose, inverse
Bit matrix	2D array of bits	+, *, transpose, inverse
Point	Record: x, y	+, translate, distance
Triangle	1D array of points	rotate, contains point?
Rectangle	Record: top, left, bottom, right	intersect?, contains point?
Bitmap	2D array of color type	rotate, flip, resize
Sprite	Record: x, y, bitmap	move, check collision
Color	Record: red, green, and blue components	to_grayscale, to_hsl, luminance
Stack	Record: stack pointer, storage array	push, pop, element count, is full?
Stepper motor	Record: step count, driver state	do clockwise step, reset step count
CPU instruction	std_logic_vector	decode, get opcode, is arithmetic?
Uniform random distribution	Record: max, min, seed, state	draw a sample, set seed
Date	Record: year, month, day	increment, +, −, to_months
Time	Record: hour, minute, second	increment, +, −, am?, pm?
AES state	2D array of bytes (4 × 4)	shift rows, mix columns
AES key	std_logic_vector	expand key, encrypt

Now let us take a closer look at one of the examples from table 13.4. A bit matrix is a two-dimensional array of bits. This kind of matrix is used, for instance, in error detection and correction (EDC) applications, and it is a good candidate for implementation as an ADT for several reasons. First, it is generic and likely to be reusable across different projects. Second, it is easy to draw a line between application code and the ADT code. Third, there is no reason for the application code to include the details of how matrix operations are performed. Finally, the underlying data representation is easy to choose—a two-dimensional array of a logic type. In our example, we will use type bit for the matrix elements.

Listing 13.2 shows an implementation for the bit matrix ADT using a package. The package declaration (lines 2–13) constitutes the public interface to the package. The type bit_matrix_type (line 4) is the underlying data structure for the ADT. Lines 7–12 introduce the subprograms that implement the operations for instances of the ADT. Ideally, this is all the information a client of the package would need to see. Therefore, it is highly important to choose clear and communicative names. Moreover, if there are any surprises in the behavior of a routine or any limitations on its inputs, then they should be documented in comments before the function declaration. The package body (lines 15–22) provides the implementation of the ADT operations.

Listing 13.2 Package implementation for an integer matrix ADT

```
1    -- Abstract data type for bit matrices and corresponding operations.
2    package bit_matrix_type_pkg is
3        -- Underlying data structure for the ADT
4        type bit_matrix_type is array (natural range <>, natural range <>) of bit;
5
6        -- ADT operations
7        function "+"(A, B: bit_matrix_type) return bit_matrix_type;
8        function "-"(A, B: bit_matrix_type) return bit_matrix_type;
9        function "*"(A, B: bit_matrix_type) return bit_matrix_type;
10       function identity_matrix(size: integer) return bit_matrix_type;
11       procedure fill_with_zeroes(matrix: out bit_matrix_type);
12       ...
13   end;
14
15   package body bit_matrix_type_pkg is
16       function "+"(A, B: bit_matrix_type) return bit_matrix_type is ...
17       function "-"(A, B: bit_matrix_type) return bit_matrix_type is ...
18       function "*"(A, B: bit_matrix_type) return bit_matrix_type is ...
19       function identity_matrix(size: integer) return bit_matrix_type is ...
20       procedure fill_with_zeroes(matrix: out bit_matrix_type) is ...
21       ...
22   end;
```

Now let us look at the client code. Without an ADT, we could expect to see something similar to the following example in the application code. The statements that perform the operation (in this case, a matrix multiplication) are mixed with the code defining the reset, enable, and clock conditions. This kind of code is hard to understand, verify, and check for correctness. It suffers from too much detail, inconsistent level of abstraction, and excessive nesting:

```vhdl
type bit_matrix_type is array (natural range <>, natural range <>) of bit;
signal A, B, C: bit_matrix_type(1 to 3, 1 to 3);
...
process (clock, reset)
    variable element_accum: bit;
begin
    if reset then
        C <= (others => (others => '0'));
    elsif rising_edge(clock) then
        if enable then
            for i in A'range(1) loop
                for j in B'range(2) loop
                    element_accum := '0';
                    for k in B'range(1) loop
                        element_accum := element_accum xor (A(i,k) and B(k,j));
                    end loop;
                    C(i,j) <= element_accum;
                end loop;
            end loop;
        end if;
    end if;
end process;
```

By replacing the matrix multiplication code with a call to the corresponding ADT operations, the application code becomes much clearer:

```vhdl
process (clock, reset) begin
    if reset then
        fill_with_zeroes(C);
    elsif rising_edge(clock) then
        if enable then
            C <= A * B;
        end if;
    end if;
end process;
```

Of course the same operation still needs to be implemented inside the package; however, this new code organization provides all the benefits discussed in the previous section. We have separated a complex piece of code into two simpler pieces. As a result, the application code is more obviously correct, and the ADT code is more focused and easier to verify.

In most applications, it is possible to find an opportunity to use an ADT. Try to get into the habit of actively searching for abstractions as you code.

13.5   Other Recommendations on Using Types

Many recommendations for using types were provided in the previous sections of this chapter. Here we reiterate some of the recommendations and add a few more to the list.

Create a Type When Your Application Asks For It   In a typical VHDL design, you will be creating a lot of types. This is actually a good thing—it means that you are using the strong typing system to your favor. The more you rely on types, the better the compiler can help you by detecting mistakes and preventing operations from being used with incorrect values.

If you need to use a type in more than one design unit, as is often the case with an entity and its testbenches, then the type must be in a package. Treat this as part of the job. Do not be economic with types, packages, or files. In many cases, moving a type to a package is the first step in discovering a meaningful abstraction.

Choose the Simplest Type That Does the Job   What is the best type for solving a problem: a type that does every imaginable operation or a type that does only what you need and nothing else? More functionality is not always good. A larger number of operations makes a type more confusing and gives you more chance to shoot yourself in the foot.

When you create your own types, implement only the functionality you need. Avoid speculative generalizations. When you choose a predefined data type, choose the simplest type that solves your problem. Do you really need an integer, or would a boolean suffice? Do you really need an std_logic_vector? Sometimes an enumeration would be more appropriate.

Create a Type to Facilitate Changes and Reduce Code Duplication   Sometimes you will find a number of objects declared with the same type and the same constraints:

```
-- In one file:
signal register_A: std_logic_vector(31 downto 0);
signal register_B: std_logic_vector(31 downto 0);
signal register_C: std_logic_vector(31 downto 0);
```

```
-- In another file:
signal register_X: std_logic_vector(31 downto 0);
signal register_Y: std_logic_vector(31 downto 0);
signal register_Z: std_logic_vector(31 downto 0);
```

In such cases, consider creating a subtype. If the subtype is used in more than one file, then it must be declared in a package. In this way, if the type needs to change, then we can make the change in a single place instead of searching for occurrences in all project files.

```
subtype register_type is std_logic_vector(31 downto 0);

signal register_A: register_type;
signal register_B: register_type;
signal register_C: register_type;

signal register_X: register_type;
signal register_Y: register_type;
signal register_Z: register_type;
```

In the original code, the use of a named constant instead of the literal value 31 would help make changes more localized, but the solution using a subtype has some advantages: it documents the fact that the objects are somehow related, it makes each declaration less verbose, and it prevents many of the problems with duplicate code that were presented in section 3.4.

Create a Record, an Array, or an ADT for Objects Frequently Used Together   Another way in which a custom type can help remove duplication is by allowing us to treat multiple objects as a single entity or perform an operation on several objects at a time. If we create a record type or an ADT, then we can write a routine and pass the related values as a single parameter. If we create an array, then we can access its elements inside loops or using array slices.

Without an appropriate data type, each element must be treated as a separate object:

```
x_pos := to_float(0);
y_pos := to_float(0);
z_pos := to_float(0);
...
x_pos := x_pos + x_velocity;
y_pos := y_pos + y_velocity;
z_pos := z_pos + z_velocity;
```

```
x_velocity := x_velocity + x_acceleration;
y_velocity := y_velocity + y_acceleration;
z_velocity := z_velocity + z_acceleration;
```

If we create a custom type (in this case, a record with x, y, and z fields), then we can treat the group as a single object:

```
pos := NULL_POS;
. .
pos := pos + velocity;
velocity := velocity + acceleration;
```

This is the clearest way to document that the objects are related. It also helps reduce the number of parameters in our routines.

Use Constants and Functions to Define Initial and Common Values   For scalar types, if we do not specify an initial or a default value for an object, it will get the leftmost value of the type. For composite types, each element is initialized according to this same rule. If we want a different initial value for our custom types, then we can declare a constant or function to provide values that make more sense:

```
constant TIME_ZERO: clock_time_type := (hours => 0, minutes => 0, seconds => 0);
constant COLOR_WHITE: color_type := (red => 255, green => 255, blue => 255);
constant SCREEN_CENTER_POS: screen_position_type := (x => 400, y => 300);
...
variable current_time: clock_time_type := TIME_ZERO;
variable background_color: color_type := COLOR_WHITE;
...
signal player_pos: screen_position_type := SCREEN_CENTER_POS;
```

If we need to initialize objects of variable size, then we can write an initialization function:

```
function identity_matrix(size: integer) return real_matrix is
    variable matrix: real_matrix(1 to size, 1 to size);
begin
    for i in 1 to size loop
        for j in 1 to size loop
            matrix(i, j) := 1.0 when (i = j) else 0.0;
        end loop;
    end loop;

    return matrix;
end;
```

Then, in the client code, we can call this function where the object is declared:

```
constant IDENTITY_MATRIX_3x3: real_matrix := identity_matrix(3);
```

If the initialization function is declared in the same package where we want the constant to be, then we must use a deferred constant: in the package declaration, we only declare the constant name and its type. Then we provide an actual value with a function call in the package body. For an example, see section 14.1.

Use a Consistent Bit Order for Multibit Values   When you create multibit values or types, choose an ordering for the bits and use it consistently throughout the design. Because we are used to seeing the most significant digit written on the left, the recommended bit order is (N downto 0).

Note that this does not mean that all arrays in your design should use descending ranges. This rationale works for multibit values but not necessarily for other array types, such as the words in a memory or the elements in a FIFO or buffer. In other cases, use the ordering that makes the most sense for your design.

To Convert from an Enumeration, Consider Using a Lookup Table   Enumeration types usually have a small number of values because they must all be named, and each value usually has a specific meaning. There are many ways to write a conversion function from an enumeration to a different type, including a big *case* statement covering all the possible values. However, a more compact solution is to declare an array whose indices are of the enumeration type and whose elements are of the converted type. In the following example, there are three operation types. If we want to convert them to another type (e.g., std_logic_vector), then we can create an array type and declare a constant array to do the conversion:

```
type operation_type is (fetch, decode, execute);

type operation_encoding_type is array (operation_type) of std_logic_vector;
constant ENCODED_OP: operation_encoding_type := (
    fetch =>"001", decode =>"010", execute =>"100"
);
```

Then in the client code, we could write:

```
signal operation: operation_type;
signal operation_register: std_logic_vector(2 downto 0);
...
if rising_edge(clock) then
    operation_register <= ENCODED_OP(operation);
end if;
```

Although the previous example uses the constant array directly, the same technique could be used inside a type conversion function.

Use an Enumeration Type to Limit the Possible Choices   When a parameter in the code can assume only a limited number of values and they are not a contiguous integer sequence, consider creating an enumeration to restrict the possible values to those that the code is prepared to accept. For instance, if a model will work properly only with three frequency values, use an enumeration type instead of an integer or floating-point number:

```
type valid_frequency_type is (FREQ_100_MHz, FREQ_150_MHz, FREQ_200_MHz);
```

This approach guarantees that your code will never be passed a value that it is not prepared to handle correctly.

14 VHDL Data Objects

The previous two chapters focused on introducing the VHDL type system, including the predefined, standard, and custom types that are available for use in our designs. Now we turn our attention to the entities that use such types in the code—*data objects*, or simply *objects*. This chapter discusses the main characteristics of the four object classes—*constant*, *variable*, *signal*, and *file*—and presents guidelines for choosing between them.

14.1 Constants

A *constant* is an object whose value cannot be changed. Constants have several uses in VHDL: they can be used as subprogram parameters, as loop parameters, as generics, or simply to represent a fixed value in a region of code. *Constant* is one of the four object classes in VHDL; the others are *variable*, *signal*, and *file*. Because it is the simplest of all four, it is a good idea to use a constant whenever possible.

In a declarative code region, we can create a constant with a *constant declaration*:

```
constant CONSTANT_NAME: type_or_subtype_name := value_expression;
```

This declares the constant and sets its value at the same time. A constant created this way is called an *explicitly declared constant*. In other programming languages and in the general programming sense, such constants are simply called *named constants*.

In some cases, we would like to declare a constant without specifying its value in the same place. This can be useful, for instance, when the constant is declared in a package and we would like to use functions defined in the package body to calculate its value. In this case, we can omit the value in the package declaration and then specify it in the package body. This kind of constant is called a *deferred constant*:

```
package matrix_pkg is
    ...
```

```
   -- A deferred constant declaration:
   constant IDENTITY_MATRIX_3x3: matrix_type;
end;

package body matrix_pkg is
   function identity_matrix(size: integer) return matrix_type is ...

   -- Full constant declaration corresponding to the deferred constant:
   constant IDENTITY_MATRIX_3x3: matrix_type := identity_matrix(3);
end;
```

14.1.1 Why Use Constants?

There are several benefits to using constants routinely in the code. This section discusses the main reasons to use named constants and the benefits of this approach.

Use Named Constants to Eliminate "Magic Numbers"

"Magic numbers" are literal values that appear in the code without a clear explanation, such as:

```
status <= status xor "10010001";

if count = 96 then ...
elsif count = 144 then ...
```

Sprinkling magic numbers throughout the code is terrible programming practice. They make the code enigmatic because it is not clear what the numbers stand for. They force the reader to look for a comment or read the surrounding statements. When the same number is used in more than one place, it makes the code harder to modify, and the changes become less reliable, as described in the next topics.

Luckily, magic numbers are easy to fix: just declare a named constant and use it in every place you use the value with the same meaning.

Use Constants to Make the Code More Readable and Self-Documenting When a magic number is used, it is hard to know what it stands for. It is also hard to tell with confidence whether the code is correct:

```
if error_code = 3 then ...    -- Bad example: what is error code 3?
```

It is not reasonable to expect that every reader will know what error_code 3 stands for. It is also an unnecessary mental burden: besides all the information we *really* need to keep in our heads to understand a program, we are taxing the reader's memory with frivolous recall tasks. If we replace the magic number with a named constant, then the code becomes more than readable—it becomes obvious:

```
constant OVERFLOW_ERROR_CODE: error_code_type := 3;
...
if error_code = OVERFLOW_ERROR_CODE then ... -- Good example, the condition is now obvious!
```

This approach also reduces the need for comments and external documentation.

Use Constants to Remove Duplicate Information   Whenever there are two sources for the same piece of information, we run the risk of contradicting ourselves in the code. If the same value is repeated over and over, then each occurrence is a duplicate copy of the same information. This violates the Don't Repeat Yourself (DRY) principle (section 4.8) and invites all the problems associated with code duplication (see section 3.4). When we move the value to a constant, it becomes the single source of truth for that piece of information.

Use Constants to Make the Code Easier to Change   If we need to change a value that is used in several places, then we have to hunt for all the occurrences of that value. This is not always easy. The values may be spread across several files. Sometimes the same value is used with different purposes, so we need to be extra sure before we make any change. At other times, we use variations of the original value, such as the number minus one or the number minus two, making the search for literal values even harder. Had we used a named constant to represent the value, the change could be made instantly and in a single place.

Use Constants to Make Changes More Reliably   Besides making changes easier, named constants allow us to change the code more reliably. Because the information is represented in a single place, there is no risk of forgetting to change one of the multiple occurrences. We also eliminate the risk of changing an occurrence that had a different purpose in the code and should not have been touched.

14.1.2   Recommendations for Using Constants

Make a Habit of Replacing Magic Numbers with Named Constants   As explained in the previous section, one of the most common uses for named constants is to replace literal values in the code. This is one of the easiest tricks that every programmer should have up their sleeves; doing this by habit will have a tremendous impact on the quality of your source code.

Be suspicious of every literal number or value that you write in the code. Whenever you find yourself about to write a literal, stop and think whether it should be a named constant instead. While you read the code, check every expression, statement, and declaration for occurrences of magic numbers. When you find one, replace it with a named constant. The declarative regions of an architecture, a package, or a subprogram are places where we would expect to see several constant declarations. Be suspicious of those regions if they only contain signals or variables.

The notable exceptions to this rule are the literal numbers 0 and 1. When used in loops or ranges, they usually represent a limit that is highly unlikely to change. Of course this exception does not apply to zeros and ones that appear inside vectors of bits; any bit-string literal is also a potential magic number. In fact, even single-bit values are magic numbers when they have some encoded meaning. In the following example, each bit in a control register defines the direction of an IO pin. A bit '1' in register io_port_direction means that the corresponding pin in is used as an output and vice versa. In the code, this is implemented using comparisons with the "magic values" '1' and '0':

```
if io_port_direction(i) = '1' then ...      --- The pin acts like an output
elsif io_port_direction(i) = '0' then ...   --- The pin acts like an input
```

The literal values '1' and '0' could be represented by named constants to document this information in the code:

```
if io_port_direction(i) = PIN_DIRECTION_OUTPUT then ...
elsif io_port_direction(i) = PIN_DIRECTION_INPUT then ...
```

How fanatic should we be about rooting out literals from the code? It depends on the likelihood that the number will ever have to change, but this is hard to estimate. We do not make a change only because we absolutely have to; we often do it because it is beneficial to the code. It also depends on how much the number is used in distant places in the code. In general, if a value is used in a single place and its meaning is obvious, then there would be little benefit in replacing it with a constant. In any case, remember that what is obvious to you today may be unclear for someone else or even for yourself a few weeks in the future.

Consider the Alternatives to Using a Constant   Named constants are a huge improvement over magic numbers, but in some occasions, there may be a better solution.

Instead of creating your own encoding scheme using named constants and integers or bit vectors, consider using an enumeration instead. This allows the compiler to perform deeper type checking and prevents the use of invalid values. Even if eventually each enumeration value must have an encoding in terms of bits, consider creating the enumeration type and a conversion function that returns a vector of bits for each value in the enumeration.

Instead of defining several constants in a "configuration package" for a design entity, consider using generics instead. This eliminates unnecessary dependencies between the entity and the package. It also communicates the information through a proper, visible channel in the entity interface. It also makes it possible to create two instances of the same entity with different configurations.

When you create a generic constant in an entity to specify the size of one of the ports, check whether it is possible to use an attribute of the port instead. If all you need is to iterate over the port elements, you can use the 'range attribute of the object. When you declare several ports with the same range constraint, check whether it would be better to use a subtype.

Declare a Constant in the Tightest Scope Possible   If a constant is used only in an architecture, then do not declare it in a package; do it in the architecture declarative part instead. The same recommendation is valid for constants in a process or subprogram: always choose the tightest scope possible. This keeps the information visible only where it is needed and puts the declarations closer to where they are used. In any case, remember to keep the constant declarations where they make sense. Configuration values, for instance, are often declared in a higher level module and then passed down to lower level modules, where they are used.

Avoid Grouping Unrelated Constants in a Package   Occasionally you will find a design where lots of unrelated definitions are grouped in a single package. These packages act like dependency magnets, attracting all kinds of unrelated stuff. In the end, they make it impossible to reuse any part of a design except by cut and paste. In project terminology, this kind of package is often called a "God Package" (because it knows all). It appears as a consequence of haste and unplanned growth in a design.

In general, it is best to keep in a package only cohesive, logically related pieces of information. If necessary, break down the package into smaller chunks with well-defined purposes. Also check whether it is possible to use proper communication channels (generic constants) or declare the constants in a tighter scope, such as in an architecture body.

14.2   Variables

A *variable* is an object whose value can be changed instantly. Conceptually, a variable only has a type and a current value. This means that variables are much simpler than signals, which have driving values, effective values, resolution functions, and projected waveforms. For this and several other reasons, it is a good idea to use a variable in cases where you have a choice between a signal and a variable. Section 14.4 elaborates on the differences between signals and variables and provides practical guidelines for choosing between them.

In the declarative part of a process or subprogram, we can create a variable with a *variable declaration*:

```
variable variable_name: type_or_subtype_name := initial_value_expression;
```

When we declare a variable, we must provide a name, a type, and optionally an initial value. In any case, even when an initial value is not provided, a variable always gets initialized in VHDL. Unlike in other programming languages, there is no risk that the variable will have an indeterminate value if it was never explicitly assigned. The default initial value depends on the variable type; for scalar types, it is the leftmost value of the type. For example, in enumerations it is the first enumeration value, whereas for the subtype natural it is the value 0. Further details about how objects get their initial values are presented in section 14.7.

The type or subtype must include any missing constraints needed to fully specify the variable dimensions. For example, if we use an unconstrained array subtype, then the variable declaration must provide all the undefined index ranges:

```
subtype register_type is unsigned(31 downto 0);

-- OK, the subtype is fully constrained:
variable register_1: register_type;

-- Error, type unsigned is unconstrained:
variable register_2: unsigned;

-- OK, object is fully constrained at declaration:
variable register_3: unsigned(31 downto 0);
```

It is possible to declare multiple variables at once by providing a list of names separated by commas. In this case, all variables will have the same type as well as the same initial value (if one is specified):

```
variable variable_1, variable_2, ... : type_name := initial_value_expression;
```

We can change the current value of a variable with a *variable assignment* statement. When this statement is executed, the variable assumes the new value instantly. There are no side effects anywhere in the code, and no history is kept about the previous values of the variable. Variable assignments are covered in section 11.1.1.

The object class *variable* has one subclass, *shared variables*, which can be accessed by more than one process. Shared variables are a case apart and are best discussed separately. In this section we discuss only normal, nonshared variables, so some of the characteristics presented here are valid for nonshared variables only. Shared variables are discussed in section 14.5.

The main characteristics of variables in VHDL are:

- **Variables are intended for local use in a process or subprogram**, whereas signals are intended for interprocess communication. Keeping this in mind will help you choose between variables and signals in most situations.
- **Variables can be used only in sequential code—namely, in processes or subprograms.** This is in line with their intended use. The main reason behind this limitation is that variables do not provide the mechanism to be used safely and with determinism outside of sequential code. If two processes could access the same variable, then they could corrupt its data or produce results that depended on their relative order of execution.
- **Variables assume a new value immediately on assignment.** After a variable assignment, the new value can be read and used in the next line of code (technically, in the next statement). Contrast this with a signal assignment in a process, in which the new value is available only after the process suspends execution.
- **(Nonshared) Variables can be declared only in sequential regions of code.** In other words, a nonshared variable can be declared only in the declarative region of a process or subprogram. A signal can be declared only in concurrent regions, but it can be used in sequential code as well.
- **In a process, variables are initialized only once at the beginning of a simulation.** Variables declared in processes are initialized when the process is elaborated in the beginning of a simulation. They are never "reset" or reinitialized; when the process loops back to the first statement, the values of all variables are kept. For this reason, if we read the value of a variable inside a process before assigning to it, then the compiler will infer storage elements to keep its previous value.
- **In a subprogram, variables are initialized every time the subprogram is executed.** For this reason, variables inside subprograms do not imply storage elements because there is no need to remember their previous values.

Variables are essential when writing sequential code. Because they get updated immediately on assignment, variables can be used as scratchpad values or to hold partial results in a computation. Sequential algorithms are often specified as a series of assignments, in which each step of the algorithm assumes that the previous step has completed and any new values are available for use. This makes variables the instrument of choice for implementing sequential algorithms.

Just like signals, variables declared in a process can also imply storage elements. If there is any execution path in a process where the value of a variable is read before it is assigned, then it may imply storage. If the assignment is made inside a test for a clock edge, then it implies a register. If not, the assignment implies a latch.

Because they are much simpler than signals, variables are much more efficient in simulations. Section 14.4 elaborates on the differences between variables and signals, and it provides practical guidelines for choosing between them.

14.3   Signals

A *signal* is an object intended for communicating values between processes or design entities. Signals have several characteristics that make them appropriate for modeling wires, nodes of combinational logic, or storage elements in a circuit. A signal includes the time dimension that is absent in variables: it has past, present, and future values. Signals have a well-defined update mechanism that allows deterministic simulation even when many processes and concurrent statements execute in parallel.

Signals may be created in the declarative region of an *entity*, *architecture*, *package*, *generate*, or *block*. To create a signal we use a *signal declaration*, shown here in a simplified form:

signal `signal_name` : `type_or_subtype` `:= default_value_expression` ;

single name or comma-separated list   optional

The type or subtype must include any missing constraints necessary to fully specify the signal dimensions. For example, if we use an unconstrained array subtype, then the signal declaration must complete all the undefined index ranges:

```
subtype address_type is unsigned(31 downto 0);

-- OK, the subtype is fully constrained:
signal address_1: address_type;

-- Error, type unsigned is unconstrained:
signal address_2: unsigned;

-- OK, object is fully constrained at declaration:
signal address_3: unsigned(31 downto 0);
```

It is possible to declare multiple signals by providing a list of names separated by commas. In this case, all signals will have the same type as well as the same default value if one is specified:

```
signal signal_1, signal_2, ... : type_name := default_value_expression;
```

A signal declaration allows the specification of a resolution function. If a signal is intended to have multiple sources, then it must be a resolved signal. There are two ways to declare a resolved signal. One is to declare it with a resolved type (see section 12.4). The other is to use an unresolved type and then specify a resolution function when the signal is declared. In the following example, the function called resolve_mvl4 is associated with signal single_wire_bus. This resolution function will be called automatically by the simulator whenever there are multiple active drivers for the signal:

```
-- A resolved signal declared with an unresolved type and a resolution function:
signal single_wire_bus: resolve_mvl4 u_mvl4;
```

When we translate a circuit diagram to a VHDL description, we naturally use signals for the wires. However, this is only one of the possible uses for a signal. Signals can also be used to implement combinational logic in a dataflow style, using expressions and concurrent signal assignments. Because this kind of assignment is reexecuted every time one of the referenced signals changes, any change on the inputs causes the outputs to be reevaluated automatically. Signals can also imply storage elements, so they can model system state or memories (although variables can serve this purpose as well). Finally, signals can be monitored for changes using *events*. In synthesizable models, we use events in a strict way: to detect signal edges and imply edge-based storage elements. In simulation, we can use events as a versatile mechanism to trigger the execution of processes.

The main characteristics of signals in VHDL are:

- **Signals are intended for interprocess communication and connections between design entities.** This determines where signals can be declared and accounts for their elaborate value update mechanism. Signals enable deterministic behavior even when there are many processes and concurrent statements executing in parallel.
- **Signals can be used in sequential or concurrent code.** Both concurrent and sequential statements can read and assign values to signals.
- **Signals can be declared only in the declarative part of concurrent code regions or inside packages.** This is in line with their main purpose as communication channels. If a signal were declared in a process or subprogram, then it would be visible only in that region and would not serve as a communication channel. A signal declared in a package is called a *global signal*; it can be used as an unstructured way to exchange information between design entities.

- **Every port in an entity is a signal.**
- **In a process, a signal is never updated immediately on assignment.** A signal assignment always schedules a value for a future time. Because simulation time does not advance while a process is executing, signals keep their original values until the process suspends execution. For this reason, whenever we read the value of a signal in a process, we will observe the value it had when the process started executing.
- **In synthesizable code, multiple assignments to the same signal in a process behave as if only the last assignment were effective.** Or, put another way, they behave as if each assignment overwrote the previous one. This is also true for simulation code when the assignments do not have an *after* clause.
- **Signals can be monitored for changes with** *events*. Events and sensitivity lists are used to model sequential and combinational logic in synthesizable code or trigger the execution of processes in simulations.

Some of these characteristics warrant further explanation. The first clarification is about the value observed when we read a signal inside a process, which is always the value that the signal had when the process resumed execution. Because simulation time does not advance while a process is executing, and because all signal updates are scheduled for a future time, it follows that we will always read the original value of a signal even after making assignments to it. In the following example, signal b gets the value that signal a had before it got incremented. The end result is that b is always one cycle behind a, which often surprises newcomers to VHDL.

```
process (clock) begin
    if rising_edge(clock) then
        a <= a + 1;     -- New value a+1 is scheduled for the next infinitesimal time unit
        b <= a;         -- b gets the original value of a, not the incremented value
    end if;
end process;
```

Another common source of confusion is the topic of multiple assignments to the same signal. There are two common cases: we can make multiple assignments from within the same process or from different processes. We will see each one in turn; however, before we start, it is useful to review the concept of a *driver* in VHDL (introduced in section 5.6.2).

A *driver* is a source of values for a signal. In simulation, a driver is a data structure containing future transactions to be applied to a signal. In synthesis, a driver may or may not correspond to the output of a gate. When a signal assignment exists inside a process, that process implies one (and only one) driver for that signal. This is an important distinction: drivers are associated with processes, not with signal declarations. If the signal is an array, then each element gets its own driver.

In a simulation, it is possible to specify a series of values for a signal if they are scheduled for different times in the future. Each of these transactions is queued in the driver and will be applied to the signal as simulation time advances. In synthesis, however, it is not possible to specify arbitrary time delays, so all assignments behave as if they were scheduled for the next infinitesimal time unit.

Now back to the subject of multiple assignments to a signal. The first case to consider is when multiple assignments are made from within the same process. Each process implies at most one driver for each signal, so all assignments to a signal will share the same driver. But in synthesis, assignment delays are ignored, so all assignments behave as if they were scheduled for the next infinitesimal time unit. Because a driver can hold at most one transaction for any given time, each assignment overwrites the previous value that existed in the driver. The net result is that only the last executed assignment inside a process has any effect. In the following example, when the process finishes executing, the driver for signal a will have only the value 3 in its queue, scheduled for the next infinitesimal time unit.

```
process (clock) begin
    if rising_edge(clock) then
        a <= 1;  -- The value 1 is scheduled for the next infinitesimal time unit
        a <= 2;  -- Value 2 overwrites the previous value 1
        a <= 3;  -- Value 3 overwrites the previous value 2
    end if;
end process;
```

The second case is when different processes or concurrent statements make assignments to the same signal. Because each process has its own driver, it is possible that they will source different values for the same signal.

Before we write code where this situation is possible, we should stop and ask ourselves what we are trying to model. If in the modeled circuit each bit should be driven from a single gate, which is the usual case, then multiple assignments from different processes are an error in the code. In contrast, if this is intended by the designer, then the solution is to use a resolved signal. In this case, multiple assignments from different processes are passed to the resolution function, which returns a single value that will be assumed by the signal.

In the following example, there are three drivers for signal single_wire_bus: one from the concurrent signal assignment (line 3) and one from each process. This is not an error because it is the intended behavior for the circuit, and the signal was declared with a resolved type (std_logic, in line 1).

```
1    signal single_wire_bus: std_logic;
2    ...
3    single_wire_bus <= 'H';  -- weak pull-up
4
5    process (all) begin
6        if condition_1 then
7            single_wire_bus <= '0';  -- forcing low
8        else
9            single_wire_bus <= 'Z';  -- high impedance
10       end if;
11   end process;
12
13   process (all) begin
14       if condition_2 then
15           single_wire_bus <= '1';  -- forcing high
16       else
17           single_wire_bus <= 'Z';  -- high impedance
18       end if;
19   end process;
```

If the signal were of an unresolved type, such as std_ulogic, then the compiler would throw an error. It is important to notice that changing the type of a signal to a resolved type is not a way to fix an error of multiple drivers detected by the compiler. If we do this in a design where the signal is not supposed to model a high-impedance bus, then the compilation will pass, but the design will not have the intended behavior. This is one of the advantages of using unresolved types by default (such as std_ulogic): if you inadvertently make multiple assignments to a signal, then the compiler will let you know it immediately.

To summarize: in a synthesizable model, if multiple assignments are made to the same signal within the same process, only the last executed assignment will be effective. If multiple assignments are made from different processes, first check that this is the intended behavior; then use a resolved signal if necessary.

14.4   Signals versus Variables

The previous sections presented the main characteristics of signals and variables. Here we contrast the two and provide practical guidelines for choosing between them.

Because signals and variables have essentially different purposes, in most cases, the intended use for an object will determine its class. However, in some other cases, the choice will be based on a particular feature of signals or variables. Table 14.1 summarizes the main characteristics of signals and variables and shows them side by side to facilitate a direct comparison. The variables considered in this table are nonshared variables. Shared variables are discussed separately in section 14.5.

Table 14.1 Main differences between signals and variables

Signal	Variable
• Intended for communication between processes or to transport values between design entities.	• Intended for local use in a process or subprogram.
• In an assignment, the new value is always scheduled for some future time (may be an infinitesimal time).	• In an assignment, the variable assumes the new value instantly.
• The value read in a process is the value that the signal had when the process resumed. Any changes are postponed until the process suspends.	• The value read in a process is the current value of the variable. Any changes can be read in the next statement.
• In synthesizable code, multiple assignments in a process behave as if only the last assignment were effective.	• In synthesizable code, the result is as if every assignment updated the variable.
• Can be used in both sequential and concurrent code.	• Can be used only in sequential code.
• Changes in the value of a signal can be monitored with events.	• Changes in the value of a variable have no observable side effects.

We can illustrate the process of choosing between signals and variables with a generic example. Our starting point is a block diagram that decomposes the system functionality across a number of modules with well-defined purposes and interfaces. When we translate this diagram to a VHDL description, we will use signals for the wires between the modules. The functionality of each module will be described in an architecture body, using processes, concurrent statements, and component instantiations. To connect these elements, we will declare signals in the architecture declarative part. Inside each process, we may use variables to calculate values using sequential algorithms or store any state local to the process. Usually the calculated values, the process state, or some value based on them will be copied from variables to signals. Finally, inside any subprogram we write, we only declare variables. We can still use signals inside a subprogram, but they must be declared outside and are usually passed as parameters.

Table 14.2 summarizes these and other guidelines to help choose between signals and variables. The last recommendation in the table deserves further explanation. There are several reasons to favor using a variable in cases where either a variable or signal would be possible. First, variables are much simpler than signals. This has advantages for both the programmer and simulator. As readers of the code, we can be sure that when a variable assignment is executed, the variable will be updated immediately with the value given in the assignment expression. When the target of the assignment is a signal, we can only be sure of its actual value after we factor in any implicit or explicit delays, its resolution function, and its force or release status.

Table 14.2 Choosing between signals and variables

Use a signal ...	Use a variable ...
• To transport values between processes or between design entities.	• To calculate values in a sequential manner.
• For values that need to be used in more than one process.	• For values used in a single process.
• When you need an object that is visible in an entire architecture.	• When you need an object that is visible only inside a process or subprogram.
• For values that need to be used in concurrent statements or in concurrent code regions.	• In algorithms where you need the values to be updated immediately (sequential or iterative algorithms).
• To model the wires of a circuit diagram.	• As scratchpad, temporary, or partial values in a calculation.
• When you need to monitor an object for events.	• When simulation efficiency is important (for example, to simulate a large number of storage elements).
• To control when a process gets executed in a simulation.	• If the object is of an access or protected type.
	• When the choice is otherwise indifferent and there is no compelling reason for using a signal.

Variables are also much more efficient than signals in simulation. This is not a decisive factor in every case, but it makes a difference when we are simulating a large number of objects, such as in a memory model. For one vendor, the difference in the amount of memory required to run a simulation can be as high as two orders of magnitude to simulate a 16 Mbit memory.[1] Signals are represented with complex data structures that may require in the simulator many dozens of bytes per modeled bit.

Finally, the fact that a variable can only be used inside a process is a good thing. We should always declare objects in the tightest possible scope. There is no need to advertise the existence of an object to an entire architecture if it is only used inside a process. In this way, there are fewer objects for us to juggle mentally while we try to understand the architecture. It also places each declaration closer to where it is used, improving readability.

One final point to discuss is the use of variables versus signals for synthesizable code. Contrary to popular belief, there is no direct relation between the class of an object (signal or variable) and how it gets synthesized in hardware. Both signals and variables can imply storage, either level-based (latches) or edge-based (flip-flops). The generated hardware will not be faster or slower just because we used a signal or variable. The timing of a synthesized circuit has nothing to do with delta delays or the fact that variables are updated immediately in a simulation. As long as the code using variables and the code using signals specify the same behavior and produce the same output values, there is no reason to expect that the synthesized hardware would be any different.

14.5   Shared Variables

Now that we have seen in detail the characteristics of signals and variables, we are prepared to discuss a special kind of variable. From the four object classes, variables are the only one with a subclass. A *shared variable* is a variable that can be accessed by multiple processes.

A shared variable, like any other variable, does not have past or future values and gets updated immediately. However, unlike nonshared variables, which can only be declared in processes and subprograms, a shared variable must be declared in a place where it is accessible by multiple processes—namely, in an *entity, architecture, package, generate,* or *block*. Moreover, because shared variables must be of a protected type, they have all the characteristics of such types: they cannot be given an initial value at declaration; they do not have predefined assignment, equality, and inequality operations; and the only way to access their data is via function calls. Finally, shared variables are not intended for synthesis, but their features make them interesting for simulations and testbenches.

In the declarative region of an *entity, architecture, package, generate,* or *block,* we can create a shared variable with a *shared variable declaration*:

```
shared variable variable_name: protected_type_name;
```

The fact that a shared variable must be of a protected type provides safety against two basic problems: it prevents data corruption when two processes try to update the variable at the same time, and it prevents inconsistent reads while the variable is being updated by some other process. However, it does not protect the code against logic errors: deadlocks, livelocks, and nondeterminism are still possible. If you choose to use protected variables, then be mindful of those typical concurrency problems. Usually it is not worth using a shared variable where a signal or nonshared variable could be used. Save them for the cases where it would be impossible or too awkward to use signals or normal variables.

Both shared variables and protected types were not originally intended for synthesis. Therefore, even if your synthesis tool offers some support, it is probably a good idea to avoid them because the resulting models may be nonportable and nondeterministic. If you need to exchange information between processes in a synthesizable model, the most common and safe way is by using signals.

Now that we have provided enough warning about the caveats of using shared variables, let us see some of their benefits. In simulations and behavioral models, shared variables offer interesting advantages. The communication mechanism provided by shared variables provides an alternative way of structuring a design, which may be simpler, more efficient, and more modular in some cases.

Consider the case of writing a behavioral model for a register file with multiple read and write ports. A behavioral model is a description that has no implied architecture and is not necessarily synthesizable. If we were to model the register file as a synthesizable component, then we would have to create a design entity, instantiate it in our design, declare the necessary interconnection signals, and drive these signals according to some protocol. In contrast, if we model the register file as a shared variable in a package, then all we need to do is include the package in a design. Every process will be able to use the register file simply by calling its methods. There is no need for extra signals, component declarations, or instantiations.

Shared variables are common in advanced simulations and testbenches. Because multiple processes can use the same shared variable, we can organize a testbench in a simpler and more modular way. This can make each process shorter, more specialized, and easier to reason about. Because shared variables are easier to include in a design, they are a good choice for functionality that must be shared across multiple modules, such as tracing, logging, and collecting statistics.

Listing 14.1 shows a common use for a shared variable in testbench code. The design under verification (DUV), called `div_by_3`, divides an unsigned input called `dividend` by the constant value three (lines 17–18). The testbench applies a predefined set of stimuli and then uses functional coverage to evaluate the test effectiveness. *Functional coverage* is the practice of sampling specific points in a design to check whether all the important values and ranges were tested (functional coverage is introduced in section 20.6). Each sampled signal or value is called a *cover point*. In the example, the input of the divider is selected as a cover point. To keep track of the values presented to the divider, we declare a shared variable called `cover_point` in line 14.

The testbench is divided into several short processes with well-defined responsibilities. Process `init_coverage` (lines 20–24) initializes the cover point variable, specifying that we want to monitor values from 0 to 999, grouped into 100 ranges of ten values each. Each range is called a *coverage bin* in verification terminology. Process `sample_coverage` (lines 26–30) does the sampling, updating the shared variable `cover_point` whenever a new input value is tested. Process `apply_and_verify_stimuli` (lines 32–38) generates the inputs to the DUV and performs assertions to check that the output values are as expected. Finally, process `report_coverage` (lines 40–44) presents the coverage results. Note that three of the four processes use the variable `cover_point`, which justifies the need for a shared variable.

Listing 14.1 Use of a shared variable in a testbench with functional coverage. The same variable is used in several processes.

```
1   library ieee;
2   use ieee.numeric_std.all;
```

```
3    use work.coverage_pkg.all;

4

5    entity div_by_3_tb is
6    end;

7

8    architecture testbench of div_by_3_tb is
9        signal duv_input, duv_output: unsigned(31 downto 0);
10       signal stimuli_done: boolean;

11

12       -- The point we want to monitor for coverage is declared as a shared variable so
13       -- that separate processes can initialize it, use it, and check coverage results.
14       shared variable cover_point: cover_point_type;
15   begin

16

17       duv: entity work.div_by_3 port map (
18           dividend => duv_input, quotient => duv_output);

19

20       init_coverage: process begin
21           -- Initialize cover point with 100 bins of 10 values each
22           cover_point.add_bins( min => 0, max => 999, num_bins => 100 );
23           wait;
24       end process;

25

26       sample_coverage: process begin
27           wait on duv_input;
28           -- Increment bin count corresponding to the current test value
29           cover_point.accumulate( duv_input );
30       end process;

31

32       apply_and_verify_stimuli: process begin
33           for i in 0 to 1_000 loop
34               ... -- code to apply stimuli and check for errors
35           end loop;
36           stimuli_done <= true;
37           wait;
38       end process;

39

40       report_coverage: process begin
41           wait until stimuli_done;
42           cover_point.print_coverage;
43           wait;
44       end process;
45   end;
```

The example of listing 14.1 will be revisited when we discuss coverage-driven constrained random verification in section 20.7.

14.6   Files

As seen in section 12.2, file *types* describe the format of files that exist in the machine running a simulation. To manipulate those files, we create instances of file types, called *file objects*. Each file object is an interface to a specific file in the host machine.

Because file types, file objects, and file operations interface with files in the machine running a simulation, they are not intended for use in synthesizable models. Consequently, VHDL file operations cannot be used to define the initial contents of models such as ROM or RAM memories. Any support for using data files with synthesizable models is extraneous to the language. If such a solution exists in a certain tool suite, then it is provided by the tool vendor.

There are two ways to use files in VHDL. The first is to read and write the values of VHDL objects directly to a file, in a format that is chosen by the used simulator and is therefore nonportable. This approach uses only operations that are built into the language, thus not requiring any package. The second method is to read and write data in a textual, human-readable format. This approach uses the text file type declared in package std.textio.

Let us look at an example of the first case. Suppose we want to store in a file instances of a record type. This record is composed of two elements: an integer number and an std_logic_vector. This vector could be input data for a testbench, and the integer could be an expected output value for a module under verification. The record type could be declared as:

```
type file_element_type is record
    stimuli: std_logic_vector(31 downto 0);
    expected_output: integer;
end record;
```

The next step is to declare a file *type*, describing the format of any file that stores instances of this record type. This type will tell the simulator how to interpret values read from the file.

```
type stimuli_file_type is file of file_element_type;
```

Note that a consequence of the way VHDL handles files is that only one type of object can be stored in each file, and this is specified when we declare the file type.

Next, we can instantiate a file *object* to interface with a specific file in the host file system. Here we have two choices: if we provide all the information required to open a file, then it will be opened automatically at elaboration. If we omit this information, then we will have to perform a call to file_open later in the code. The following examples illustrate the two cases. The value read_mode used in the second example specifies what kind of operation will be performed in the file; the valid choices are read_mode, write_mode, and append_mode.

```
-- Option #1: Create a file object and open the associated file in read mode:
file stimuli_file: stimuli_file_type open read_mode is "stimuli.txt";

-- Option #2: Just create the file object; do not open any file:
file stimuli_file: stimuli_file_type;
...
-- Later, in sequential code, we open the file by calling file_open:
file_open(stimuli_file, "stimuli.txt", read_mode);
```

File types and file objects can be declared in most declarative regions of the code, including *processes*, *subprograms*, *architectures*, and *packages*. When we declare a file type, several operations are created automatically for objects of that type. These operations include the procedures file_open, file_close, read, write, and flush, which allow us to manipulate the file, and the function endfile, which allows us to check whether the end of a file was reached.

Listing 14.2 shows a process that writes two VHDL objects to a file. The example uses only built-in file IO capabilities, that is, it does not use package textio. Therefore, the generated file is not human-readable and is probably nonportable. The write operations are in lines 16 and 17. When we are done writing to the file, we call file_close (line 19) to release the resources associated with the file in the file system. Whether it is necessary to call file_close depends on where the file object is declared and whether other parts of the code will try to access the same file. For instance, if the file is declared in a subprogram, then it is not necessary to call file_close because this procedure is implicitly called at the end of the subprogram. To be on the safe side, it is good practice to close a file after we are done reading or writing.

Listing 14.2 Writing VHDL objects to a file (in a tool-dependent and non-textual file format)

```
1    -- Write two records to a file named "stimuli.txt" in a tool-dependent format
2    process
3        -- A type describing the format of each object in the file
4        type file_element_type is record
5            stimuli: std_logic_vector(31 downto 0);
6            expected_output: integer;
7        end record;
8
```

```
 9        -- A type for the file itself
10        type stimuli_file_type is file of file_element_type;
11
12        -- An object of class 'file' whose format is given by the two types above
13        file stimuli_file: stimuli_file_type open write_mode is "stimuli.txt";
14    begin
15        -- Write two records to the file
16        write(stimuli_file, (x"ff00ff00", 1));
17        write(stimuli_file, (x"55aa55aa", 2));
18
19        file_close(stimuli_file);
20
21        wait;
22    end process;
```

Listing 14.3 shows a process that reads a file of the same format generated in listing 14.2 and prints its contents to the screen. The read operation is performed in line 21. Note that the values read from the file are copied directly to a VHDL object (the variable current_element), and no conversions are necessary. The procedure write as used in line 24 is not related with the stimulus file; it simply outputs the object value to the screen.

Listing 14.3 Reading a file and printing its contents (in a tool-dependent and nontextual format)

```
 1 -- Read file "stimuli.txt" and print its contents (file contents are in a tool-dependent format)
 2 process
 3    -- A type describing the format of each object in the file
 4    type file_element_type is record
 5        stimuli: std_logic_vector(31 downto 0);
 6        expected_output: integer;
 7    end record;
 8
 9    -- A type for the file itself
10    type stimuli_file_type is file of file_element_type;
11
12    -- An object of class 'file' whose format is given by the two types above
13    file stimuli_file: stimuli_file_type open read_mode is "stimuli.txt";
14
15    -- A variable to hold the value of each element read from the file
16    variable current_element: file_element_type;
17 begin
18
19    while not endfile(stimuli_file) loop
```

```
20        -- Read a single record from the file
21        read(stimuli_file, current_element);
22
23        -- Print the values of the record fields
24        write(output,
25            to_string(current_element.stimuli) & " " &
26            to_string(current_element.expected_output) & LF
27        );
28    end loop;
29
30    file_close(stimuli_file);
31
32    wait;
33 end process;
```

The approach used in the examples of listings 14.2 and 14.3 is simple to use because the objects can be read and written directly, without any kind of conversion. However, the files are vendor-specific, nonportable, and not readable by humans. There is also the restriction that the entire file must contain only one type of object.

To overcome these limitations, we can use the types and operations defined in package textio to read and write files as lines of text. In this approach, we forgo any structured access to the file and assume the responsibility for parsing it correctly. This approach is more flexible, and the resulting files are portable and human-readable. On the downside, the code is a little more verbose and error-prone.

Listing 14.4 shows the examples of listings 14.2 and 14.3 converted to use text files. For writing information to a file, the general approach is to declare an intermediate variable of type line, write information to this variable using write operations, and then write the entire line to the file at once. In the example, we first declare variable current_line in line 4. Next, we write information to this variable in lines 6, 7, 10, and 11 and write the line to the file in lines 8 and 12. We use a similar approach for reading information from the file: we read the file one line at a time (line 27) and then extract information from that line one object at a time (lines 28 and 29).

When writing information in text format, it is important to insert separators between the individual values so that the read operations can distinguish between individual pieces of data (and also to make the file more readable by humans). In lines 7 and 11, the integer number is left-justified in a field of at least three characters. The extra spaces do not cause any problem for the read operations because they skip over leading whitespaces in the line.

Listing 14.4 Reading and writing text files using package `textio`

```
1 -- Write two lines to file "ascii_stimuli_out.txt" in text format (human-readable)
2 process
3     file stimuli_file: text open write_mode is "ascii_stimuli_out.txt";
4     variable current_line: line;
5 begin
6     write(current_line, bit_vector'(x"ff00ff00"));
7     write(current_line, 1, field => 3, justified => left);
8     writeline(stimuli_file, current_line);
9
10    write(current_line,  bit_vector'(x"55aa55aa"));
11    write(current_line, 2, field => 3, justified => left);
12    writeline(stimuli_file, current_line);
13
14    file_close(stimuli_file);
15
16    wait;
17 end process;
18
19 -- Read all lines from file "ascii_stimuli_in.txt" in text format (human-readable)
20 process
21     file stimuli_file: text open read_mode is "ascii_stimuli_in.txt";
22     variable current_line: line;
23     variable number: integer;
24     variable stimuli: std_logic_vector(31 downto 0);
25 begin
26     while not endfile(stimuli_file) loop
27         readline(stimuli_file, current_line);
28         read(current_line, stimuli);
29         read(current_line, number);
30
31         write(output, to_string(number) & " " & to_string(stimuli) & LF);
32     end loop;
33
34     file_close(stimuli_file);
35
36     wait;
37 end process;
```

On a final note, although the `write` procedure used in line 31 is used to output text to the screen, it is in fact a file operation. The name `output` denotes a special text file associated with the standard output stream provided by the simulator. Because every file type has a predefined write operation for its element type, we can call write(output, some_string) to print a string to the screen. In this way, we can bypass the chores of declaring a variable of type line for outputting simple messages.

14.7   Object Initial Values

We close this chapter with a discussion of how objects get their initial values in the beginning of a simulation. The topic of initialization and reset of synthesizable models will be covered in section 19.5.

A variable declaration may optionally include an expression called the *initial value expression* for that variable. This expression is evaluated every time the declaration is elaborated, and then its value gets assigned to the variable.

If the variable declaration does not provide such an expression, then a *default initial value* is applied, which depends on the variable type. For scalar types, it is the leftmost value of the type. For composite types, each individual element gets the default value from its type. For access types, it is the value null. Here are some examples of variables initialized with the default values for their types:

```
type parity_mode_type is (no_parity, even_parity, odd_parity);
variable parity_mode: parity_mode_type; -- Initial value is no_parity
variable carry_bit: std_logic;          -- Initial value is 'U'
variable syllable: string(1 to 3);      -- Initial value is (nul, nul, nul)
variable current_line: line;            -- Initial value is null
variable count: integer;      -- Initial value is the smallest representable
                              -- integer (e.g., -2,147,483,648)
```

The place where a variable is declared determines when it gets initialized. Variables declared in processes are initialized when the process is elaborated. This happens only once, at the beginning of a simulation, and this is the reason that variables in processes can imply storage elements: when a process gets reactivated, its variables are not reinitialized. In contrast, variables declared in subprograms are reinitialized every time the subprogram is executed. For this reason, variables in subprograms cannot imply storage.

For signals, the initialization mechanism is a bit more involved. A signal declaration may also include an expression, in this case called a *default expression*. When evaluated, this expression gives the *default value* for that signal (note the subtle difference from variables, for which an *initial* value expression provides an *initial value*). Just like for variables, if a default expression is not specified, then the default value will be taken from the type of the signal. For a signal of type std_logic or std_ulogic, it is 'U', which is the first value in the enumeration.

Every signal has a default value, either given from a default expression or taken from its type. However, this value may or may not correspond to the initial value of the signal when simulation starts. Remember that a signal may have multiple drivers. During initialization, each driver gets the default value from the associated signal. Then all sources are passed to the resolution function, and its return value gets attributed to the signal.

In any case, the initial value may never be observed in a waveform, at least not for more than one infinitesimal time unit. Because all processes are executed once before the simulation begins, the signal may be assigned a new value before simulation time advances. Note that this applies only to signals that model combinational logic; in contrast, signals that represent registers should contain their given initial values when simulation starts.

Contrary to other programming languages, in VHDL, it is good practice *not* to initialize signals and variables at their declarations in synthesizable code. There are a few reasons for that. First, when we run the synthesizable code in the simulator, it allows us to see which signals were never assigned any value, which is helpful for debugging. Types based on `std_ulogic` will initialize to `'U'`, making it easy to spot uninitialized signals in a waveform. The operations defined in the standard packages are prepared to handle uninitialized values and propagate them so we can track any initialization problem (or lack of initialization) until its origin. Second, strictly speaking, initial values are not supported for synthesis. However, in practice, synthesis tools will try to honor initial values when they make sense and the implementation technology allows. For example, in many FPGAs, initial values are used to define the power-up condition of signals and variables modeling registers. Finally, synthesizable models usually include a reset signal anyway, which provides a proper initialization mechanism. For all those reasons, it is a good idea not to provide initial values out of habit when we write declarations in synthesizable code.

We will get back to the subject of initialization in section 19.5. In any case, keep in mind that an initial value is not a substitute for a proper initialization or reset signal. Unless you have a good reason, it is good practice to always have a reset signal that brings the system to a known default state.

Part V   Practical Coding Recommendations

15 Excellent Routines

The routine may be the most important invention in Computer Science, but many VHDL developers behave as if it did not exist. A survey of half a million lines of code in open source projects revealed that half of them did not use any subprograms.[1]

Does this mean that functions and procedures are not as useful in VHDL as in other programming languages? Not at all. Most likely, many designers are missing opportunities to reduce system complexity, improve readability, and avoid duplicating code. This chapter presents guidelines to help you decide when to create a subprogram and practical advice on how to write routines that will make your code easier to understand and maintain.

15.1 What Is a Routine?

A routine is a sequence of statements that have been wrapped together and given a name. Other programming languages may call a routine by different names, with the same or slightly different meaning. Most of the time, however, routines, subroutines, subprograms, functions, procedures, and methods mean essentially the same. The term *routine* is usually favored by software developers. The term *subprogram* is the denomination given in the VHDL LRM for both functions and procedures. In this book, the terms *routine* and *subprogram* are used interchangeably.

A routine is a procedural abstraction: a programmer calls a subprogram, trusting it to perform a task and ignoring any implementation details. If the abstraction is created properly, then the routine can be used as a black box, based solely on its name, parameters, and documentation.

It is a common misconception that routines are useful only for code that must be reusable and generic. Routines are, above all, a great tool for organizing our source code. It makes sense to create them even when they have a fixed behavior or when they are used only once in the code.

15.2   Why Use Routines?

In the strictest sense, routines are optional. We can write complex code without routines, and we can write large projects ignoring them completely. If we do so, however, our code will be harder to understand, debug, and improve. This section presents the most compelling reasons to use routines in VHDL designs.

To Reduce Code Complexity   Our code does not have to be complex only because it does many things, but it will be unnecessarily complicated if it does many things without a clear structure and organization. Routines help reduce complexity in two ways: by helping us focus on one part of the problem at a time, and by providing a less detailed view of any complex task.

These two benefits are best explained with an example. Figure 15.1a shows a relatively complex task involving a sequence of three operations. Those familiar with cryptography will recognize the operations as three steps from the Advanced Encryption Standard (AES). Because there are no boundaries or delimiters between the steps, all data objects share the same scope and are visible by any statement. As a consequence, operations can interfere with one another, forcing us to read the entire code to be sure of its behavior.

```
-- sub bytes
for i in 0 to 3 loop
    for j in 0 to 3 loop
        state_2(i,j) :=
            sbox(state_1(i,j));
    end loop;
end loop;

-- shift rows
for i in 0 to 3 loop
    for j in 0 to 3 loop
        state_3(i,j) :=
            state_2(i,(j+i) mod 4);
    end loop;
end loop;

-- add round key
for i in 0 to 3 loop
    for j in 0 to 3 loop
        state_4(i,j) := state_3(i,j) xor
            round_key(i,j);
    end loop;
end loop;
```

```
function sub_bytes(state: state_type)
        return state_type is
begin
    -- Code for operation 'sub bytes'
end;

function shift_rows(state: state_type)
        return state_type is
begin
    -- Code for operation 'shift rows'
end;

function add_round_key(state: state_type)
        return state_type is
begin
    -- Code for operation 'add round key'
end;

...

state_2 := sub_bytes(state_1);
state_3 := shift_rows(state_2);
state_4 := add_round_key(state_3);
```

(a) A complex operation performed
    in a big block of sequential code.

(b) The same operation broken down into
    three simpler routines.

Figure 15.1

Using routines to organize a complex task.

In figure 15.1b, the task has been broken down into three functions: sub_bytes, shift_rows, and add_round_key. Of course it is still necessary to implement the code inside each routine, but the new organization has two main benefits. First, each routine can be understood in isolation. Because each routine is smaller, is more focused, and has less variables, each of them is easier to understand than the original task.

The second benefit is that the code using the routines provides a clear overview of the main task. With a quick glance at the routine calls, it is possible to identify the overall sequence of operations and tell what each step does. It is also easier to see that the output of each routine is used as an input to the next. In case we still need more information, we can examine the code details by looking inside one routine at a time. By breaking down a complex operation into subprograms, the code becomes easier to work with and understand.

To Avoid Duplicating Code Duplicate code is at the root of many problems in software development. An application that has similar code in two or more places is harder to maintain. For each modification, we must remember to update all the similar places in the code. It is only a matter of time until we forget to update one of those places, introducing bugs that may be hard to detect. The code is also longer and harder to read; we must read similar sequences of statements over and over, paying close attention to any small differences. Duplicate code is also a design problem: it is a failure to decompose a system into nonoverlapping modules.

Extracting duplicate code to a routine and calling it from all other places is the most effective technique to remove code duplication. To illustrate some benefits of this technique, let us see an example. The following code converts x and y coordinates to memory addresses for accessing a video framebuffer. The logic to calculate a memory address from a pair of coordinates is duplicated in the three lines:

```
-- Convert x and y coordinates to memory addresses in the framebuffer
write_address <= x_in(9 downto 2) & y_in(9 downto 2);
read_address <= x_out(9 downto 2) & y_out(9 downto 2);
next_address <= x_next(9 downto 2) & y_next(9 downto 2);
```

If this conversion logic needs to change, then we will have to look for occurrences of the same expression all over the source code, possibly in several different files. One way to prevent this problem is to encapsulate the conversion logic inside a function:

```
function memory_address_from_x_y(x, y: in bit_vector) return bit_vector is
begin
    return x(9 downto 2) & y(9 downto 2);
end;
...
```

```
write_address <= memory_address_from_x_y(x_in, y_in);
read_address <= memory_address_from_x_y(x_out, y_out);

next_address <= memory_address_from_x_y(x_next, y_next);
```

The new version has more lines of code because of the function declaration, but it offers several advantages:

- The knowledge about how to calculate an address from a pair of coordinates is located in a single place. If this logic needs to change, then we only need to modify one line of code.
- Because the function name is self-explanatory, the client code (the code that calls function memory_address_from_x_y) is self-documenting and easier to understand. The original comment becomes redundant and unnecessary.
- The client code can safely ignore the details of how to transform a pair of coordinates to a memory address and concentrate on higher level functionality. This allows us to think closer to the problem domain and saves us from dealing with low-level implementation details all the time.

Whenever you see sections of code that look similar, think of a way to remove this duplication. In many cases, the best way is to move the common code to its own routine.

To Make the Code More Readable   Sometimes the best way to make a sequence of statements easier to read is to hide them behind a routine with a descriptive name. Suppose you are skimming through some code and run into the following lines:

```
-- Check whether my_array contains a 0
number_to_check := 0;
element_found := false;
for i in my_array'range loop
    if my_array(i) = number_to_check then
        element_found := true;
    end if;
end loop;
...
```

It is easy to read the statements and figure that the code searches for the value 0 in an array. However, in the context of a larger task using this operation, it is irrelevant how this processing is done; it only matters that the code works as expected. If we encapsulate the code snippet in a function, then the client code becomes much clearer:

```
array_contains_zero := array_contains_element(my_array, 0);
```

As a bonus, the code is now self-documenting. We can remove the comment at the top because it says the same as the function call.

To Make the Code Easier to Test   Certain aspects of the system behavior may be buried deep inside a design. For instance, an important calculation may be implemented with several individual statements, inside a process that is a small part of an architecture. Testing this behavior is hard because we do not have direct access to its inputs and outputs; we may have to put the circuit through a long sequence of states to reach the condition we want to test. We may also have to check the result indirectly if the calculated values are inaccessible from outside the module.

If we encapsulate this piece of behavior inside a function, the design will be not only clearer but also easier to test. By declaring the function in a package, we have direct access to its inputs and outputs, and the code can be tested with a simpler testbench. Because creating testbenches requires significant effort, anything we can do to make testing easier improves the odds that the code will be properly tested.

To Make the Code Easier to Change   Sometimes we need to write code knowing that it is highly likely to change in the future. For instance, we may use a temporary algorithm that does not offer the required performance, only because our first goal is to make the system produce the correct values. Or we may write a nonsynthesizable version because it is needed to test other parts of the circuit. In such cases, the client code will be more stable if we wrap the volatile code inside a routine.

15.3   Functions versus Procedures

VHDL is among the languages that make a distinction between two kinds of routines. The first type, called a *function*, is similar to a function in mathematics: it processes one or more input values and produces exactly one output value. The other kind of subprogram, called a *procedure*, encapsulates a group of statements that communicate with the calling code in two possible ways: via input and output parameters or by direct manipulation of objects declared outside the procedure. Figure 15.2 shows one example of each. The two examples are functionally equivalent.

Peter Ashenden, in the *Designer's Guide to VHDL*,[2] makes the following distinction: a procedure is a generalization of a statement, whereas a function is a generalization of an expression. Hence, a procedure can be used in place of a statement, whereas a function can be used only where an expression is allowed.

This section elaborates on the distinction between functions and procedures and then discusses the reasons for choosing one over the other.

```vhdl
function sum_function(vect: integer_vector) return integer is
    variable sum: integer := 0;
begin
    for i in vect'range loop
        sum := sum + vect(i);
    end loop;
    return sum;
end;
```

(a) A function.

```vhdl
procedure sum_procedure(vect: in integer_vector; sum: out integer) is
begin
    sum := 0;
    for i in vect'range loop
        sum := sum + vect(i);
    end loop;
end;
```

(b) A procedure.

Figure 15.2
The two kinds of subprograms available in VHDL: function and procedure.

15.3.1   Functions

A *function* encloses a sequence of statements that operate on zero or more input values and produce exactly one output value. From the client's point of view, the actual execution of the statements is secondary: what really matters is the value that the function calculates and returns.

A function call is used as part of an expression; it cannot be used as a stand-alone statement. When the expression is evaluated, the function is executed, and its return value is inserted in place of the function call. Because it is used in the middle of expressions, a function call fits nicely in the client code, leaving it uncluttered; it is not necessary to declare auxiliary data objects just to receive the result of a calculation. However, the fact that a function call is an expression imposes some limitations: a function cannot use *wait* statements, cannot be used as a stand-alone statement, and must produce its result in a single simulation cycle. Function parameters can only be of mode *in*, and their classes must be *constant, signal,* or *file* (variables are not allowed).

VHDL defines two kinds of functions: *pure* functions, which always return the same value when called with the same input values; and *impure* functions, which may return different values even though the input values are the same. An impure function is less restrictive on the objects it may access: besides the function parameters, it can access data objects that are visible at the place where the function is declared. For instance, when an impure function is declared in an architecture, it can read the corresponding entity ports; when it is declared in a process, it has access to the process variables; and when it is declared in a package, it has access to global data objects declared in the package.

15.3.2 Procedures
A procedure encloses a sequence of statements that do not return a value. Unlike functions, whose main purpose is to produce an output value, procedures are called for their behavior and for their effects on other objects (e.g., on the procedure parameters or signals and variables declared outside the subprogram).

A procedure call cannot be used in an expression; it must be a stand-alone statement. It can, however, calculate as many output values as we wish, as long as they are communicated via output parameters. This means that, in practice, procedures can do the same job as functions. The only downside is that passing the output values may be cumbersome, if it forces the client code to declare auxiliary objects to receive the results, or sneaky, when data is passed through objects not specified in the parameter list.

Procedures do not have as many limitations as functions, and they can be used to implement modules with flexible interfaces. Procedure parameters can be of mode *in*, *out*, or *inout*, and their classes can be *constant*, *variable*, *signal*, or *file*. Because of their flexibility, procedures can be used almost like an entity or component and are an interesting alternative for the implementation of combinational circuits.

15.3.3 Choosing between Functions and Procedures
Before we look at specific guidelines for choosing between functions and procedures, let us review the main differences between the two kinds of subprograms (table 15.1).

Table 15.1 Distinctive characteristics of functions and procedures

Characteristic	Functions*	Procedures
Where and how it is used	In concurrent or sequential code. Always in an expression.	In concurrent or sequential code. Used as a standalone statement.
How to send values to the subprogram	Via *input* parameters.	Via *input* or *inout* parameters; via other objects visible to the subprogram.
How to return values from the subprogram	Via a single return value (required).	Via one or more *output* or *inout* parameters; via visible objects writable by the subprogram.
Designator (name)	An identifier or an operator (e.g., "+", "-", "and", "or").	An identifier.
Timing issues	Cannot include *wait* statements; must be evaluated in a single simulation cycle.	Can include *wait* statements, which cause the calling process to suspend.

*Some characteristics apply to *pure* functions only.

As shown in the table, some characteristics are exclusive of functions or procedures; when we need one of these characteristics, we are forced to use the construct that supports it. For example, if the calculation must take place inside an expression (such as the initial value in an object declaration), then we must use a function. In contrast, if we need to use *wait* statements or calculate multiple output values, then we must use a procedure.

There are, however, many situations where it is possible to use either kind of subprogram. Here we discuss the main reasons for choosing one over the other in such cases.

The main advantage of using functions is that they fit perfectly in expressions in the client code. Functions can be used in a manner that is indistinguishable from predefined operators; they do not require the declaration of auxiliary objects to receive the output values, resulting in less clutter and code overhead. In the following example, the value of signal `seven_seg` is calculated directly from `num_lives_p1` and `num_lives_p2`, using the functions `bcd_from_int` and `seven_seg_from_bcd`. We can write the three function calls in a single expression without any need for auxiliary data objects:

```
seven_seg <= seven_seg_from_bcd( bcd_from_int(num_lives_p1) & bcd_from_int(num_lives_p2) );
```

Had we used procedures, the same operation would require auxiliary data objects to hold the partial results:

```
signal num_lives_p1_bcd: std_logic_vector(3 downto 0);
signal num_lives_p2_bcd: std_logic_vector(3 downto 0);
...
convert_int_to_bcd( num_lives_p1, num_lives_p1_bcd );
convert_int_to_bcd( num_lives_p2, num_lives_p2_bcd );
convert_bcd_to_seven_seg( num_lives_p1_bcd & num_lives_p2_bcd, seven_seg );
```

Another advantage of using functions is their restricted interaction with the program state and external objects. This may look like a disadvantage, in the cases where such restrictions prevent us from using a function. In all other cases, however, it is a big advantage: when you call a function, you may be sure that it will not change its parameters or leave any side effects in the program state. When we read the client code, we know that there is nothing hidden behind a function call: all it does is provide a value, which depends only on its inputs. This makes the code much easier to reason about; we do not need to read the function implementation to know that the program state is safe.

In contrast, the main advantage of using procedures is their greater flexibility. A procedure can produce any number of output values, including zero. A procedure has access to the data objects that are visible where it is declared; it can also read and write to files, which is not allowed in functions because file operations count as side effects.

Procedures have several characteristics that make them well suited for testbenches. Because they allow signal parameters and *wait* statements, procedures can be used to produce stimuli and control the passage of time. Their ability to produce text output and perform file I/O is also useful in testbench code.

The disadvantages of using procedures are the opposite of the advantages of using functions. When they are used simply to calculate an output value, procedures make the client code more verbose because they cannot be inserted in expressions. When you see a procedure call, you cannot be sure about what it causes to the program state without inspecting its implementation.

As a general rule, functions are the recommended kind of subprogram wherever they can be used. If the goal of a subprogram is to calculate one value, then use a function even if it requires some rewriting on the client code. In contrast, if the subprogram does not return a value, if it absolutely must return more than one value, or if it is called for the effects it has on other objects, then use a procedure. Table 15.2 summarizes the guidelines for choosing between functions and procedures.

15.4 Writing Great Routines

Now that we have a strong grasp of the main reasons to create a routine and the key factors in choosing between functions and procedures, we can move to the next part of the problem: how to create high-quality routines that improve the quality of our source code.

Table 15.2 When to use a function or procedure

Use a function ...	Use a procedure ...
When the main goal of the subprogram is to produce a value.	When the main goal of the subprogram is to execute statements or change the values of existing objects.
When the subprogram must return a single value.	When the subprogram does not produce a value, or when it produces more than one value.
When the subprogram call is part of an expression.	When the subprogram call is a standalone statement.
For data conversion, logic, and arithmetic operations.	For operations involving the passing of time (using a *wait* statement).
To overload predefined operators (e.g., and, or, "+", "-").	When you need side effects (e.g., file operations). *
To implement mathematical formulas.	To implement algorithms involving the passage of time, or to group a sequence of statements.
Whenever you can, to keep the client code simple.	Whenever you need more flexibility than provided by pure functions.*

*An impure function could also be used in these cases.

In *Clean Code*, Robert Martin summarizes in one sentence most of what it takes to write good routines: "choosing good names for small functions that do one thing."[3] We will discuss these and other guidelines in this section.

Have a Single, Well-Defined Purpose   A good routine should do one thing, and do it well. A function that does only one thing is small and easy to understand. It takes only the necessary parameters and is trivial to test. Keeping routines focused also makes them more robust and less likely to change in the future. In contrast, a function that does many things has many possible reasons to change, and each change has a higher risk of breaking something else.

One problem with enforcing this rule is to know exactly what "do one thing" means. One way to check whether a function does only one thing is to try to describe it clearly in one sentence. If the sentence has the words "and" or "or," then the function probably does more than one thing. Another way is by looking at the level of abstraction of the routine statements. If all the statements or subprogram calls are one level below the name of the routine, then the routine is doing one thing.[4]

Boolean parameters in the form of "flag arguments" are a strong indication that a function does more than one thing. In the following example, the argument named value_is_bcd (line 8) steers the function through two distinct paths. One of the disadvantages of this kind of code is that we need to declare variables for both possible paths (lines 11–14). As a result, the function is littered with more variables than would be necessary to accomplish each task individually.

```
1    -- Return a vector for driving a series of 7-segment displays.
2    function unsigned_to_seven_segment(
3        value: unsigned;
4        number_of_digits: integer;
5        -- When true, treat the input value as a BCD number where
6        -- every 4 bits hold one digit from 0 to A. When false,
7        -- treat the input number as an unsigned integer.
8        value_is_bcd: boolean
9    ) return std_logic_vector is
10       ...
11       variable segments: std_logic_vector(number_of_digits*7-1 downto 0);
12       variable unsigned_digit: unsigned(3 downto 0);
13       variable bcd_quotient: unsigned(value'range);
14       variable bcd_remainder: unsigned(3 downto 0);
15   begin
16       if value_is_bcd then
17           ... -- code treating the input value as BCD digits
18       else
```

```
19          ... -- code treating the input value as an unsigned integer
20      end if;
21
22      return segments;
23  end;
```

The proper way to fix this code is to divide it into two functions. The function caller probably knows whether it provides an unsigned or a BCD value, so it might as well call the correct version of the function.

As a final remark, remember to use the principle of strong cohesion to check whether the function does only one thing. If you can separate the routine in two different tasks, and each one uses an independent set of data, then it is probably doing two things. For routines, it is almost always possible to reach the highest level of cohesion—namely, functional cohesion (see "Types of cohesion" in section 4.5). Don't settle for anything less.

Make It Small   There are many advantages to keeping routines small. Short routines are naturally less complex; they are easier to understand, easier to test, and more likely to focus on a single purpose. Short routines are also more likely to be correct because they have fewer places for bugs to hide. They are also easier to read because we can see the entire routine at once.

There are no hard and fast rules on how long a function should be. Besides sheer length, other factors indicate when a function is growing too large, such as complexity, number of variables, and number of indentation levels. Most of the time, it is better to use these metrics rather than the number of lines of code.

Having said that, there is a good practical reason to keep your routines short. When you are trying to understand part of a program, your ability to understand it is proportional to how much of it you can see at a time. For this reason, many authors recommend keeping functions within the length of a page; with today's technology, this corresponds to approximately 50 lines.

Routine length is one of the metrics commonly enforced by static analysis tools. Depending on the programming language, maximum values of 20, 40, or 60 lines of code (LOC) are common. For VHDL, SciTool's *Understand* defaults to a maximum length of 60 lines. For C++, VerifySoft's *Testwell* recommends 40 lines or less. For Ruby, a significantly more expressive language, the *Roody* linter defaults to a maximum of 20 lines per method.[5]

There is no problem in exceeding the 60-line limit occasionally if your routine calls for it. However, keep this number in mind and, as you approach it, start thinking of ways to break the routine into smaller tasks. At any rate, going over 150 LOC is asking for trouble. The Institute for Software Quality states that routines longer than that have been shown to be less stable, more subject to change, and more expensive to fix.[6] A study by Richard Selby and Victor Basili showed that routines with 143 lines or less are 2.4 times less expensive to fix than larger routines.[7]

The last point we need to address is how to keep our routines short. The most effective way is to prevent them from growing too large in the first place: when a routine approaches your chosen length limit, start looking for ways to move part of its functionality to a new routine. A good way to do this is to remove any duplicate code. Keep an eye out for expressions or sequences of statements that look similar and extract them to their own routines. Another effective way to reduce a routine is to decompose it into a series of shorter subprograms. Use abstraction: find a sequence of statements that have a clear purpose and move them to their own routine. Loops, especially nested ones, are natural candidates. The inner code is often at a different level of abstraction and performs a well-delimited subtask. Move the inner code to its own function, and pass the loop iterator as a parameter.

Keep a Uniform Abstraction Level in the Routine   Your code will be easier to follow if each operation leads naturally into the next, in a clear and natural order. It is hard to keep this flow if the code mixes essential operations with low-level details. As you read the code, every time you run into a series of intricate low-level statements you need to stop, analyze them, and mentally chunk them into a meaningful operation.

To avoid this, all operations in the routine should have approximately the same importance; in other words, the entire routine should contain only tasks at the same level of abstraction. Listing 15.1 shows a routine that violates this principle. The routine has a fairly complex job: it calculates the value of a pixel at the given coordinates for use in an on-screen display. This routine has a number of good characteristics: it is well documented, it has a good name, it is not too long, and it builds on other routines to perform its task. The problem, however, is that parts of the routine are at a much lower level of abstraction.

Listing 15.1 A routine with mixed levels of abstraction

```
1    -- Calculate the value of a pixel for the OSD (on-screen-display) at the
2    -- given coordinates, looking up the character code in the text buffer,
3    -- and then character bitmap in the font ROM.
4    function get_osd_pixel_at(x, y: natural) return pixel_type is
5        ...
6    begin
7        -- Find the ASCII code of the character at x, y
8        char := get_character_at_pixel(x, y);
9        code := get_ascii_code(char);
10
```

```
11      -- Calculate the pixel position in the font bitmap, using
12      -- only the lowest-order bits of the x, y coordinates
13      x_uns := to_unsigned(x, x_uns);
14      y_uns := to_unsigned(y, y_uns);
15      x_offset_uns := x_uns(2 downto 0);
16      y_offset_uns := y_uns(3 downto 0);
17      offset.x := to_integer(x_offset_uns);
18      offset.y := to_integer(y_offset_uns);
19
20      -- Get the pixel from the character bitmap
21      bitmap := get_character_bitmap(code);
22      pixel := bitmap(offset.x)(offset.y);
23
24      return pixel;
25  end;
```

Tasks such as get_character_at_pixel and get_character_bitmap (lines 8 and 21) are major steps in the computation. In contrast, lines 13–18 are doing bit twiddling and perfunctory type conversions. The inconsistent level of abstraction disrupts the reading flow and makes the routine harder to follow by littering it with unnecessary details. To avoid this, make sure that all operations in a routine have approximately the same level of abstraction.

Use Few Indentation Levels   A routine that uses many levels of indentation is probably more complicated than it needs to be. The excessive indentation also leaves less space for actual code on each line. Limiting nesting to about three or four levels is a good guideline that makes your routines simpler and easier to read. See "The Problems with Deep Nesting" (section 10.1.1) for the details.

Return Early When It Improves Readability   The practice of structured programming says that each software module should have a single entry and a single exit points. This guideline was created to make the code more intellectually manageable, in a time when the *goto* statement was thriving. Today, programming languages do not allow jumping to the middle of a routine; however, it is still possible to return from several different places.

   In certain cases, however, returning from the middle of a routine can make it easier to understand. One example is when the routine has already accomplished its goal—for instance, finding an element in an array. Another example is at the top of a routine, where you need to check input parameters and then return special values or abort the execution. This kind of return performed at the top of a routine is often called a "guard clause." Getting those special cases out of the way early helps you concentrate on the normal processing that comes after them.

An example of this guideline is shown in figure 15.3. On the left side (figure 15.3a), we see a modified version of function sqrt from ieee.math_real, implemented using nested *if-else* statements. On the right (figure 15.3b), the code has been flattened by using guard clauses at the beginning. The guard clauses handle all the special cases, so that when we arrive at the normal processing performed inside the *while* loop, we can focus solely on it.

The kind of code shown in figure 15.3a has a tendency to appear in regions with many conditionals. In software, frequently occurring bad ideas and practices are called *anti-patterns*; the shape of code shown in figure 15.3a is known as the *arrow anti-pattern*. Keep an eye out for it.

To solve this problem, use early returns or extract the inner part of conditional and iterative structures to their own functions. Be mindful, however, that an excessive number of return points is also confusing. When we are debugging and the return value is different from what we expected, we need to look harder to find out which statement returned it. Use multiples returns only when they improve readability.

```
function sqrt(x: real) return real is
  ...
begin
  if (x < 0.0) then
      assert false report "sqrt: x < 0 in"
          severity error;
      sqrtval := 0.0;
  else
      if x = 0.0 then
          sqrtval := 0.0;
      else
          if (x = 1.0) then
              sqrtval := 1.0;
          else
              oldval := inival;
              newval := (x/oldval + oldval) / 2.0;

              while (abs(newval-oldval) > err) loop
                  oldval := newval;
                  newval := (x/oldval + oldval) / 2.0;
              end loop;

              sqrtval:= newval;
          end if;
      end if;
  end if;

  return sqrtval;
end;
```

```
function sqrt(x: real) return real is
  ...
begin
  if x < 0.0 then
      assert false report "sqrt: x < 0"
          severity error;
      return 0.0;
  end if;

  if x = 0.0 then
      return 0.0;
  end if;

  if x = 1.0 then
      return 1.0;
  end if;

  oldval := inival;
  newval := (x/oldval + oldval) / 2.0;

  while (abs (newval-oldval) > err) loop
      oldval := newval;
      newval := (x/oldval + oldval) / 2.0;
  end loop;

  return newval;
end;
```

(a) Nested code with a single return.   (b) Flattened code with multiple returns.

Figure 15.3
Use of guard clauses to reduce nesting.

Give the Routine a Good Name   Choosing good names is the better part of writing understandable code. Meaningful names make it possible to understand the code long after it has been written. When you choose names that are meaningful, clear, and unambiguous, you make your life a lot easier, not to mention the life of all other programmers who work with your code.

Good naming is so important in programming that the next chapter is dedicated entirely to this topic. Here we provide only a quick overview of the subject; for a more elaborate discussion, see chapter 16.

Ideally, the name of a routine should describe everything it does and pack all the information for it to be used correctly. It is a big mistake to think that this is an unattainable goal; in most cases, it is possible to choose names that are perfectly descriptive, with little effort. In the few cases where finding a good name is hard, this is often the sign of an underlying design problem; when we do not understand a problem well, choosing good names is always hard.

As developers, we have all the freedom in the world to choose good names. We have at our disposal words and concepts from the problem domain, from the solution domain, and from our Computer Science knowledge. Yet too often we choose names that are ambiguous, meaningless, or vague. Here are some guidelines to help you avoid this trap.

The first guideline is a matter of attitude. Take the job of naming routines seriously. Do not choose names casually or rush through the process. It is probably a good idea to try out a few or several names before you commit to one of them. To test how your names are working, look at the client code that uses the routine. The routine call should read naturally and unmistakably.

Once you have chosen a name, do not be afraid to change it for a better one when it occurs to you. As our knowledge of the solution improves, better names will naturally come up. Changing a name for a better one only takes a moment but provides benefits from then on, every time the code is read. Today's source code editors and IDEs can rename subprograms and other identifiers automatically, so the modifications cost nothing and are free of mistakes.

Make the name of the routine as long as necessary to explain everything it does. Your goal is to make the use of the routine self-evident. If a routine's main purpose is to return an object, then name it after the return value. If a routine's main purpose is to perform an action, then name it after everything it does. Use strong verbs, and avoid wishy-washy words such as "do," "perform," or "process."[8] Fully descriptive names make the client code easier to understand and easier to check for correctness.

Always prefer meaningful, unabbreviated, correctly spelled words to compose your names. Do not create names that need to be deciphered and require us to memorize their real meaning. Instead, use self-evident and unambiguous names. Every programmer reading the code should be able to infer the same meaning from the same name;

abbreviations and cryptic prefixes or suffixes work against this goal. Finally, use pronounceable names; this allows you to explain the code to another programmer using plain English.

15.5   Parameters

Before we can present the remaining guidelines about the use of routines, we must review a few concepts about subprogram interfaces and parameters.

A routine is a module; as such, it should be treated as a black box. The only thing the client code should need to know about a routine is its interface. Essentially, the interface of a subprogram is composed by its name, parameters, return value (in the case of a function), and documentation. Although there is more to a routine than that, these are the main factors that make a routine easy to use.

The creator of the routine should do everything to make its use self-evident, and a good part of it is the choice of parameters. Name, type, and number of parameters are important factors in making a routine more usable. The following topics discuss the use of parameters in VHDL and provide a series of guidelines for using them effectively in your routines.

15.5.1   Formal versus Actual Parameters

Parameters are the pipes that carry values from the client code to inside a routine and vice versa. In Computer Science and VHDL terminology, there is a distinction between two kinds of parameters: *formal* and *actual*. The difference is one of point of view: formal parameters refer to the objects declared in the routine interface and used inside the routine, whereas actual parameters refer to the values provided by the client code. Formal parameters are the objects inside the parentheses of a routine declaration. Routine parameters are specified in a *parameter interface list*, or *parameter list* for short. Figure 15.4a shows the declaration of a function called sum. This function has two formal parameters: augend and addend.

Actual parameters, in contrast, are supplied by the caller of the routine. In other words, actual parameters are the values or objects that appear inside the parentheses when the routine is called. In figure 15.4b, variable a and the literal number 2 are actual parameters in the call to function sum. Actual parameters are also called *arguments*.

Another way to put the distinction between formal and actual parameters is to say that formal parameters are placeholders, whereas actual parameters are concrete objects and values. When the routine is called, formal parameters inside the routine are associated with the corresponding actual parameters in the client code.

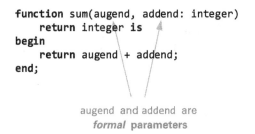

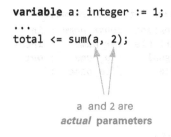

augend and addend are
formal parameters

a and 2 are
actual parameters

(a) Formal parameters in the declaration of a routine.

(b) Actual parameters in a routine call.

Figure 15.4
Formal versus actual parameters.

15.5.2 Interface Objects

Each formal parameter in a routine is a declaration of an *interface object*. For a function, interface objects provide constant values that can be used in the routine statements. For a procedure, interface objects can move values in and out of the subprogram.

The simplest declaration of an interface object has only two pieces of information: a name and a type. However, if we want to fully configure an interface object, then its declaration may have up to five parts:

1. A parameter **class**: *constant, variable, signal,* or *file*;
2. An **identifier**: the name of the formal parameter;
3. A parameter **mode**: *in, out,* or *inout*;
4. The object **type** or **subtype**;
5. A **default value**: a static expression used to initialize the object when no actual value is passed in the subprogram call.

Figure 15.5 shows a parameter list that declares four interface objects of different classes and modes. The *parameter mode* (*in, out,* or *inout*) defines the direction in which the values flow into and out of the routine. The *parameter class* defines how formal and actual parameters will be associated, and how the interface object can be used *inside* the routine. It is not necessarily the same as the class of the actual parameter; for example, a signal in the client code can be associated with a constant parameter inside the routine. *Default values* are used to initialize the interface object when the routine is called without a corresponding actual parameter. In this way, we can write routines with optional parameters.

Of the five parts, only the name and type are mandatory. If a mode is not specified, then it will default to *in*. If a class is not specified, then it will default to *constant* for mode *in* and to *variable* for modes *inout* and *out*. Therefore, if we do not specify either a class or a mode, then the parameter will default to *constant* and *in*. As we will see in a

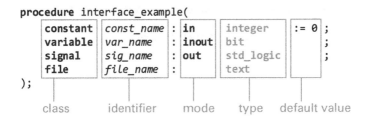

Figure 15.5
The five parts of an interface object declaration.

moment, this is a sensible default that allows the routine to work as expected in most cases because most parameters are in fact of mode *in*, and any expression can be used as an actual value for a formal parameter of class *constant*.

An important point to note is that not all combinations of class, mode, and default value are valid. Here we discuss their main restrictions.

The first constraint is that the class of a formal parameter restricts its mode. If the class is *constant*, then its *mode* must be *in*. This is easy to remember because it would make no sense to pass constant values back to the client code. In contrast, if the class is *variable* or *signal*, then all three modes (*in*, *out*, *inout*) are allowed. This also makes sense because procedures can be used to read, write, or modify both signals and variables. Finally, parameters of class *file* have no mode.

The second constraint is that the class of a formal parameter restricts what can be supplied by the client code as an actual parameter. Formal parameters of classes *signal*, *variable*, and *file* can only be mapped to actuals of the same class. In contrast, formals of class *constant* can be associated with any expression provided by the client code. Fortunately, a signal name, a variable name, a constant name, and a literal value are all examples of expressions, so they can all be used when the formal parameter is a *constant*. Also, *constant* is the default class for *in* mode parameters; therefore, if you do not specify a class, then your subprograms will work in all those cases. Figure 15.6 shows which actual parameters can be associated with each class of formal parameters in a subprogram call.

The third constraint is that the kind of subprogram restricts which modes can be used: if the subprogram is a function, then all parameters must be of mode *in*. If it is a procedure, then all three modes (*in*, *out*, and *inout*) can be used, subject to the other restrictions. Finally, default values can only be used with parameters of mode *in* and cannot be used with class *signal*.

These are the main limitations on the combination of parameter classes and modes; however, there are many other restrictions on what can be done with the parameters inside a subprogram. For the complete reference, see section 4.2.2 of the VHDL Language Reference Manual.[9]

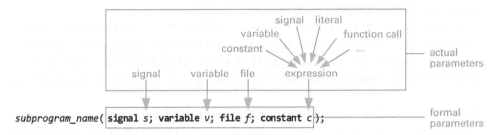

Figure 15.6
Valid associations from actual to formal parameters.

The key takeaway from these rules is that class *constant* is the most flexible for a formal parameter: it allows the actual parameter to be a signal, a variable, a constant, a literal, or any expression. If you have never worried about specifying parameters classes in subprogram declarations, then this is probably the reason.

15.5.3 Subprogram Calls

In a subprogram call, actual parameters provided by the client code are associated with the formal parameters used inside the routine. How this mapping is done depends on the class and mode of the formal parameter.

For classes *constant* and *variable*, parameter passing is done by copying. The objects on the caller side are never touched while the routine is executing. If the parameter is of mode *in* or *inout*, then the formals are initialized with the values of the actual parameters. Inside the routine, all the manipulation happens only on the formal parameters. When the routine finishes execution, for parameters of mode *inout* or *out*, the value of the formal parameter is copied to the object associated with the actual parameter.

For the class *signal*, parameter passing is done by reference. If the parameter is of mode *in* or *inout*, then the actual signal is associated with the formal parameter at the beginning of the subprogram. If the parameter is of mode *inout* or *out*, then the driver of the actual signal is associated with the driver of the formal parameter at the beginning of the subprogram (for more about drivers, see section 5.6.2). An interesting effect of this pass-by-reference mechanism is that if you read a signal at different places in the subprogram, then it is possible to see different values if the reads are separated by a *wait* statement.

Formal parameters and local variables in subprograms are always initialized anew every time the subprogram is called. This has an interesting implication for synthesis: variables declared in subprograms never model storage elements because they are always assigned an initial value on every call.[10]

15.5.4 Guidelines for Using Parameters

In software, errors in the interface between modules are one of the most common causes of bugs. The main reasons include excessive interface complexity, ambiguous design, incomprehensible interfaces, and lack of documentation. This section presents guidelines to help you avoid this trap by creating routines with high-quality interfaces.

Take Few Parameters   Routines that take more than a few parameters have several disadvantages. First, a large number of parameters is a strong indication that the logic inside the routine is complicated. Second, it may be a sign that routine is doing more than one thing. Third, the large number of parameters increases the coupling between the client code and the routine.

Before writing a routine with more than three parameters, first make sure it is doing only one thing. If you can identify two or more smaller tasks that use only part of the input arguments, then create separate routines and pass each one only the required information.

Another effective way to reduce the number of parameters is to combine some of them into a single object, as described in the next guideline. A wrong way to reduce the number of parameters is to read objects declared outside the routine, bypassing the parameter list. Passing input values implicitly does not reduce coupling and does not make the routine less complicated.

Consider Passing an Object of a Record Type Instead of Separate Parameters   When a routine has a long list of parameters, it is possible that some of them are part of the same conceptual unit. In such cases, consider whether the arguments could be grouped into a single object of a record type or array.

In the following example, the function `rectangles_intersect` nominally takes eight arguments; however, it is easy to see that each group of four arguments actually describes a single abstract entity:

```
function rectangles_intersect(
    rect_1_top, rect_1_left, rect_1_bottom, rect_1_right: coordinate_type;
    rect_2_top, rect_2_left, rect_2_bottom, rect_2_right: coordinate_type
) return boolean;
```

If we formalize the abstraction as a record type, then the function can be rewritten with only two parameters:

```
type rectangle_type is record
    top, left, bottom, right: coordinate_type;
end record;
```

```
function rectangles_intersect(rect_1, rect_2: rectangle_type) return boolean;
```

Use Assertions to Check Input Parameters   If your routine could malfunction under certain combinations of input values, then make sure it is protected against them. Test the input values using assertions, and raise an error if an invalid input is detected. In the following example, the matrix product can only be calculated when the number of columns in matrix *A* equals the number of rows in *B*. Instead of relying on a comment to warn the users of the routine, a better idea is to use an assertion:

```vhdl
function "*"(A, B: matrix_type) return matrix_type is
    variable C: matrix_type(1 to width(A), 1 to height(B));
begin
    -- Product A × B is defined only if the number of columns in A
    -- equals the number of rows in B.
    assert width(A) = height(B)
        report "Invalid matrix dimensions in '*'" severity failure;
    ...
    return C;
end;
```

Use an assertion to detect any condition under which a function might not work. Choose a descriptive message in the assertion text; it is a good idea to include the name of the routine and the values of the offending parameters. This way you will be notified immediately if something unexpected happens, and you will have precise information to help you fix it.

Document Interface Assumptions about Parameters   VHDL makes it possible to enforce many kinds of restrictions on input parameters. If your routine works only with a subset of integer numbers, then declare the parameter using a range. If your routine works only for a few discrete values of an input, then consider using an enumerated type.

In some cases, however, it is not possible to use a range to restrict the input values. One example is when the inputs are floating-point or fixed-point numbers; another example is when the routine cannot handle certain specific values, such as in some mathematical formulas. In such cases, document any assumptions about the input parameters using a comment in the function header.[11] As a general rule, prefer to pack any important information directly into the parameter name; when this is not possible, document it in a header comment associated with the routine or with the parameter.

Use Named Association to Clarify Subprogram Calls   Routine calls should be self-evident in the client code; the reader should not have to look at the subprogram declaration or implementation to understand the purpose of a subprogram call. In some cases, however, the routine name alone is not enough to convey all the information. If the routine has more than a few parameters, it may be hard to remember or figure out the role of each argument in the subprogram call. In other cases, the routine interface may be unclear, but we may not be allowed to change it (e.g., when it is part of an external or third-party library).

In such cases, it may be a good idea to use named association instead of positional association to make the client code more readable. In the following example, it is not possible to guess the meaning of the literal values 2 and 3:

```
output_matrix <= delete(input_matrix, 2, 3);
```

Using named association, we can make the intent of the code more obvious:

```
output_matrix <= delete(input_matrix, delete_row => 2, delete_column => 3);
```

This is an application of a well-known principle in usability engineering: reduce the cognitive load on the users' brains by putting more information in the world and requiring them to keep less information in their heads. Applied to programming, this means that code that keeps more information on the screen and puts less of a burden on our memory is always better to work with. For more about this topic, see "Recognition Versus Recall" in section 3.6.1.

Use Default Values to Abbreviate Subprogram Calls   When a routine is called frequently with the same value for an actual parameter, or when it is possible to provide a value that makes sense most of the time, consider providing a default value for the parameter. In the following example, the parameters round_style, num_guard_bits, and denormalize default to round_nearest, 3, and true, allowing the subprogram calls to omit them:

```
function add(
    left_op, right_op: float;
    constant round_style: round_type := round_nearest;
    constant num_guard_bits: natural := 3;
    constant denormalize: boolean := true
) return float;

...

sum_1 <= add(op_1, op_2);
sum_2 <= add(op_2, op_3);
sum_3 <= add(op_2, op_3,
    round_style => round_nearest,
    num_guard_bits => 5,
    denormalize => false
);
```

For this technique to work with positional association, parameters with default values must come last in the parameter list. The reason is that it is not possible to "skip" an actual parameter with positional association—in positional association, actual parameters are expected to appear in the exact same order as the formal parameters. If the parameters with default values come last, then they can be omitted in the subprogram call.

Omit the Class and Mode When They Are the Same for All Parameters   When all parameters have the same class and mode in a subprogram, and the default class and mode (*constant* and *in*) are acceptable, leaving them out makes the declaration much shorter and improves readability. In the following example, the unnecessary keywords *constant* and *in* are included in the interface list:

```
function rectangles_intersect(
    constant rect_1: in rectangle_type;
    constant rect_2: in rectangle_type
) return boolean;
```

The following declaration is equivalent to the previous one, but it omits the default class and mode, making the function header more concise:

```
function rectangles_intersect(rect_1, rect_2: rectangle_type) return boolean;
```

Explicitly Define All Classes and Modes When They Differ between Parameters   When the parameters have different classes and modes, it is better to be explicit and specify the class and mode for every parameter. If we omit some of the classes or modes, then the reader will be forced to remember (or learn) the default rules. Even if we know the rules, determining the actual class and mode each time we read the declaration requires an extra effort.

Figure 15.7a shows a procedure definition with the default class and mode specifications omitted. It may take the reader a while to determine the class of the parameter packet or the direction of parameter rx_clock.

Figure 15.7b, in contrast, explictly defines all classes and modes, even when the defaults would be enough. In this way, the reader finds all the information directly in the declaration. Because the interface object declarations are longer, consider using one parameter per line in such cases.

We can apply the same reasoning for functions. Because the mode is always *in*, it is always safe to omit it. When all parameter are *constants*, the class can also be omitted. However, when one of the parameters in a function is a signal, it is best to explicitly state the class of all interface objects.

```
procedure receive_packet(                    procedure receive_packet(
    active_edge: edge_type;                      constant active_edge: in edge_type;
    signal rx_data: in bit;                      signal rx_data: in bit;
    signal rx_clock: bit;                        signal rx_clock: in bit;
    packet: out packet_type                      variable packet: out packet_type
);                                           );
```

(a) Default classes and modes omitted.   (b) All classes and modes explicitly declared.

Figure 15.7
Two equivalent definitions for a procedure.

Put Parameters in Input-Modify-Output Order   To provide consistency to the readers and programmers who use your routines, establish a convention of ordering the parameters by their mode. The order input-modify-output is easy to remember because it follows the order of operations inside a routine: reading data, changing it, and outputting the results.[12]

One common exception is for abstract data types (ADTs). Usually, an object of an ADT comes first in the parameter list regardless of its mode. This reinforces the idea that the operation is performed *on* the abstract object. The subsequent parameters then follow the conventional order.

Consider Using Enumerations for Sparse Parameter Types   When a parameter is of a scalar type but only some values are allowed, consider defining an enumeration type for the parameter. This makes the code safer because it is not possible to pass invalid values to the routine; it also makes the code shorter because it is not necessary to validate the input parameters or provide default clauses.

In the following example, the key length is specified as an integer parameter, but only three values are actually used. This makes it necessary to check the input for invalid values:

```
function num_lookup_rounds(key_length: in natural) return natural is
begin
    assert (key_length = 128 or key_length = 192 or key_length = 256)
        report "Invalid key length in num_lookup_rounds" severity failure;

    case key_length is
        when 128 => return 10;
        when 192 => return 12;
        when 256 => return 14;
        when others => return 0;
    end case;
end;
```

In the following rewritten version, the integer parameter was replaced with an enumeration, and it is not possible to pass an invalid value. In consequence, it is not necessary to verify that the input is valid or provide a when others clause:

```
type key_length_type is (KEY_LENGTH_128, KEY_LENGTH_192, KEY_LENGTH_256);

function num_lookup_rounds(key_length: in key_length_type) return natural is
begin
    case key_length is
        when KEY_LENGTH_128 => return 10;
        when KEY_LENGTH_192 => return 12;
        when KEY_LENGTH_256 => return 14;
    end case;
end;
```

Avoid Flag Parameters   Flag parameters are arguments that can assume a limited number of values and cause the routine to perform different actions depending on their value. In the following example, add_sub is a flag parameter that choses the kind of arithmetic operation performed in the routine:

```
function add_sub(data_a, data_b: in signed; add_sub: in std_logic)
    return signed is
begin
    if add_sub = '1' then
        return data_a + data_b;
    else
        return data_a - data_b;
    end if;
end;
```

Flag parameters should be avoided because they only exist if a routine does more than one thing. Passing flag parameters is also an undesirable form of coupling—namely, control coupling (for more about control coupling, see section 4.4.1).

The proper way to fix a function like this one is to divide it into separate routines. If the function has a boolean parameter, then split it into two routines: one for when the flag is true, and the other for when it is false. If the two routines share some of the statements, then extract the common code to another subprogram. The client code should make the decision and then call the appropriate routine.

Avoid Side Effects   A routine has side effects if it accesses an object that is not in its parameter list or if it interacts with the outside code by means other than its parameters or return value. Pure functions cannot have side effects; this is why they always produce the same output when called with the same arguments. Procedures and impure functions, in contrast, may behave differently each time, even when called with the same arguments.

Side effects are a powerful and useful tool; they are also indispensable for certain kinds of operations, such as text output and file I/O. In general, however, they have many disadvantages and should be kept to a minimum.

Subprograms with side effects are deceptively simple because their interfaces do not provide an accurate idea of what they accomplish. They are also harder to understand. Because they breach encapsulation, we cannot think locally while we read the subprogram; we need to consider all other places where the shared objects may be modified. Any external object that is read or written in the subprogram could be working as an invisible parameter.

Side effects also make the subprogram less predictable. Results may depend on previous history or the order of execution of subprograms. The subprogram may also break inadvertently under modification if we change a module that touches the shared objects. Finally, side effects make a subprogram harder to test and reuse. The routine cannot be tested in isolation; at least it will be necessary to instantiate or emulate all the shared objects.

The appeal of using side effects is that they allow the creation of uncluttered interfaces. We must weigh this benefit against all their disadvantages. If you absolutely must use them, then keep them extremely simple and document all the side effects in the routine header.

Use Routine Parameters and Result Values to Make Dependencies Obvious   When you have a series of subprograms that perform a sequence of transformations on the same data, organize your code so that all the exchanged data is visible, and the output of one subprogram is the input to the next.[13] To achieve this, we must create routines that are completely orthogonal, do only one thing, and receive only the necessary data to do their job.

In the following example, the procedures communicate via shared objects that do not appear at their interfaces. Therefore, is not possible to say which procedures operate on which part of the data. The routines are also nonorthogonal because each one affects global data that is used by the others. It is also unclear whether the order of the calls is relevant:

```
ip_address := remote_host_ip(packet);
find_remote_host;
update_tcp_state;
update_connections_list;
```

To make the dependencies more obvious, we can replace the three procedures with functions. In the following rewritten version, all input data is passed to the function via parameters, and all output data is passed through return values:

```
ip_address := remote_host_ip(packet);
remote_host := get_host(ip_address);
tcp_state := update_tcp_state(remote_host, packet);
update_connections_list(remote_host, tcp_state);
```

In this new implementation, the dependencies are obvious and enforced by the program structure; it is not possible to call the routines in the wrong order. We cannot call get_host before we have ip_address, and we cannot call update_connections_list before we have remote_host and tcp_state. Also, there is no shared state, and all input and output of data are explicit. The variables have an extremely short span and are active for only a few lines. Finally, the code is easier to debug because each intermediate step and each output value can be used as checkpoints.

Programs that rely on calling functions in a given order are usually fragile. Prefer parameters and result values over shared data objects, and try to arrange the routine dependencies so that the return value of one is the input value to the next. This transforms the coupling between the routines to data coupling, the most desirable form of coupling.

Use Unconstrained Arrays When Possible   If a formal parameter is an unconstrained array (an array without a fixed range), then the subprogram will take the size and direction from the actual parameter. This means that you can write subprograms that are entirely generic with respect to the shape of their array parameters.

To write this kind of subprogram, we need to query information about the parameter inside the routine. One way of doing this is to use the array attributes 'range and 'length, as shown in the following example:

```
function mirror(vector: std_logic_vector) return std_logic_vector is
    variable mirrored_vector: vector'subtype;
begin
    for i in vector'range loop
        mirrored_vector(i) := vector(vector'length - i - 1);
    end loop;

    return mirrored_vector;
end;
```

In the example, the attribute 'range is used to loop through the vector elements, and 'length is used to calculate the offset for indexing into the assignment source. The return value also takes its range and size from the actual parameter, using the 'subtype attribute that returns the fully constrained subtype of the input called vector. As a result, the subprogram can be used with vectors of different sizes:

```
mirror(b"1111_1100_1010_0000");    -- returns 0000_0101_0011_1111
mirror(b"1111_0000");              -- returns 0000_1111
```

Do not be more restrictive with ranges than necessary. This will make your subprograms easier to use and reuse.

For Reusable Subprograms, Normalize Parameter Ranges and Directions   Because the range of actual parameters is defined by the client code, a subprogram should not make assumptions about their size or direction. If the arguments are expected to have the same length, then it is best to check this with an assertion. Also, if the subprogram has more than one array argument, then remember that their ranges may be different. It is not safe to use indices from one object when referring to elements in a different object: if the ranges do not match, then the indices could be invalid.

In the following example, function vectors_are_equal should return true when the two std_logic_vector parameters vector_1 and vector_2 are equal. However, the client code is providing two objects with incompatible ranges. If the function does not account for this possibility, then it could cause an index out of range error if it tried to access C1 using the indices from C2:

```
constant C1: std_logic_vector(31 downto 24) := b"1100_1010";
constant C2: std_logic_vector(1 to 8) := b"1100_1010";
...
-- This will cause an 'index out of range' error
-- if the function does not normalize its parameters:
result := vectors_are_equal(C1, C2);
```

The solution is to normalize the two arguments so that they have a uniform indexing scheme. This can be done using aliases or auxiliary variables. In the following example, the parameters vector_1 and vector_2 are normalized to the same ascending range starting at one.

```
function vectors_are_equal(vector_1, vector_2: std_logic_vector) return boolean is
    alias vector_1_normalized: std_logic_vector(1 to vector_1'length) is vector_1;
    alias vector_2_normalized: std_logic_vector(1 to vector_2'length) is vector_2;
```

```
begin
    assert vector_1'length = vector_2'length
        report "length mismatch in vectors_are_equal" severity failure;

    for i in vector_1_normalized'range loop
        if vector_1_normalized(i) /= vector_2_normalized(i) then
            return false;
        end if;
    end loop;

    return true;
end;
```

We could achieve the same effect using variables instead of aliases:

```
function vectors_are_equal(vector_1, vector_2: std_logic_vector) return boolean is
    variable vector_1_normalized: std_logic_vector(1 to vector_1'length) := vector_1;
    variable vector_2_normalized: std_logic_vector(1 to vector_2'length) := vector_2;
begin
    ...
```

If your subprograms are intended to be reusable, or if you want to be on the safe side, then remember to normalize the ranges of your array parameters.

15.6   Where Should Routines Go?

We close this chapter with a discussion of the appropriate places to declare a routine. Subprograms can be declared in a package, in an entity, in an architecture, in a process, or even inside another subprogram. In practice, most subprograms are declared in a package or in the declarative region of a process or an architecture.

There are various factors to consider when choosing the right place for a subprogram. If the routine will be called from multiple places, then we need to declare it in a scope accessible to all units that will use it. We want to keep declarations close to where they are used, but we also want to avoid making processes and architectures too long. We want to hide complexity, but we want to see the subprogram statements easily when necessary. The following are the advantages and disadvantages of the most common places for declaring subprograms.

The most localized place to define a subprogram is in the declarative region of another subprogram, but this is little used in practice. Realistically, the first place to consider is the declarative region of a process. From the point of view of encapsulation and information hiding, this is a good place because it prevents all other processes from using the routine. However, the main reason to write a subprogram inside a process is because it will also have *write* access to signals declared in the architecture and to the entity ports. This is only possible if the subprogram is declared in a process.

However, declaring a procedure in a process does not reduce the amount of code inside the process unless there was a lot of duplication. Even a short routine declaration could make the process too big to fit in one screen or to view the sensitivity list and the process statements at the same time. In short, the only valid reason to declare a subprogram in a process is when you need write access to signals or ports, but remember to keep the process to a reasonable length.

The second common place to declare a subprogram is the declarative region of an architecture. Unlike subprograms declared in processes, a subprogram declared in an architecture does not have *write* access to the architecture signals; it only has *read* access. However, it can be used by any process or concurrent statement in the architecture body. For this reason, if the subprogram is used in several places in the architecture, then the architecture declarative region is a good place for it.

The third and final common place for a subprogram is inside a package. The main advantage of this approach is that the subprogram can be called from any design entity that uses the package. However, the subprogram will not have visibility to signals or variables declared outside the package. If you follow the recommendation of minimizing side effects, then this should not be a problem in most cases.

Routines declared in packages are also easier to test; they can be tested directly by including the package in a testbench, calling the routine with the correct input values, and checking the outputs against their expected values. This is significantly easier than testing routines declared in processes or architectures, which most of the time can only be tested indirectly.

A minor disadvantage of using packages is that it may feel like overkill to create a package containing a single subprogram. In such cases, you may want to wait until there are at least two subprograms or other associated declarations to justify creating the package. However, try to resist the temptation of creating "miscellaneous" packages with incohesive functionality.

One last point about subprograms declared in packages is that we have a choice of whether to make them visible outside the package. If we want the subprogram to be visible by other design units, then we should declare the subprogram header in the package header. In contrast, if the routine is used only inside the package, then the subprogram should be declared only in the package body. This provides better information hiding by preventing other design units from accessing routines that are only relevant to the package's internal implementation.

16 Excellent Names

We use names to refer to the parts of a design. If we choose names that are clear, accurate, and descriptive, then the pieces will fit together logically, and the design will be self-evident. If we choose names that are too cryptic, meaningless, or ambiguous, then the system will be nearly impossible to reason about.

Understanding a large piece of code is never easy, but good naming makes a real difference. Unfortunately, most developers were never taught how to choose proper names. This chapter provides a frame of reference to help you recognize bad names and many guidelines to help you create great ones.

16.1 How Long Should a Name Be?

All other things equal, short names are better than long ones. Unfortunately, when we remove or omit part of a name, we never keep all other things equal. We might be removing a word that clarified an object's role in the code. We could be making the name less pronounceable and thus harder to memorize and use in a conversation. We could be using an abbreviation that makes perfect sense for us but will take other developers some time to decipher.

We use names to identify each object and operation in our code. A good name should provide enough detail to communicate what an object is or what an operation does. Names that are too short do not convey enough meaning; they force the reader to look into the surrounding code and recall things from memory all the time. If you need to memorize what a name really means, then the name has failed its purpose of describing the entity it represents.

However, we do not want our names to be too long. Names that are too long are also harder to read and obfuscate the structure of the code. Long names also leave less space for statements and other operations in a line of code. Moreover, some tools may concatenate long lists of names to create unique identifiers or construct commands; if the names are too long, then they will be impossible to read.

So how can we tell if a name is too long or too short? The best measure is not how many characters a name has but whether it provides all the necessary information and has all the characteristics of a good name. If a name does not describe an object fully and accurately, then it is too short. If the name does not tell everything that a routine does, then it is too short. If the name is cryptic or does not read naturally because it was truncated, then it is too short. In contrast, if the name is so long that it makes an expression hard to read, then it is too long. If we cannot fit more than two or three names in a line of code, then they are probably too long.

There is a limit to how long a name can be before it becomes too cumbersome, but it is more elastic than most developers think. As a rule of thumb, names up to three or four words or 20 characters rarely need to be shortened. At around 30 characters, the names may start to feel uncomfortable and require some attention. However, in many occasions, this attention is in the form of a restructuring of the code rather than artificially compressing the names.

The rest of this section presents recommendations to make sure that your names strike a good balance between the amount of information, readability, and length.

Make a Name as Long as Necessary   Creating names that are too short is a habit that most programmers must break to avoid delaying the work of other developers in the team. The first lesson to learn is: choosing names that are clear and unmistakable is a top priority on your list. Making them very short is not. It is a poor tradeoff to make a name shorter by removing meaningful information from it or by making it harder to read.

In most cases, you should not worry about the length of a name. Let other factors such as making the name more informative, clear, pronounceable, and unambiguous decide how long it should be. A long name that spells out all the important information is better than a short but cryptic one. A long name that fully describes an object is better than a short name that needs a comment to be understood.[1]

Favor Clarity over Brevity   A name is clearer if it uses natural language words and describes all the important information about an object or a routine. A name is less clear if it needs to be deciphered, is unpronounceable, or uses nonstandard abbreviations. When you have to choose between a clear name and one that is short, always choose clear.

A common belief, especially among less experienced developers, is that every reader will need some time to "get used to the code" anyway, and figuring out what the names really mean is only part of the process. This is, of course, nonsense—there is no reason to make learning the code any harder than it already is. If you can give your readers a head start by providing all the information neatly packed into the names of you objects and routines, then by all means do it.

Use Abbreviations Judiciously Programmers seem to have a natural urge to abbreviate words or phrases that would be perfect names for an object or a routine. Some will make a deliberate effort to abbreviate every name in the code. Please try to break this habit if you have it! Clear, unabbreviated, and pronounceable names should always be your first choice. The problem with abbreviations is that they are harmful to many of the good qualities of a name. They make the name less pronounceable, more ambiguous, and harder to spell correctly. They are one more thing to be deciphered and remembered. As a general rule, avoid abbreviating a name unless there is a real need and this would bring a clear and significant gain.

For the cases where an abbreviation is justifiable, here are some guidelines. The first is to prefer standard abbreviations, the ones that can be found in a dictionary (including dictionaries of computer or technical terms). Thus, `stats` is a good abbreviation for statistics, `ack` is a good abbreviation for acknowledge, and `arg` is a good abbreviation for argument. We can also use standard or well-established abbreviations and acronyms from the hardware field, such as `clk`, `rst`, and `req`, as long as they cannot be mistaken for anything else. What does a name with a suffix `_int` mean? Is this an internal signal, integer value, or interrupt? The few characters saved are not worth the loss in clarity and the confusion we inflict on our readers.

The second recommendation is to avoid abbreviations that save only one or two letters of typing. Abbreviating `key` as `ky` and `result` as `rslt` does not justify the loss in readability. In each case, we would be exchanging a perfectly pronounceable and natural word for a cryptic name for the savings of one or two letters. For the same reason, we should use abbreviations only for names that appear frequently in the code. If a name appears only occasionally, then the savings in space do not justify the loss in readability.

The third recommendation is to keep a project-wide dictionary of abbreviations. If we want our code to look more uniform, then we need to make sure that everyone is using the same abbreviations. This dictionary could be a text file, a word processing document, or some other sort of database. In any case, it should be maintained alongside the source code, in the revision control system. Even abbreviations that are obvious or well known should be in this dictionary.

Another important recommendation is to use abbreviations consistently. Do not abbreviate the same word differently in several parts of the code. Avoid abbreviating a word in one name and then using it unabbreviated somewhere else.

Finally, when you remove unimportant words from a name, make sure you keep the part of the name that tells what an object "is." For example, if a variable represents the number of processed packets, make sure its name includes one of the words `number`, `count`, or `num`. Simply naming it `packets` would be inaccurate: the variable does not hold the packets themselves; it keeps a count of them. Similarly, if a constant holds a register address, then make sure its name includes the words `ADDRESS`, `ADDR`, `INDEX`, or `OFFSET`, for instance. Simply naming it `REGISTER` would be inaccurate.

Choose the Length of a Name in Proportion to Its Scope One way to make names as short and meaningful as possible is to exploit scope. Each name needs to be precise enough to identify an entity in the context where it appears. If an object is used in several modules, then the name must be descriptive enough to identify it wherever it is used. In contrast, if an object is used only in a few lines of code, then it can have a much shorter name.

Let us illustrate this with an example. Suppose that a net is used across different hierarchy levels. At the bottom level, we need to provide little information to make a name unique. In the example of figure 16.1, the innermost entity is called `fifo`. Inside this entity, the name `element_count` has enough information to be understood unmistakably.

Going up one level in the hierarchy, inside the `output_controller` block, we may have to add more information to make the name unique. The output controller may be responsible for several other low-level units, and `element_count` may be vague or ambiguous. What is an element in this context? By naming the signal `fifo_element_count`, we eliminate any ambiguity.

Finally, at the top level, the name `fifo_element_count` may not be descriptive enough. The top-level entity could control two FIFOs: one for input and the other for output data. Therefore, we may have to further qualify the name. In this case, `output_fifo_element_count` could be a good choice.

We used signals in a design hierarchy for this example, but the same idea applies to any named entity in the code. A constant in a package that is used by several design units will tend to have a longer name than a constant that is local to an entity or a function. A single-letter name such as `i`, `j`, or k works fine as the index of a short loop, but it should never be used in other occasions.

This relationship between the length of a name and its scope has two implications. The first is that one of the best ways to make names short is to structure our code using small scopes. If we keep our routines short and our design units small and focused, then there will be less need for longer names. The second implication is that if we give an object a short name, then we are signaling to the reader that this object is not intended for use outside of a small region of the code.

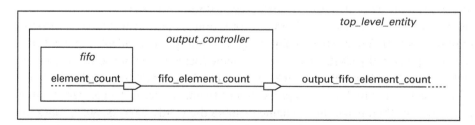

Figure 16.1
The length of a name should be proportional to its scope.

16.2   General Naming Guidelines

This section presents guidelines that help create better names for any kind of named entity in the code.

Choose Meaningful Names   The name of an object or routine should contain all the important information about it. Names such as a, x, l, and iv are meaningless and force us to look somewhere else in the code to understand what they represent. Names such as `tcp_header_length` and `interrupt_vector_address` convey all the information to communicate what the object represents. Meaningful names put more information in the code and allow us to keep less information in our heads.

The more information we pack into a name, the more meaningful it is. When deciding what to put in a name, our first concern should be to describe as precisely as possible what something is or does. In the case of an object, it should tell us immediately what the object represents or contains. In the case of routines, the name should describe everything the routine does.

Choosing meaningful names also means to avoid empty or meaningless words. Be wary of words that only seem to have some meaning, such as `data`, `value`, or `info`. In almost every case, it is possible to find a more descriptive word.

Take Your Time   Names are too important to be chosen casually. Choose good names, and your code will be easy to write and understand. Choose names that are ambiguous, incomplete, or inaccurate, and nothing will make sense.

Some programmers behave as if names were not as important as statements in the code. In this view, statements are "real" code, and names are just second-class entities used by the statements. This is fundamentally wrong. Names make a huge difference in readability, and readability is one of the best proxies for quality in a program. A well-chosen name shows that the author spent enough time thinking about the role of a variable or routine. Moreover, code is read much more often than it is written. Choosing good names takes time, but it pays off in the long run.

Instead of using the first name that crosses your mind, get into the habit of evaluating a few candidates. Check their pros and cons in your head, and type the names on the screen to get a feel for them in the code. Iterate a few times, and move forward only when you have found a name that accurately describes the object or routine.

Don't Be Afraid to Change a Name for a Better One   Imagine you have been working on a project for a few weeks. When would you be able to choose the best names? In the beginning, before you started implementing the project? Or at the end, when you are most intimate with the design?

We should do our best to choose an accurate and meaningful name from the start, when we declare an entity. However, we should also leverage the knowledge gained by implementing the system. Get into the habit of renaming data objects and routines whenever you think of a better name. Every time you change a name to reflect your improved knowledge, you are improving the quality of your code.

Novice programmers tend to avoid changing names for fear of breaking the code. This is one of the worst habits you could pick up as a programmer. There are at least two ways to avoid this trap. The first is to use automated renaming tools, available in modern IDEs or source code editors. The second is to have a good set of automated testbenches that tell you when you break something. By running the tests after every modification, we ensure that the design functionality is unchanged.

Finally, never keep bad names on the grounds that you or other programmers "are already used to them." The meaning of a name should never have to be memorized; you should be able to read it directly from the code. Besides, after you put the code aside for a while, you will not remember what those names meant anyway.

Use Clear, Natural Language Words

Code that reads well is easier to understand. Using natural language words helps the code read naturally because there is no need to translate each name into something else. Read the following statements, and try to feel how the code looks obvious. The assignment statement may be read as "gets" or "gets the value of."

```
next_count <= (current_count + 1) when count_enabled else current_count;
```

```
...
```

```
if current_position = LAST_POSITION then
    next_state <= done;
end if;
```

Clear and natural names make correct code look obvious and wrong code stand out. Another advantage is that you do not need to remember how you *chose* to write a name because the name *is* the correctly spelled word.

Use Pronounceable Names   If you choose not to use natural language words, then at least use names that are pronounceable. This makes it easier to use the name when you need to explain the code to someone or when you are reading the code silently in your head.[2] Following this rule, THRESH is a better abbreviation than THRSLD, and STIM is a better abbreviation than STML. However, none of them is a better name than the unabbreviated words THRESHOLD and STIMULUS.

In *The Elements of Programming Style*, Kernighan and Plauger propose a simple test to assess the clarity of a program.[3] If another programmer can understand your code when you read it aloud over the telephone, then it is clear enough. Otherwise it should be rewritten. This test has been called "the telephone test" for code readability. Choosing names that are unpronounceable and need to be spelled out would cost you several points in the telephone test.

Use Searchable Names   Another reason never to use single-letter names is that they are nearly impossible to search for in the source text.[4] This is a common need: we often need to see where a variable or routine is used to understand how it fits into the program. We also need to search for the names if we want to change them. These operations will be hard if you use excessively short names such as a, b, y, and q.

Use Names from Your Problem and Solution Domains   A good way to create names that are short and meaningful is to use terms that have a lot of meaning built into them. We can draw inspiration from the problem domain or solution domain[5] (see section 4.2.1 for the definitions).

Problem domain names relate to objects and operations at a higher level of abstraction. A typical application should use many names from its problem domain; otherwise it may be a sign that we are working at too low of an abstraction level. For example, in a CPU design, one would expect to see instructions, opcodes, and ALU operations. If most operations are performed directly on names such as data_bus and address_bus, we are probably missing an opportunity to work at a higher level of abstraction.

In contrast, solution domain names include Computer Science terms and basic hardware structures. If a design uses FIFOs, stacks, or arbiters, then we should see those names appear in the code.

When we have a choice between the two, it is usually better to pick the name that is closer to the problem domain. For example, suppose we need to create a function to calculate the inverse of a matrix. If we create a variable to represent the return object, then it could be called return_matrix (which speaks to the solution domain) or inverse_matrix (which speaks to the problem domain). Picking the one that speaks to the problem domain packs more meaning into the name.

Choose Names That Cannot Be Misunderstood   Before committing to a name, examine it carefully and try to imagine whether it could be interpreted differently from what you intended. Sometimes the name seems obvious to you, but how does it look to a fresh reader? Always try to imagine how the name would be seen from the perspective of someone who is not as intimate with the code as you are. For example, what does the name SAMPLES_MAX mean? Is it the maximum number of samples or the maximum value in an array of samples? In the first case, `SAMPLE_COUNT_MAX` would be less ambiguous. In the second case, `SAMPLES_VALUE_MAX` is more explicit.

Choose Names That Do Not Require a Comment If after writing a declaration you feel the need to add an explanatory comment, check whether the comment could be avoided by using a more communicative name. In most cases, the name of an object or a routine should be enough for someone else to use it with confidence. If the name of an object does not tell what it holds, then choose another name. If the name of an object or a routine is not enough to communicate how it should be used, then look for a better name and check for deeper problems in the code.

Avoid Mental Mappings Good names tell you immediately what something is or does. Bad names require you to translate a name into something meaningful before you can really work with it. This unnecessary mental burden could be easily avoided by choosing clearer names.[6]

As an example, if you create an enumeration type for an automatic gate and decide to name your states `S1` for closed, `S2` for opening, `S3` for opened, and `S4` for closing, then you have imposed on the reader the mental burden of memorizing this relationship. Other developers will need to translate the names to their real meanings before they can think about the problem. This is a waste of mental capacity. Luckily this is easy to fix: instead of arbitrary names, choose meaningful names that have an obvious connection with the entity they represent. Meaningless names, inaccurate names, and abbreviations all require translation or mapping before they can be used. A much better approach is to name things what you call them in real life.

Be Wary of Names Including "And" or "Or" Names with the conjunctions *and* or *or* are usually a sign that an entity is trying to do more than one thing. Review the code and see whether it should be two separate entities instead. In some cases, it is enough to use another name at a higher level of abstraction, denoting a concept that encompasses the two lower level entities.

Consider Including Physical Units in Names VHDL offers *physical types*, which are a handy way to represent physical quantities or measurements associated with units and scale multipliers (see "Physical types" in section 12.2.1). If for some reason you are not using them (e.g., because they do not allow the desired precision or range), then you can make the units explicit by including them in the name. For instance, if an entity uses a generic constant for a clock frequency, then `CLOCK_FREQUENCY_IN_MHZ` or `CLK_FREQ_MHZ` are better names than `CLOCK_FREQUENCY` or `CLK_FREQ`. This is an example of using names that cannot be misunderstood.

16.3   Guidelines for Naming Data Objects

The rules we have seen so far are general and apply to any named entity in the code. This section presents rules that are specific for data objects such as signals, variables, and constants.

Fully and Accurately Describe the Entity That the Object Represents   Every data object represents a piece of information used in the code. This information may have a parallel in the real world, such as the position of a switch, or it may exist only in the code, such as the state of an FSM. In any case, the most important factor in naming a data object is to choose a name that fully and accurately describes the entity it represents.[7]

There are many factors to choosing an accurate and descriptive name. First, it should contain all the information needed to associate the name with the entity it represents. Second, it should describe as precisely as possible what information the object holds. Third, it should be detailed enough that any developer can use the object correctly in the code. Finally, the name needs to be unambiguous in its context and immune to misinterpretation.

We have already touched on these guidelines in the previous section, but let us examine a few examples more specific to data objects. Suppose we are designing some network hardware and need to keep track of the number of received packets. What would be a good name for an object holding this value? Would the name `packets` be a good idea? Probably not. This name has at least two big problems. First, it does not tell the whole story: we are not counting just any kind of packets; we are interested in a specific set of packets (the ones we have received), so this information should be in the name (unless this is the only kind of packet in the design). The second problem is that the name is missing precisely the word that would tell what information the object represents: the *number* of packets. In fact, the name `packets` is misleading because it suggests that the object stores the packets themselves. A better name would be `number_of_received_packets`, `received_packet_count`, or `num_received_packets`.

As another example, suppose we are designing the status register of a CPU. In this register, each bit is a CPU flag, and bit number 7 is the *overflow* flag. Instead of using the literal number 7 in the code, we should use a constant to hold this value. What would be a good name for this constant? `OVERFLOW_FLAG` sounds like a good name, but is it accurate? To answer this question, we should ask ourselves what information the constant really represents. In this case, it is certainly not the state of a flag; it is a bit position. More precisely, it is the position of the bit representing the overflow flag. Therefore, a better name would be `OVERFLOW_BIT_POSITION`, `OVERFLOW_FLAG_INDEX`, or `OVERFLOW_BIT_POS`. All those names tell us precisely what kind of information the constant holds.

All this may sound rather obvious, but failing to name an object after what it really represents is a common mistake. After choosing a name, remember to check whether the name accurately describes the value that the object holds.

Name Things What You Call Them   When choosing a name, a good way to start is to write down the same words that you would use to describe the object to someone.[8] Thus, to name a constant that holds the clock frequency in megahertz, a good name would be `CLOCK_FREQUENCY_IN_MHz`. To represent the number of elements in a FIFO, we could use `number_of_elements_in_fifo` or `fifo_element_count`. These names require no effort from the reader and are almost impossible to misinterpret. Later, you can use other techniques to make the names shorter, such as removing any unnecessary words or exploiting object scope.

Use a Noun or Noun Phrase Name   Data objects represent entities manipulated in the code. Just like objects in the real world, they should be named with a noun or noun phrase. Sometimes a simple noun is enough to identify the object uniquely within its scope: depending on the context, `accumulator`, `root`, `radicand`, `paddle`, `ball`, or `framebuffer` could all be good names. At other times, we need to provide additional context to properly identify an object. In such cases, the name will be a compound noun, such as `cypher_text`, `elapsed_time` or `inverted_matrix`.

For scalar objects (objects that can hold only one instance of a value), use singular names. For array types, which can hold many elements of the same type, use a plural name or a name that implies a collection. A plural name is usually enough to indicate that an object holds a collection of elements; most of the time, suffixes such as `_vector` or `_array` are redundant and unnecessary. Examples of good plural names are `registers`, `display_segments`, or `switches`. Examples of names that indicate a collection include `register_file`, `input_fifo`, or `event_queue`.

Avoid Repeating the Object Type in Its Name   When you declare an object, you must provide an object class, a type, and a name. Repeating the class or type into the object's name is adding redundant information.

This practice has several disadvantages. First, it makes the name longer without adding meaningful information. Second, the object type is often added in an abbreviated form, making the name less pronounceable. Third, it is not really necessary: VHDL is a strongly typed language, and the compiler makes all necessary checks to ensure that object types are used correctly. Finally, if we feel tempted to use this technique, then it may be a sign that the code is too long or confusing or the names are poorly chosen.

In some cases, however, an exception is justifiable. When two objects represent the same entity but differ only in type, it is helpful to use a suffix or prefix to differentiate between them. A common practice is to use the same base name for both objects and add a suffix indicating the type of each. For example, if an input signal is of type `std_logic_vector` but we need to use it as an unsigned value, then we could name them `operand_slv` and `operand_uns` to make it clear that they represent the same entity. However, this practice should be used only when necessary and never by default.

Put Computed-Value Qualifiers at the End of a Name   A typical program has many data objects that hold computed values. For instance, we often need to calculate the sum, average, maximum, or minimum of a series as part of a larger algorithm. In such cases, it may be a good idea to use the same base name for all the related values and a suffix to indicate the computed-value qualifier.[9]

This is more of a style issue than a hard-and-fast rule, but it can help make the code more readable. Steve McConnell, the author of this recommendation, cites two main advantages. First, the most important part is at the beginning of a name, so it is more prominent and gets read first. Second, by sticking to this convention, we avoid inconsistent naming schemes where an object is named `packet_lenght_average` (with the qualifier at the end) and another one is named `max_packet_length` (with the qualifier at the beginning). Following this rule makes such names easier to choose and produces more consistent code.

Avoid Nondescriptive Names Such as *temp, aux, value,* or *flag*   Indistinct names such as `temp`, `aux`, `value`, or `num` should be avoided because they do not reveal anything about the entity that the object represents. Such empty names may happen in code for several reasons. Sometimes the programmer thinks that the object is not important enough to deserve a proper name. At other times, the developer wants to dodge the effort of choosing a better name. It is also possible that the developer is not completely sure about the object's real purpose in the code. Whatever the case, this is a bad sign.

Do not take the easy way out by choosing a generic name. If you feel tempted to use one of the names in this list, then stop and see whether you can think of a name that better communicates what the object represents. For example, suppose we need to create a function to calculate the sum of the elements in an array. The routine will probably use an intermediate variable to hold the partial sum. Instead of using `temp` or `value` to keep this partial value, we could use more meaningful names such as `partial_sum`, `running_total`, or `accumulated_value`.

Other names that *appear* to have some meaning on the surface but need to be viewed with suspicion include `input`, `output`, `data`, and `flag`. In almost every case, it is possible to find a better, more descriptive name for the object.

Avoid Numerals in Names[10]   It is common to find code such as the following example, in which data objects have numerals embedded in their names.

```vhdl
signal pending_interrupt_0: boolean;
signal pending_interrupt_1: boolean;
signal pending_interrupt_2: boolean;
signal pending_interrupt_3: boolean;
signal interrupt_vector_0: integer;
signal interrupt_vector_1: integer;
signal interrupt_vector_2: integer;
signal interrupt_vector_3: integer;
...
if pending_interrupt_0 then
    isr_address <= interrupt_vector_0;
elsif pending_interrupt_1 then
    isr_address <= interrupt_vector_1;
elsif pending_interrupt_2 then
    isr_address <= interrupt_vector_2;
elsif pending_interrupt_3 then
  isr_address <= interrupt_vector_3;
end if;
```

In this example, the eight individual signals are actually two arrays in disguise. Using hard-coded numerals in the names makes the code harder to generalize, which in turn makes it more verbose. If we declare the signals as arrays, then we can use a loop and rewrite the assignments using array indexing:

```vhdl
signal pending_interrupts: boolean_vector(0 to 3);
signal interrupt_vectors: integer_vector(0 to 3);
...
for i in pending_interrupts'range loop
    if pending_interrupts(i) then
        isr_address <= interrupt_vectors(i);
    end if;
end loop;
```

Besides making the code more compact, this approach also makes it possible to move the code to a function that takes two unconstrained array parameters. Writing a function for working with the original names would require eight parameters, and it would only work with exactly this number of arguments.

Another use of numerals in names is to differentiate between objects. For instance, to overload the operator "/" for a custom data type, we could write the function:

```vhdl
function "/"(operand_1, operand_2: custom_data_type) return custom_data_type;
```

In this example, the numbers were added only as a way to differentiate between the two parameters. The problem with this approach is that the numbers do not contribute any meaning to the names. Almost always we can find more descriptive names if we put a little effort into it. In this example, we could use `left_operand` and `right_oper-and` to make explicit which one is which. Another solution would be to use the proper arithmetic terms `dividend` and `divisor`.

In other cases, when the objects are really indistinct, it may be acceptable to use numerals in the names. The reader should be allowed to interpret this as a sign that the objects are interchangeable. In any case, remember to consider using an array, and always try to come up with better names first.

Use Prefixes and Suffixes Judiciously Prefixes and suffixes are an effective way to convey additional information about an object. We can use them to indicate that an object has a special role in the code or differentiate between similar objects. However, like abbreviations, prefixes and suffixes are detrimental to many qualities of a good name and of code in general. They make a name less pronounceable. They add one more level of encoding, requiring the reader to decipher a name instead of reading it. They are one more convention that needs to be learned and documented. They clutter the code, reducing the visibility of the most important part of a name—the words that convey its meaning. In many cases, they are just redundant information.

It is funny that prefixes and suffixes are recommended on the grounds of making a name more readable, when in truth they tend to produce abominations such as `s_i_cnt_o_r_l`. If we spend more characters with redundant prefixes and suffixes than with the meaningful part of a name, then there is something wrong with our naming conventions. *Readability* means the ease of understanding the code structure and operation. When we pack a lot of information in a name, it makes the name more informative but not necessarily more readable. If the code needs to be deciphered before it can be understood, then this does not help with readability at all.

The reasoning behind some naming conventions that recommend redundant suffixes is that if we know that a signal is, say, an integer, then we will not try to assign an `std_logic_vector` to it. If we know that an object is a signal, then we will not use it as the target in a variable assignment. This is redundant because the compiler will not let us do that anyway. In practice, this kind of mistake could exist in the code only for a brief period, until we hit the compile button. This kind of convention does not make a design safer, and it does not prevent us from committing logical errors.

So how can we tell the good prefixes and suffixes from the bad ones? This is easier than it might seem. Good suffixes have semantic meaning and provide information that could not be inferred from the object type or class alone. Table 16.1 shows some examples of semantic suffixes that can be added to signal names. These are good uses

Table 16.1 Examples of semantic suffixes that do not add redundant information

Suffix	Meaning
_n	Added to a signal to indicate that it is active low.
_z	Added to a signal to indicate that it is part of a tri-state bus.
_reg	Added to a signal to indicate that it is the output of a register.
_asyn	Added to a signal to indicate that it has not been synchronized.

for suffixes because this information is not readily visible in the code. Besides, they can help spot errors and make the code easier to verify.

In contrast, bad suffixes are redundant and simply restate information that is readily visible in the object declaration. This includes the object type, the object class, or a port mode. Although this information is not always useless, it is at least redundant. The compiler enforces all the rules regarding object types, classes, and modes and will not compile our code if we make this kind of mistake. If we need to be reminded of an object's type constantly, then there may be something wrong with our code. Most likely we are failing to choose descriptive names for our objects.

There is, however, one justifiable exception. When two objects represent the same entity but differ only in type, it is helpful to use a suffix or prefix to differentiate between them. A common practice is to use the same base name for both objects and add a suffix indicating the type of each. For example, in a memory model, we may need both an std_logic_vector and an integer object to represent a read address. In this case, we could call them read_address_slv and read_address_int to make it clear that they represent the same value. In any case, save this practice for when it is really needed and avoid doing this by default anywhere else.

Guidelines for Naming Boolean Objects   Boolean variables and boolean signals are a powerful way to organize our code and document our intent. We can use boolean objects to break down a complicated test into smaller parts, document the test in a conditional statement, or make the code using the objects more readable. Here are some guidelines that help create good names for boolean objects.

The first guideline is to name a boolean object as an assertion—a statement claiming that something is true. According to this recommendation, pixel_is_transparent, fifo_is_full, and output_is_ready are all examples of good names.

Another alternative is to name boolean objects thinking about how they read as the condition of an *if* statement. According to this recommendation, end_of_file, fifo_full, or input_available are good naming examples. We can see how they look natural when used in a conditional statement:

```
while not end_of_file loop ...
```

```
if fifo_full then ...
```

```
if input_available then ...
```

Finally, as a general rule, prefer positive names. In other words, avoid the word *not* or the prefix *un-* in the name. They make the test harder to read when the condition must be negated:

```
if not fifo_not_full then ...
```

```
if not unnormalized then ...
```

Code like this is not unlike Orwell's *not un-* formation, a mental vice common in unclear speech.[11] "A not unblack dog was chasing a not unsmall rabbit across a not ungreen field...."

16.4   Guidelines for Naming Routines

A computer language allows us to describe a system using our own words, in addition to the predefined language constructs. In this vocabulary, data objects are nouns, and routines are verbs. Routines are used to manipulate objects or change the system state; therefore, they should be named after the actions they perform. This section presents guidelines to help you choose great names for your routines.

Name It from the Viewpoint of the User   A routine is a procedural abstraction: a developer must be allowed to call the routine, trusting it to perform a task and ignoring any implementation details. This allows the client code to work closer to the problem domain. It also has two implications on how we should name the routine.

The first implication is that a routine should not be named after its implementation. Instead, the name should describe its higher level goal.[12] This approach helps ensure that the client code makes no assumptions about how the routine performs its task, giving us freedom to change the implementation if needed.

As an example, consider a routine used in a game to check whether two objects have collided. In computer graphics, 2D images or animations are called *sprites*, and a common way to test for a collision between them is to check whether their bounding rectangles intersect. Thus, a possible name for this routine could be `rectangles_intersect()`. However, this name reveals implementation details about how the operation is performed and does not refer to the terms used in the application. To fix these problems, we should choose a name that uses terms from the problem domain, such as `sprites_collide()`.

The second implication is that when you name a routine, you should always think about how it will read in the calling code. This is part of a programming practice called *design from the client side*. When evaluating alternative choices for a name, test them in actual code to see how it looks, and try to imagine typical use situations. Modern IDEs make it easy to rename a routine, so you can experiment with several names, in your own code, in very little time.

If a Routine's Main Purpose Is to Return a Boolean, Then Name It after an Assertion or a Question   Many functions encapsulate a test that returns *true* or *false*. Such routines are called *predicate functions*. Sometimes the task may seem almost too simple to deserve a routine, but this practice offer two main benefits: it makes the calling code clearer, and it removes duplication when the test is performed more than once.

A common technique for naming this kind of function is to name them after an assertion—a statement claiming that something is true. Here are some examples of functions named following this guideline:

```
function pixel_is_transparent(pixel: pixel_type) return boolean;
function rectangles_intersect(rect_1, rect_2: rectangle_type) return boolean;
function possible_to_increment(digit: bcd_digit_type) return boolean;
function array_contains_element(arr: array_type; elem: object_type) return boolean;
```

We can test how well those names work by reading them as the condition of an *if* statement. See how the code looks obvious, even though we have no idea of how the functions are implemented:

```
if pixel_is_transparent(x, y) then ...
if sprites_intersect(paddle, ball) then ...
if array_contains_element(allowed_keys, pressed_key) then ...
if possible_to_increment(hours_digit) then ...
if rectangles_intersect(rectangle_1, rectangle_2) then ...
if rectangle_contains_point(rect, point) then ...
```

A possible variation is to name the function after a question rather than an assertion. This approach also works, but it does not read well in English all the time:

```
if is_pixel_transparent(x, y) then ...
if can_be_incremented(hours_digit) then ...
```

If a Routine's Main Purpose Is to Return an Object, Then Name It after the Return Value   The distinctive feature of a function is that it returns a value that can be used in expressions. Naming a function after its return value feels natural to the client programmer and reads well in the middle of expressions. Here are some examples of function names following this recommendation:

```
function cosine(operand: real) return real;
function minimum(values: integer_vector) return integer;
function bounding_box(rect_1, rect_2: rectangle_type) return rectangle_type;
function transpose(matrix: matrix_type) return matrix_type;
function height(matrix: matrix_type) return integer;
function identity_matrix(size: integer) return matrix_type;
function gray_code(binary_value: std_logic_vector) return std_logic_vector;
```

A large part of the functions you will find in VHDL will be *type conversion functions*. Such functions accept one argument of a given type and return an equivalent value in another type. To name type conversion functions, choose one of the following guidelines.

The first possibility is to be explicit and name the function using the template `new_type_from_original_type`:

```
function real_from_float(float_value: float) return real;
```

The main advantage of this convention is that it makes it easy to confirm visually that the code is correct. In the following example, note how the part of the name describing the new type (real) is close to the assignment target, and the original type (float) is close to the object being converted:

```
sample_real <= real_from_float( sample_float );
```

Another alternative is to use the template `original_type_to_new_type`. This scheme is more common in practice, but compared with the first approach, it has the disadvantage of making the assignments look twisted:

```
sample_real <= float_to_real( sample_float );
```

A third approach is to rely on the function overloading mechanism of VHDL and omit the original type from the function name:

```
sample_real <= to_real( sample_float );
```

This last convention is used in most of the VHDL standard libraries, and it works well in practice. It is short, easy to remember, and easy to understand.

If a Routine's Main Purpose Is to Perform an Action, Then Name It after Everything It Does Some routines are called for their effects on the system state: they may change objects passed as parameters or the environment in which they are executed. For such routines, choose a name that describes at a higher level of abstraction everything the routine does.

To name this kind of routine, start with a verb. Choose a strong verb, such as ini-tialize, disable, or add. Avoid meaningless or weak verbs such as do, perform, or process. Use the imperative form of the verb, and write as if you were giving a command to the routine.

When the verb is enough to describe the operation, it is not necessary to add a noun. For instance, in a game routine, initialize(ball) and disable(enemy_ship) may be perfectly clear. It would be redundant to name them initialize_ball(ball) and disable_enemy_ship(enemy_ship).

When the verb does not provide enough detail about what the routine does, add a noun to make the meaning perfectly clear. In the names update_coordinates(sprite) and update_bitmap(sprite), the noun adds meaningful details, and it is required to differentiate between the two routines. However, simply repeating the argument type in the routine name would be unnecessary: update_sprite(sprite) is again on the redundant side. Subprogram overloading makes it unnecessary to repeat the object type in the routine's name. As long as the parameter types are different, the compiler will be able to choose between routines with the same name. Therefore, it is not necessary to create different names for routines that differ only on the type of object they accept. As an example, consider the case of three routines named update_pixel_color(pixel), update_line_color(line), and update_rectangle_color(rectangle). As long as the meaning is the same, we can reuse the same name for all the routines: update_color(pixel), update_color(line), and update_color(rectangle).

Use your judgment to decide how much information to provide in the routine name. When in doubt, however, err on the side of providing more information rather than less. You can always rename the routine later if you decide that a name is too long or detailed.

If a Routine Is Hard to Name, Then Check for Any Design Problems Routines that do one thing and have a well-defined purpose should be easy to name. If you find yourself struggling to name a routine, then check for any design problems in the code. If all the names you can think of use weak verbs such as *do, perform,* and *process,* then it may be a sign that the routine's purpose is not completely clear. If you find yourself using names joined by connecting words such as *and* or *or,* then either the routine does not have a single purpose or it is named at the wrong level of abstraction. Names can tell a lot about the quality of a design. Sometimes bad names help make a design problem evident. Conversely, looking for ways to improve a name can lead to reorganizations that make the code clearer.

16.5   Guidelines for Naming Other VHDL Constructs

Naming Entities   Design entities are the primary hardware abstraction in VHDL. They provide the building blocks of a design, much like physical components used to build a hardware system. Therefore, it makes sense to name entities like we name things in the real world: using nouns or noun phrases.

A simple or compound noun is acceptable when it evokes an entity that any reader would recognize. This includes the names of well-known hardware elements (such as and_gate, register_bank, or adpcm_codec) and chip names or part numbers (such as hct00 or mc14532b). It also works well for top-level entities, which are often models of things that exist in the real world (such as alarm_clock or pong_game).

For entities that are not universal or well-known functions, such as most blocks in a custom design, a good recommendation is to use an agent noun—a noun derived from an action, such as *driver* from *drive* or *encoder* from *encode*. This guideline yields names such as parity_generator, signature_detector, or uart_transmitter. Agent nouns are preferable to nouns formed with the suffix *–ion* (as in parity_generation or signature_detection) because the former is focused on the entity that performs the action rather than the action itself.

As always, avoid using names that are cryptic, are meaningless, or do not provide enough detail. Names such as e, ent, circuit, design, or protocol are totally meaningless. With little effort we can find names that are much more communicative. Cryptic names, such as nnum1b or shf3mode, are indecipherable to anyone except those most intimate with a design. Additionally, remember to avoid nonstandard abbreviations.

Certain kinds of entities have specific roles in a solution. In these cases, a naming convention such as adding a semantic suffix to the entity name can help set them apart from other entities in the design. One example is to add the suffix _tb or _testbench to all entities containing testbenches. Another example is to use the suffix _top to identify entities that can be used as the top level in a design hierarchy.

Table 16.2 shows some examples of names that could be improved, along with a description of the corresponding problems and suggestions of improved names.

Naming Architectures   An architecture body describes the internal operation or structure of an entity. Because it characterizes and describes the implementation of an entity, it makes sense to name it with an adjective or adjectival phrase. Typical examples

Table 16.2 Examples of entity names that could be improved

Name that could be improved	Problems	Improved name
add	Named after a verb; could be a good name for a routine.	adder
parity	Named after an output value; could be a good name for a function.	parity_generator
signature_detection	Name in passive form (with suffix -*ion*).	signature_detector
system	Too generic.	alarm_clock_top

include behavioral, structural, or pipelined. There are different schools of thought on naming architectures; here we will survey the most common ones.

The first approach is to create architecture names prescriptively. This includes practices such as adding a standard suffix to the entity name (e.g., adder_a or adder_arch) or simply repeating the entity name as the architecture name. This approach can save you some time because you do not need to put much thought into choosing a meaningful name. It also makes the connection between entity and architecture obvious. Nevertheless, you may be missing a good opportunity to provide useful information; moreover, this approach does not work when an entity has multiple architectures.

The second approach is to choose from a list of predefined names. A common practice is to use the level of abstraction at which the architecture is implemented. Typical names following this convention would be behavioral, rtl, dataflow, structural, and gate_level.

The third approach is to name the architecture with a description or distinguishing feature of the implementation. Examples of names that describe an implementation include two_stage_pipeline, recursive, and combinational. Examples of names that are a distinguishing feature of the implementation include with_timing, ideal, and area_optimized.

We can choose the convention that makes more sense in our workflow, but as a general rule, it is better to choose names that have more meaning packed into them. Remember to follow the common naming guidelines and avoid names that are meaningless, enigmatic, or vague. Finally, do not use names that would be better suited for an entity. In other words, do not attempt to describe in the name of an architecture what the corresponding entity does.

Naming Files   To keep a design organized, each file must contain a single conceptual unit. In VHDL, this guideline gives rise to two different conventions. The first is to put a primary unit and the corresponding secondary unit in the same file. In this approach, a package declaration and a package body would go in a single file. An entity

Table 16.3 Example of a file naming convention

File contents	File name
One entity and one corresponding architecture	*entityname*`.vhd`
One entity	*entityname*`_ent.vhd`
One architecture	*entityname_architecturename*`_arch.vhd`
One configuration	*entityname_configname*`_cfg.vhd`
One package and the corresponding package body	*packagename*`_pkg.vhd`
One testbench (when there is a single testbench per entity)	*entityname*`_tb.vhd`
One testbench (when there are multiple testbenches per entity)	*entityname_testname*`_tb.vhd`

declaration and its architecture body would also go in a single file. In these cases, it is a good idea to give the file the same name as the primary unit.

The second approach is to put each compilation unit in its own file. Following this convention, a package declaration, the corresponding package body, an entity declaration, and its architecture body should all be placed in separate files.

There are advantages to both approaches. Putting the primary and secondary units together reduces the number of files and eases the navigation across files while editing the code. It also keeps related information closer together. However, giving each unit its own file may reduce unnecessary recompilation because when we analyze a primary unit, all the secondary units that depend on it must be reanalyzed. For example, if a package declaration and its body share the same file, then any change in the package body will require that all units using the package be recompiled. This could be avoided by putting the package body in a separate file from the package declaration.

You may want to experiment with both approaches and choose the one that works best in your workflow. In any case, it is important that the name of a file be related with the units it contains. A file naming convention is strongly recommended. Table 16.3 shows an example of a file naming convention than can be used in VHDL projects.

When you create your naming conventions, remember to choose the lowest common denominator between valid VHDL names and valid file names in your operating system. For instance, avoid characters that are not valid in VHDL names, such as spaces and dashes, and do not rely on capitalization to distinguish between files.

Naming Processes Every process can be named with a label before the *process* keyword. This label is not mandatory, but it provides two benefits: it documents the process intent, and it can provide useful debug information. Simulators identify processes by name, helping you cross-locate simulation events with the corresponding processes in the source code.

In any case, keep in mind that not all processes need to be named; only do so if you have something meaningful to say. Processes exist at a much lower level and higher granularity than entities and architectures, and trying to give a name to each of them would add unnecessary clutter to the code. Also, an undescriptive or a meaningless name is worse than no name at all.

A process is a delimited region of sequential code within an architecture. Most processes are created to drive one or more signals, which can be considered the outputs of the process. If we think about it, this role is similar to that of an entity or a procedure. Therefore, we can use two different approaches for naming a process.

The first approach is to name it as you would name a procedure: use the imperative form of a verb or verb phrase. Examples of processes named following this recommendation include `increment_pc`, `generate_outpus`, and `count_events`.

The second approach is to name a process as you would name an entity: using a noun or noun phrase. If the process models a basic element or well-known function, then we can use a plain noun such as `latch` or `next_state_logic`. If the process models a function that is specific to our design, then we can use an agent noun such as `event_counter` or `input_sorter`.

Any of the two approaches works well, so it is not necessary to stick to a single form. Know the available choices, experiment with them, and choose the one that works better. In any case, avoid meaningless suffixes such as `_proc`, `_process`, or `_label`. They do not add any meaningful information and only clutter the code.

Naming Types   In a typical VHDL design, it is usual to create a large number of types. Choosing good names for them may be even more important than naming data objects because types are visible in a much larger scope and can affect a large extent of the code.

When we name a type, we should use a name that describes the category of objects that it represents. For example, suppose we want to create a type to represent instructions in a CPU. A good name for this type would be `instruction`—the name of the class of objects that have this type. This is in line with the names of predefined types such as `integer`, `real`, or `bit`.

However, this has an undesirable consequence: we would not be able to use the name `instruction` for our signals or variables because the name would be taken by the type. This would force us to use less desirable names (such as `instr`, which is less pronounceable) or create an artificial distinction by adding extra words to the name (such as `current_instruction`).

One way to avoid this problem is to use a standard prefix or suffix to identify user-defined types. This approach makes the type names clearly identifiable and prevents them from clashing with object names. There are a few choices on how to identify custom types, but the clearest approach is to add the suffix `_type` at the end of the name. Other common approaches include adding the suffix _t or the prefix t_ to the base name. In any case, do not abbreviate the word *type* by removing only the last letter and using the suffix _typ. You get all the disadvantages of an abbreviation for the space savings of a single letter.

A last recommendation about naming types is to use names that are problem-oriented rather than language- or implementation-oriented. When you create a subtype, avoid encoding its limits or size into the name. For instance, instead of integer_0_to_15 or word_32_bit, find out what the types really represent and name them accordingly. If you want to create a type for CPU registers, then create a register_type instead of an unsigned_32bit_type. Repeating the type information is redundant and creates an unnecessary dependency between the type and its name. Problem-oriented names also document our intent for a type. If we create a type named unsigned_32bit_type for the registers of a CPU, then we are more likely to use it for different purposes than if we named it cpu_register_type.

Naming Enumeration Types and their Values   A particular case in naming types is that of *enumeration types* (also called *enumerated types* or *enumerations*). An enumeration type is a set of named values. To declare an enumeration, we must give names to the type and each value in the enumeration.

For enumeration types, we can use the same guidelines provided for naming types in general. However, a common doubt is whether the name of an enumerated type should be singular or plural. For instance, suppose we need to name an enumeration that defines the states in a finite state machine. Should we call it state_type or states_ type? The common practice is to use the singular form, for several reasons. First, the VHDL standard libraries use the singular form in enumerations. For instance, see the type named *boolean*, which includes the values *true* and *false*, or the type *character*, which includes all valid characters. Second, an enumerated type is not different from other scalar types such as *integer* or *real*, which are also named in the singular. Third, popular programming languages that provide standard naming guidelines, such as Java and C#, explicitly recommend using singular names for enumerated types.[13] You will be in good company if you do so. Finally, we can think about how the type will read when used in a declaration. If we create an object called my_favorite_color, then this object is an instance of a color (and not of a "*colors*"). An FSM current state is a state (and not a "*states*"), and so on.

As for the names used as values in an enumeration, there are a few recommendations. Some programmers like to add a common prefix to all values of an enumeration to prevent name clashes when different types share a common value. For instance, the value idle could exist in both a cpu_state_type and a motor_state_type. To avoid name clashes, we could add the prefixes cs_ to the elements belonging to cpu_state_ type and ms_ to those of motor_state_type. Following this convention, we would use `cs_idle` and `ms_idle` in the enumerations. Another advantage of this convention is that it clearly shows that the enumeration values are related because they share the same prefix. It could also prevent name clashes with other identifiers, such as the names of data objects.

In VHDL, this practice is more of a style issue than a real need because enumeration literals are proper values and are treated as such by the compiler. In most situations, the compiler is able to disambiguate between them and select the proper type. The compiler can check, for instance, the type of the target in an assignment or the type of the other operand in a comparison. In the few cases where the names are really ambiguous, we can use other tricks. For instance, we could use *type qualification* to tell the compiler that a value belongs to a given type. In the earlier example, we could write `cpu_state_type'(idle)` or `motor_state_type'(idle)`. Alternatively, if the enumeration is declared in a package, then we could use selected names. In the example, we could write `cpu_state_type.idle` or `motor_state_type.idle`.

Finally, another stylistic issue is whether the values of an enumeration should use the same case convention as constants. In other programming languages, where enumerations are constants in disguise, it makes sense to use the same convention. In such languages, if the convention is to name constants using `ALL_CAPS`, then the enumeration values would be named `OPENED`, `CLOSING`, `CLOSED`, and so on. This is more of a stylistic issue, but for the sake of consistency and to avoid undermining the convention used for constants, in VHDL it may be a good idea to use lowercase for enumeration values. Nevertheless, it is useful to see clearly in the code when a name is an enumeration value. A common solution is to use syntax highlighting to show enumeration literals in a different font style or color.

Naming FSM States   The states in an FSM are typically represented with an enumeration type. Therefore, all the considerations for naming enumerations also apply. The only additional recommendation is to use more communicative names than `S1`, `S2`, and `S3`. For example, in the case of a garage door, we could use `opened`, `opening`, `closed`, and `closing`. In a CPU, we could use `fetch`, `decode`, and `execute`. This saves our readers the trouble of memorizing the connection between arbitrarily chosen names and what they really represent. Also, try to use a more meaningful name than `state` for the state variables. They certainly describe the state of *something*, so try to encode this information into the name unless it is really obvious.

Naming Constants   Named constants are a big help in readability and maintainability. Constants make the code more readable because instead of a mysterious number in the code (say, `1080`), we can see a description of what the value means (`VERTICAL_RESOLU-TION`). They also make the code more maintainable because if the value ever needs to change, we can do it in a single place instead of hunting for literal values in the code.

All guidelines for naming data objects also apply to constants. There is, however, one more factor to consider. In many situations, it is useful to tell them apart clearly from other objects in the code. For instance, in an expression, if we can identify which parts are constant, then we can concentrate our debugging efforts on the parts that are allowed to change. As another example, there are cases where using a constant or signal in an expression can result in significantly different hardware circuits.

To identify constants in the code, a common convention is to write their names in ALL_CAPS. This is in line with the style conventions of many popular programming languages. For instance, constants in Ruby, Python, and Java are named using ALL_CAPS. Although NAMES_WRITTEN_IN_ALL_CAPS are harder to read than names_written_in_lowercase, this convention works well for constants because they are used less frequently than variables or signals. In this way, named constants stand out without making the bulk of the code unreadable.

Naming Generics   Most generics are generic constants. For such generics, it makes sense to use the same conventions used for naming constants. However, unlike named constants, generics are part of an entity's interface, so it is especially important to name them clearly and accurately. Aim at choosing a name that provides all the information needed to use the generic correctly. If besides the name we still need to write a comment, then we have failed to choose a good name.

A recurrent problem in naming generics is to leave out precisely the part that tells what the generic holds. Take the name INPUT_BITS, for instance. For the author, it may look perfectly clear that it contains the *number* of bits in an input port. However, this is not what the name says. According to the name, this generic holds "bits"; otherwise, it is lying. To avoid any confusion, always be explicit: if the generic specifies a *number* of elements, then make sure that NUMBER, NUM, or COUNT appears in the name. If it specifies a dimension, then make sure the name includes SIZE, LENGTH, or LEN. If the module has more than one dimension, then refer to each of them by name, using, for instance, WIDTH, HEIGHT, or DEPTH.

Let us take a closer look at some examples. Suppose we are modeling a FIFO in which the size of each word and the number of stored words are configurable using generics. A terrible choice would be to use arbitrary, single-letter names such as N and M because these names do not carry any meaning. Even if the reader has some idea of what needs to be configured, these names do not give a hint as to which generic defines which dimension. We can improve the generics somewhat by using names that hint at the corresponding dimensions, such as BITS and WORDS, for instance. However, these names still fail to state clearly what they represent—namely, the *number of bits* for each word in the FIFO and the *number of words* in the FIFO. In other words, those names are lying: the constant named BITS does not contain bits, and the constant named WORDS does not contain words.

Table 16.4 Examples of generic names that could be improved

Names that could be improved	Problems	Improved names
N, M, G1, G2	The name is meaningless.	NUM_BITS, WORD_WIDTH, WORD_SIZE, DATA_WIDTH, SELECT_WIDTH, SEL_WIDTH
ADDR	The name is lying; the generic does not hold an address.	ADDRES_WIDTH, ADDR_WIDTH
BITS	The name does not provide enough information. The name is lying; the generic does not contain "bits".	NUM_BITS, BIT_COUNT, SIZE, LENGTH, WIDTH
CLOCK	The name is missing the word that tells what the object represents.	CLOCK_PERIOD, CLOCK_FREQUENCY
LOG	The name is lying; the generic does not contain a "log" object. Is this a switch to enable logging, or the name of a log file?	LOG_FILENAME, LOG_ENABLE, ENABLE_LOGGING
CONFIG	The name is meaningless.	COUNT_MODE, COUNT_DIRECTION, DEBUG_ENABLED, NUM_STAGES, NUM_CORES

Taking those names in context, an astute reader would probably figure out what they mean. However, if our code needs to be figured out, then it is not clear enough. We want any reader to know immediately and unmistakably what each generic means. Fortunately, this is easy to fix: we just need to describe more precisely what the generics represent. In this case, a good choice would be WIDTH_IN_BITS and DEPTH_IN_WORDS. These names are so clear that they are almost impossible to misconstrue, and this is all that really matters. Yes, they are longer than N and M, but this is because they carry enough information to be useful on their own. Another good choice would be WORD_WIDTH and FIFO_DEPTH. The words *width* and *depth* make it clear that the objects represent dimensions; the names are also a little bit shorter.

In summary, try to pack all the important information into generic names, and make sure they are descriptive enough to be used without resorting to comments. A reader should be able to use a generic without making any assumptions and without looking at the code inside the architecture. Finally, make sure that the name really tells what the generic holds. Do not call it BITS if it does not contain bits, and do not call it ADDRESS if it does not contain an address. Table 16.4 shows examples of generics whose names could be improved, as well as suggestions to make them easier to use.

Naming Signals Signals are the communication channels within an architecture body, carrying values between concurrent statements, processes, and instantiated

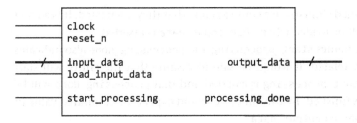

Figure 16.2
Good port names make for a module that is easier to use.

components. Signal names are especially important to describe the operation of an architecture. In a schematic, when we look at a net, we can see where it comes from and where it goes to. In code, all we have is the signal name.

Signals are also data objects, so all the recommendations for naming objects also apply. Every signal should have a clear name that fully and accurately describes the information it represents. In some cases, however, it may be useful to provide additional information about the role of a signal in the design. This can be done by attaching a semantic suffix to the signal name. Some examples include:

- using a suffix (such as _n) to indicate that the signal is active low;
- using a suffix (such as _z) to indicate that the signal is part of a tri-state bus;
- using a suffix (such as _reg) to indicate that the signal is the output of a register;
- using a suffix (such as _async) to indicate that the signal has not been synchronized.

These kinds of suffixes are important because they register our intent and add information that is not present in any other place in the code. In contrast, prefixes and suffixes that merely repeat information provided at the signal declaration (such as its type) are redundant and should be avoided.

Naming Ports Ports are the most visible part of an entity's interface, so it is especially important to give them meaningful and communicative names. Ideally, an entity's interface should provide all the information needed to instantiate and use it correctly. Although this is usually not possible using port names alone, it is a goal worth pursuing. The more effort we put into choosing descriptive port names, the easier the entity will be to use. It also minimizes the amount of guesswork required from the reader.

Let us illustrate this with an example. Suppose we need to use the entity from figure 16.2 in a design. For this example, the exact operation performed by the entity is not relevant.

Because the port names are clearly spelled out, it is relatively easy to figure out how to use the entity in our code. The names clock and reset_n indicate a synchronous design with an active-low reset. The signal named input_data is probably where we would provide the data to be processed by the entity. Because the signal load_input_data is named similarly, we could assume that they are related. It is a good bet that input_data will be loaded when load_input_data is asserted.

The symmetry of the names start_processing and processing_done also indicates that they are somehow related. It would be safe to assume that the entity will start doing its work when start_processing is asserted and that processing_done will be asserted when the job is finished. By this time, we would expect the computed value to be present at the port named output_data.

Do you feel like we are restating the obvious here? This is exactly what we should strive for. In a good design, everything turns out to be precisely what you expected.[14]

Of course, there are still details to be sorted out, but we can be pretty confident that we have grasped the overall working of the interface. Needless to say, if the signals had been named a, b, c, x, and y, then we would still be scratching our heads trying to figure out where to start.

Naming Loop Parameters   A *loop parameter* or *loop index* is an object that assumes different values during the iterations of a loop. The sequence of values is defined in the loop range. For example, in the loop statement:

```
for i in 1 to 10 loop ... end loop;
```

the loop parameter is named i, and the loop range is 1 to 10.

Loop parameters have certain characteristics that warrant special conventions. For simple loops spanning few lines of code, it is customary to use single-letter names such as i, j, and k. Although this seems to go against the recommendations of using communicative names, this is in line with choosing the length of a name in proportion to its scope. If the loops are short, then the loop header will always be within view, and there is little chance that we could mistake the loop index for something else. Also, loop parameters are commonly used as array indices and in other expressions, often several times in a single line of code. This increases the pressure for shorter names. Finally, these identifiers have been used for loop indices since Fortran, so they are an established convention. Most programmers will readily identify the names i, j, and k as loop indices.

In any case, avoid using the identifiers i, j, and k (and single-letter names in general) elsewhere in the code. Even in loops they should be avoided if the loop is longer than a few lines of code, if it is nested, or if its logic is complex. In such cases, consider using names that refer to what the loop is iterating over, such as stage, cell_number, or table_row.

17 Excellent Comments

Comments are annotations intended to make the code easier to understand and maintain. Done right, they can save precious development time and improve productivity. Done wrong, they duplicate information and may confuse our readers. This chapter discusses the many types of comments and gives advice on how to comment our code more effectively.

17.1 Comments: Good or Bad?

Comments have a legitimate role in making the code easier to understand. They communicate what code alone cannot express, such as the reasons behind an implementation choice. They convey information at a higher level than the code, reducing the need for reading and understanding individual statements. This aids comprehension and can save valuable development time.

However, excessive commenting is a bad coding practice. Comments are often duplicate information. They are frequently inaccurate and get outdated easily. Even worse, comments often work as a crutch, when the code really should be reorganized. On many occasions, when we find a heavily commented region of code, the purpose of the comments is to prepare the reader for the bad code that follows. Comments also add clutter to the code, which goes against the original goal of making the code easier to read.

For all those reasons, comments are a polemic issue. Some programmers think that comments are generally good, and the more of them, the better. Others strive for self-documenting code and try to avoid any superfluous comments. The key to settling this argument is to understand that there are different kinds of comments: some are generally good, some are almost certainly bad, and some may be good or bad depending on the context and circumstances. A programmer should know the typical kinds of comments and the problems associated with them. This chapter will give you the conceptual tools to decide when to comment your code and the practical advice for writing good comments when they are really needed.

17.2   Basic Principles

Don't Explain Bad Code—Rewrite It!   Many comments start off on the wrong foot, when a programmer realizes that the code is becoming hard to understand. To make it "clearer," the developer adds a comment to explain the convoluted logic behind the code.

This is absolutely the wrong approach. The first principle of good commenting is: *don't explain bad code—rewrite it.*[1] You should never use comments to explain tricky code. If your code is getting complicated, then simplify it until it becomes self-explanatory. This makes the comments unnecessary.

Express Yourself in Code, Not in Comments   There are many ways to register our intent in the code. As a rule, you should never settle for a comment when you can express yourself in the code proper. Among other advantages, code is always up-to-date with system behavior, whereas comments get outdated easily. Code can be debugged and verified, whereas comments can never be tested or trusted. Moreover, comments are often duplicate information. Consider the following example:

```
-- Set to '1' if you want to use a hardware multiplier, '0' otherwise
generic MULTIPLIER_CONFIG: std_logic := '0';
```

In this declaration, a comment was needed only because the generic was poorly named. If we choose a name that matches its role, then the comment becomes irrelevant:

```
generic USE_HARDWARE_MULTIPLIER: boolean := false;
```

As you code, always try to move information from comments to the code proper. A simple way to do this is to choose meaningful names for data objects and routines; this practice alone makes most comments unnecessary. Another option is to create auxiliary data objects or routines, which are not strictly necessary, but make the code easier to understand. As a general rule, you should always try to express your intent in code that is processed by the compiler, relying on comments only as a last resort.

Minimize Duplicate Information   Comments should add useful information to what the code already says. This usually means registering the author's intent or summarizing a region of code at a higher level of abstraction. If we offer more detail than necessary, then we risk duplicating information, which adds multiple sources of truth to the code and makes it harder to maintain (see sections 3.4 and 4.8 for details).

In the following example, the comment in line 4 is saying more than the corresponding declaration. This adds a second source of information to the code, which will conflict with the actual code if the value in line 2 is changed:

```
1    -- Maximum number of elements that can be stored in FIFO
2    constant FIFO_DEPTH: natural := 16;
3
4    -- True when FIFO has 16 elements
5    signal fifo_full: boolean;
```

Another common example is to add a file's name in a comment inside the file, or a routine's name in a comment describing the routine. Both will cause problems if the names need to be changed.

Keep Comments at the Level of Intent Comments should be kept at the level of intent. A comment written at the level of intent records the author's intention and explains the purpose of a section of code. Another way to state this principle is to say that that comments should tell us *why* instead of *how*. Comments that tell us *how* are often redundant with the code:

```
-- Divide pixel coordinates by the font size:
text_column <= pixel_x / FONT_WIDTH;
text_row <= pixel_y / FONT_HEIGHT;
```

This type of comment is likely to get outdated if we change the implementation. In contrast, comments that tell us *why* capture the developer's intent and withstand modifications:

```
-- Convert graphic coordinates to text mode coordinates:
text_column <= pixel_x / FONT_WIDTH;
text_row <= pixel_y / FONT_HEIGHT;
```

Keep Comments Close to the Code They Describe One way to reduce the chance of a comment getting outdated is to put it close to the code it describes. The customary place is immediately before the code. This way the comment prepares the reader for what is to follow.

A corollary of this principle is that when we have a list of declarations and want to describe each one of them, it is better to comment each declaration individually. The alternative, grouping all the comments in a big block before the declarations, puts the comments farther away from the code they describe. The chance that a developer will edit the code and forget to update the comment increases with the distance between them.

17.3   Types of Comments

17.3.1   Good Comments

Certain types of comments are approved of by most developers. Such comments are either necessary or beneficial; as long as they are not wrong or outdated, they deserve their place in the code. This section discusses the kinds of comments that do more good than harm in most situations.

Explanation of Intent   Certain information cannot be expressed in code alone. Sometimes it is useful to document the reason behind a decision, telling why you chose a particular path from the multiple choices available. At other times, you may want to explain why a special case is handled differently. See the following example:

```
1    function sqrt(x: real) return real is
2    begin
3        -- Handle sqrt(0) manually to prevent division by 0
4        if x = 0.0 then
5            return 0.0;
6        -- Return exact value for sqrt(1) to avoid rounding errors
7        elsif x = 1.0 then
8            return 1.0;
9        end if;
10
11       -- Proceed using standard algorithm for all other values
12       ...
```

In this code, lines 3 and 6 spell out why the square roots of 0 and 1 are calculated differently. No amount of code or good variable naming could explain this decision.

Another place where comments are often needed is to explain business logic—information that refers to the problem domain and needs to be used in the code. In the following example, the comment describes why the statement is necessary:

```
-- We never read from the LCD, so set its read/_write port to 0
lcd_rw <= '0';
```

This comment makes the connection between the code and the business logic without adding redundant information. Comments that explain the programmer's intent are good because they do not repeat what the code already says. Such comments are less likely to get outdated when the code undergoes low-level changes. They are also helpful in maintenance: another developer can compare what the code does with what it was supposed to be doing, making it easier to detect errors.

Summary Comments   A summary comment is an abstract of a section of code. Used correctly, this kind of comment helps maintenance and increases developer productivity because it reduces the need for reading individual lines of code to grasp their overall intent. In the following example, each comment summarizes the next five lines of code:

```
1    -- Initialize matrix with an identity matrix on the left side
2    for i in dense_matrix'range(1) loop
3        for j in dense_matrix'range(2) loop
4            dense_matrix(i, j) := '1' when (i = j) else '0';
5        end loop;
6    end loop;
7
8    -- Copy elements from the sparse matrix
9    for i in 1 to sparse_matrix.elements_count loop
10       row := sparse_matrix.elements(i).row;
11       column := sparse_matrix.elements(i).column;
12       dense_matrix(row, column) := '1';
13   end loop;
```

When summary comments are used as in the previous example, they are also called *narrative comments* because they guide the reader through the program's flow.

Summary comments are a mild form of duplicate information, but because they are at a different level of abstraction, they are not overly redundant with the code. However, as any other kind of comment, they can get outdated. When you add a summary comment, remember to treat it as part of the code and always keep it up-to-date.

17.3.2   Double-Edged Comments

Very few kinds of comments are good enough to be recommended without reservation. However, many other kinds can be useful depending on the situation and the quality of their contents. Such comments must be added judiciously, *never* by default or mandate. This section discusses the kinds of comments that can be good or bad depending on their use.

Documentary Comments   Sometimes the names we choose for our design units, routines, or data objects cannot convey all the information our readers will need. In those cases, it is useful to add a documentary comment close to their declarations.

Documentary (or documentation) comments are generally a good idea, but they fall under the double-edged category for two reasons. First, in many instances, a comment would be unnecessary if the entity were properly named. In those cases, the comment serves as a crutch for the bad name. Second, in the cases where the name already conveys the relevant information, a comment is superfluous and redundant.

The following example uses documentary comments to describe an entity's purpose and its ports:

```
-- Generate a sequence of test addresses.
entity test_controller is
    port (
        r: in std_logic;    -- The reset signal
        c: in std_logic;    -- The clock signal
        a: out std_logic_vector(31 downto 0)  -- The test address
    );
end;
```

Granted, the information provided in the comments is useful. Otherwise, how could we know that signal a contains the generated test address? The problem with this style of coding is that it puts an unnecessary burden in our brains because we need to recall from memory the relationship between those terse names and their meaning in the code. A better approach is to keep more information in the code and less in our heads. This is easy to fix: whenever you find a similar situation, just move the information from the comment into the object's name. Unlike a comment, the name accompanies the object everywhere and does not have to be memorized:

```
entity test_address_generator is
    port (
        reset: in std_logic;
        clock: in std_logic;
        test_address: out std_logic_vector(31 downto 0)
    );
end;
```

Documentary comments should be used only when it is impossible or too cumbersome to embed all the required information in an object's name. The following example makes good use of a comment for a data declaration:

```
-- Number of seconds elapsed between leaving the 'connection_established'
-- state and entering the 'connection_closed' state.
variable connection_duration_in_secs: natural;
```

When a documentary comment is used with a routine, it is called a *routine header*. Such comments may provide useful information for the user of the code, such as a description of what the function does and any conditions that may cause errors.

```
-- Returns 'base' to the power of 'exponent'.
-- Raises an error if base = 0.0 and exponent <= 0.0.
-- Raises an error if base < 0.0 and exponent has a fractional part.
function "**"(base, exponent: real) return real;
```

Another benefit of documentary comments is that they are understood by modern IDEs and automatic documentation generators. In source code editors that are aware of VHDL syntax, hovering over a name anywhere in the code pops up a window showing the associated header comment. This can save a lot of vertical scrolling between an object's declaration and the places where it is used in the code.

Reading Aids Some comments are an attempt to make the code easier to read, by calling attention to details that could be overlooked. Others mark special places in the code or try to separate distinct sections in a source file. This can be helpful if used sparingly; most of the time, however, it just adds clutter to code. Usually, they are a failure to acknowledge the real problem: the code is too long or too complicated.

An example of a reading aid that serves as a crutch for bad code is *end-of-block comments*. This technique consists in adding an endline comment after the closing keywords of a control structure or compound statement (lines 12, 14–16, and 19 in listing 17.1). This might make sense in long sections of code with deeply nested control structures. However, in short and straightforward code, they only add clutter. If a loop or conditional statement seems complicated enough to deserve this kind of comment, then consider this a sign that the code needs to be simplified. In any case, avoid writing end-of-block comments as a default practice.

Listing 17.1 Examples of end-of-block comments (not recommended)

```
 1 function shift_left(matrix: matrix_type; shift_amount: natural) return matrix_type is
 2    variable shifted_matrix: matrix_type(matrix'range(1), matrix'range(2));
 3    constant NONZERO_COLUMNS_COUNT: natural := width(matrix) - shift_amount + 1;
 4 begin
 5    -- for each matrix row
 6    for i in matrix'range(1) loop
 7       -- for each matrix column
 8       for j in matrix'range(2) loop
 9          -- copy element or fill it with zero
10          if j < NONZERO_COLUMNS_COUNT then
11             shifted_matrix(i, j) := matrix(i, j + shift_amount);
12          else -- j >= NONZERO_COLUMNS_COUNT
13             shifted_matrix(i, j) := '0';
14          end if; -- copy element or fill it with zero
15       end loop; -- for each matrix column
```

```
16    end loop; -- for each matrix row
17
18    return shifted_matrix;
19 end; -- function shift_left
```

Section headings are another common reading aid. Sometimes they mark places in the code to make them easier to spot when skimming through a source file. At other times, they separate distinct sections of the code. Listing 17.2 shows several examples of section headings, most of them unnecessary.

Listing 17.2 Excessive use of section headings (not recommended)

```
 1 --------------------------------------------------------------------------------
 2 -- Library and use clauses
 3 --------------------------------------------------------------------------------
 4 library ieee;
 5 use ieee.std_logic_1164.all;
 6 use ieee.numeric_std.all;
 7
 8 --------------------------------------------------------------------------------
 9 -- Entity declaration
10 --------------------------------------------------------------------------------
11 entity synchronous_ram is
12    --------------------------------------------------------------------------
13    -- Generic declarations
14    --------------------------------------------------------------------------
15    generic (
16        RAM_DEPTH: natural := 1024;
17        DATA_WIDTH: natural := 32
18    );
19    --------------------------------------------------------------------------
20    -- Port declarations
21    --------------------------------------------------------------------------
22    port (
23        ----------------------------------------------------------------------
24        -- Input ports
25        ----------------------------------------------------------------------
26        clock: in std_logic;
27        reset: in std_logic;
28        ...
29        ----------------------------------------------------------------------
30        -- Output ports
31        ----------------------------------------------------------------------
```

```
32         data_out: in std_logic_vector
33     );
34 end;
35
36 --------------------------------------------------------------------------
37 -- Architecture declaration
38 --------------------------------------------------------------------------
39 architecture rtl of synchronous_ram is
40     ...
```

The first thing to consider is that this technique only works if used sparingly. If the code is full of banners, as in the prior example, each of them becomes less effective. Second, there are better ways to achieve the same goal without littering the code. If necessary, try to add some vertical separation using blank lines, which are less intrusive. Moreover, modern IDEs provide an instant outline of a source file, highlighting the most relevant places in the code hierarchy. Many source code editors allow you to bookmark arbitrary points in the code in case you need to return to them often. In any case, if you feel like your source files are getting hard to navigate, then consider breaking them down into smaller files.

Markers in the Code   Markers are notes inserted in the code as a reminder that something needs to be done or as a request for other programmers to look into an issue. Markers are not supposed to be left in the finished code. In the following example, a to-do note was used to remind the developers that part of the code will need attention in the future:

```
-- TODO: The transparent color must be configurable in the future,
-- but we will use a hardcoded value for now.
if sprite(x,y) = 255 then -- transparent color
    output_pixel <= BACKGROUND_COLOR;
else
    output_pixel <= sprite(x,y);
end if;
```

To use markers effectively, we should select a predefined set of labels corresponding to the kinds of issues we want to track, making it easier to search for occurrences of a certain task. TODO, FIXME, and TBD (to be done) are common tags for pending development tasks. Modern IDEs can recognize such tags and create to-do lists automatically.

Developer notes and other markers in the code are useful tools. They only qualify as double-edged comments because they tend to accumulate if not removed diligently. If you use them, then assume the responsibility to weed them out periodically.

17.3.3 Bad Comments

Now that we have seen the kinds of comments that are generally good and those that can be good under the right circumstances, we turn our attention to comments that should not exist in high-quality code.

Explanation of the Code   Unlike comments that explain a programmer's intent, comments that explain the code itself are a bad sign. Explanatory comments show around logic that is too complicated to understand from the code alone; they underline a failure to write code that is clear enough to stand on its own. Of course, when you are maintaining someone else's messy code, you would rather have comments to help you; this is because bad code needs a lot of explanation. When writing new code, however, you should never settle for a comment when you could express yourself in code. Strive to communicate your intent using plain code, relying on comments only as a last resort.

To illustrate the problem with explanatory comments, consider the example in listing 17.3. The fact that a single assignment needs several lines of comments is a clear sign that something is wrong. The code also suffers from unnecessary conversion functions, indicating that the signal types were poorly chosen.

Listing 17.3 Explanatory comments trying to compensate for bad code

```
1 -- When the quadrant is even, the table address is the least significant phase bits.
2 -- We know that the quadrant is even when phase(PHASE_WIDTH-2) is '0'. When the
3 -- quadrant is odd, we must subtract the least significant phase bits from the total
4 -- table size.
5 sine_table_address <= phase(PHASE_WIDTH-3 downto 0) when phase(PHASE_WIDTH-2) = '0'
6    else conv_std_logic_vector(2**(PHASE_WIDTH-2)-conv_integer(phase(PHASE_WIDTH-3
7    downto 0)), PHASE_WIDTH-2);
```

To get rid of the explanatory comments, we can move information from the comments into the code proper. We can do this by creating routines, constants, aliases, and temporary signals or variables, among other techniques. The comment in lines 1–4 mentions three pieces of information: a condition ("quadrant is even"), a vector slice ("least significant bits"), and a numeric value ("total table size"). We can express this information in code using a signal, an alias, and a constant:

```
constant SINE_TABLE_SIZE: integer := 2**(PHASE_WIDTH - 2);
alias least_significant_bits is phase(PHASE_WIDTH-3 downto 0);
signal quadrant_is_even: boolean;
...
quadrant_is_even <= (phase(PHASE_WIDTH-2) = '0');
```

Then the conditional signal assignment becomes much more readable:

```
sine_table_address <=
    least_significant_bits when quadrant_is_even else
    (SINE_TABLE_SIZE - least_significant_bits);
```

When you find yourself about to write a comment such as the one in this example, stop and fix the code so that it does not need an elaborate explanation.

Superfluous or Redundant Comments Some comments do not add any new information; they just repeat the code in other words. The classic example of a redundant comment is:

```
-- increment counter
counter <= counter + 1;
```

Even when disguised as useful information, comments like this are still redundant. The following comment seems to add important information because it makes it easier to understand what the statement does:

```
-- increment column counter
cl_ctr <= cl_ctr + 1;
```

However, the only reason that it seems useful is because the signal name was too abbreviated. If we name the signal properly, then the comment becomes irrelevant:

```
-- increment column counter
column_counter <= column_counter + 1;
```

Now we can go ahead and delete the comment without remorse. Listing 17.4 shows many examples of redundant comments; none of them earns their lunch. If you find comments like these in your code, then delete them without mercy.

Listing 17.4 Examples of redundant comments

```
1    architecture example of redundant_comments is
2        -- the data bus width
3        constant DATA_BUS_WIDTH: natural := 64;
4        -- the value of PI
5        constant PI: real := 3.14159;
6        -- the clock period
7        constant CLOCK_PERIOD: time := 10 ns;
```

```
 8      ...
 9   begin -- start of architecture example
10
11         -- toggle clock every 10 ns
12         clock <= not clock after 10 ns;
13
14         -- main test process
15         main_test_process: process (clock) begin
16
17             -- on rising clock edge
18             if clock'event and clock = '1' then
19                 -- increment counter
20                 counter <= counter + 1;
21                 ...
22             end if;
23
24         end process;
25   end; -- end of architecture example
26
27   -- end of file
```

Journal Comments   Sometimes you will find a list of changes at the top of a file, as in listing 17.5.

Listing 17.5 A changes journal in form of a comment

```
-- Revision History
-- ----------------
--
-- Revision  Date      Author      Change
-- --------  --------  ----------  -----------
--     1.0   10/08/14  Bob Tables  Initial revision
--     1.1   10/22/14  Bob Tables  Counter width increased to 18 bits
--     1.2   11/05/14  Jim Hacker  Added initial values for simulation
--     1.3   11/06/14  Jim Hacker  Fixed typos
--     2.0   11/17/14  Bob Tables  Changed port names to match other modules
--     2.1   11/18/14  Jim Hacker  Added 'done' signal
--     2.2   12/13/14  Jim Hacker  Constant ADDRESS_WIDTH changed to a generic
--     2.3   12/14/14  Bob Tables  Fixed rounding algorithm (thanks to Jim Hacker)
```

In other cases, you may find the same sort of remark spread throughout the code:

```
if count > ('0' & rate(15 downto 1)) then -- fixed 01/02/15 by Jim Hacker
```

The problem with this approach is not that the comments are useless, but that there are much better tools for this job. Revision control tools such as SVN, Git, and Mercurial can keep a complete history of the codebase. They allow you to restore any version and undo any small change if you want. The tools can tell which files and lines were modified in each revision, with a complete record of date, time, and authorship.

In contrast, history comments are often incomplete, are inaccurate, and grow irrelevant over time. Who knows what else was changed besides what the author decided to report? The information contained in this kind of comment is also hard to use in practice. How would you undo a change? How do you find the changes in the code? The best advice is to avoid such comments and leave the task to your version control software.

The only occasion in which journal comments are justifiable is when you expect your readers to come across the source files completely out of context—for instance, when they are distributed without any form of revision control. Still, consider storing and distributing your code following the modern practices of software configuration management; if you give your peers access to a source code repository, then they can access any revision whenever they need to.

Commented-Out Code   Sometimes, when browsing through code, you may come upon some lines that are commented out. The feeling is puzzling; there is no way of telling why those lines are there. Are they still useful? Is it safe to delete them? Consider the following example:

```
if phi2 = '1' then
    output <= fcs_input & output_xor;
--else
--    output_xor <= gf_input xor input;
end if;
```

Why would someone leave those commented lines in the code? Maybe this was a quick fix or an experiment, or maybe someone just forgot to delete them. Because you cannot be sure, you leave them as they are, and they will keep littering the code for a long time. To prevent this, make sure there is no commented-out code before committing your work. Remember that your revision control tool can automatically keep the old code for you.

17.4   Comment Layout

Comments are supposed to help us understand the code and make it easier to maintain. A good layout scheme should not interfere with the code, and it should not get in our way as we write or modify the code. Moreover, it should not be time consuming.

This section presents considerations and guidelines about how to lay out comments around our source code.

17.4.1 Comment Syntax in VHDL

Syntactically, VHDL allows two kinds of comments: *single line* and *delimited*. A single-line comment starts with two dashes (- -) and goes until the end of the line. If the comment must span several lines, then each line needs to start with the two dashes. We are allowed to write code before (to the left of) the dashes if we want.

```
-- This is a single-line comment.
-- This is another single-line comment.
begin -- There may be code to the left of a single-line comment.
```

A delimited comment is enclosed within the /* and */ symbols. It can span multiple lines, and there may be code before the /* symbol and after the */ symbol.

```
/* This is a delimited comment starting and ending on the same line. */
/* This is a delimited comment
   spanning multiple lines. */
process /* There may be code on either side of a delimited comment. */ begin
```

Because delimited comments are a recent addition to VHDL, most of the examples you will find (and most developers out there) still use single-line comments only. This approach works well; otherwise delimited comments would have been added earlier to the language. One advantage of limiting yourself to single-line comments is that you can reserve multiline comments for quickly "commenting out" sections of code. If the region to be commented already contained delimited comments, then this would not be possible because delimited comments cannot be nested. A delimited comment ends at the first */ symbol, no matter how many opening /* symbols exist before it.

Another use for delimited comments is when you want to write more code after the comment, as in the following association list:

```
generic map (clock_frequency => 50_000 /* in kHz */, pulse_duration  => 10 /* in ms */)
```

If we wanted to comment-out the line above using a delimited comment, then it would cause a compile error because the compiler would understand that the comment ended right before the comma.

Because mixing comment styles can cause nesting problems, and because it is counterproductive to spend time switching styles on a per-comment basis, it is not a bad idea to limit ourselves to single-line comments in general. We can still use delimited comments in the special occasions mentioned earlier. Single-line comments are also convenient for short comments because they are more economic, saving us some typing and horizontal space.

17.4.2 Comment Layout Choices

The difference between single-line and delimited comments presented in the previous section is only syntactic. For our study of comments, a more important distinction is between *full-line* and *end-of-line* comments (also called *endline* comments). Full-line comments have the entire line for themselves, whereas endline comments share the line with code.

```
-- This is a full-line comment. The line consists of a comment only and no code.
/* This is also a full-line comment. */

variable minimum_value: integer;    -- This is an endline comment.
variable maximum_value: integer;    /* This is another endline comment. */
```

For our purposes, a *block comment* is any comment spanning more than one line. Listing 17.6 demonstrates the introduced terms with an example. We will use this example to comment on some good and bad layout choices.

Listing 17.6 Comment layout examples

```
 1 -- This is a block comment. It spans multiple lines and describes the code that
 2 -- follows. All lines in a block comment should be indented at the level of the first
 3 -- line of code underneath it.
 4 process is
 5     variable minimum_value: integer;    -- This is an endline comment for a data declaration
 6     variable maximum_value: integer;    -- Another endline comment for a data declaration
 7 begin
 8    if rising_edge(clock) then
 9        -- This is a full-line comment. Prefer this style to endline comments.
10        if enable then
11            data <= load_data;  -- An endline comment for a statement. This should be avoided.
12        end if;
13    end if;
14 end process;
```

The first thing to notice is the indentation. The example shows two full-line comments (lines 1–3 and 9). To prevent comments from interfering with the code, full-line comments should be indented at the level of the first statement that follows, as shown in the example. This makes it clear that the comments refer to the correct level of the code. Indenting the statements and failing to do the same with comments would ruin the outline of the code, making it hard to identify and understand any control structures.

Another important remark is the vertical position of comments relative to the code they describe. Because code is usually read from top to bottom, and because we use comments to prepare the reader for what is to follow, it makes sense to place the comments immediately before the code they describe. This also helps the reader to decide whether a block of code is relevant before reading it.

Now let us turn our attention to endline comments (lines 5, 6, and 11 in the example). At first the end of the line seems like a convenient place for a comment. After all, it is close to the declaration or statement that the comment describes. In practice, however, endline comments have several problems.

First, they are harder to format. When there are multiple endline comments, you will probably want to align them vertically so that the comments do not look ragged. This requires manual alignment using tabs or spaces. It also means that the distance from the left margin will be determined by the longest line of code. Moreover, if the comment does not fit in the space that is left, then you may have to start a new line, which will be mostly blank and waste a lot of space. During all the time you spent trying to make the comments look neat, you could have moved on to the next development task and been doing actual work.

Second, endline comments are a pain to maintain. If you rename an object, then you may have to realign all other comments around the affected line. This is especially bad because renaming is a highly recommended practice; we should not stop until we have found a name that fully and accurately describes an object, and this is hard to do on the first try. Furthermore, renaming should be done automatically by a tool to be time-effective and reliable; after a change, you should not have to revisit all the places where a name is used and rearrange the code around it. Some programmers even get this backward and think that renaming is bad because it messes with formatting. Nonsense! Instead, you should aim for layout styles that do not get in your way while you are coding and do not break under modification.

Finally, one last reason to avoid endline comments is that they leave less space for the comment text. This means that you will be hard pressed to make the comments as short as possible, instead of as clear as possible.[2]

Endline comments may work in situations where the code in the surrounding lines has approximately the same length, the comment is short, and the code is not expected to change much. This is the case with some data declarations, association lists, or enumeration types. Overall, endline comments should be used sparingly or avoided.

17.5   Commenting Guidelines

17.5.1   Layout Guidelines

Choose a Low-Maintenance Commenting Style[3]   One problem with comments is that they tend to get outdated relative to the code. We can make the maintenance job easier by using comment styles that withstand modifications. To illustrate this, let us look at an example with several problems.

```
-------------------------------------------------------------------------------
-- Entity seven_segment_display_driver                                       --
-- ===================================                                       --
--                                                                           --
-- Output some text on the seven-segment displays depending on the FSM state. --
--                                                                           --
-- Inputs      Function         | Outputs     Function                       --
-- ======      ========         | =======     ========                       --
-- fsm_state   The state of the FSM | displays    Array with 5 SSD displays  --
-------------------------------------------------------------------------------
entity seven_segment_display_driver is
    port (
        fsm_state: in state_type;
        displays: out slv_array(0 to 4)(6 downto 0)
    );
end;
```

The first thing that jumps out is how much time it took the author to format this header. The comment looks nice, but the same information could be provided with a much simpler style. Besides taking more time to write, this layout takes more time to maintain. The column of dashes on the right margin needs to be realigned whenever the text changes. The comment also repeats names found in the code, making it harder to keep up-to-date. The underlining beneath the entity name will also need to be updated if the entity is renamed. The column layout for the inputs and outputs makes everything harder to maintain: if we change one of the port names or their descriptions, then we need to realign everything on the right.

Here's a more sensible version of the same comment without any duplicate information:

```
-- Output some text on the seven-segment displays depending on the FSM state.
entity seven_segment_display_driver is
    port (
        -- The state of the FSM
```

```
        fsm_state: in state_type;
        -- Array with 5 SSD displays
        displays: out slv_array(0 to 4)(6 downto 0)
    );
end;
```

If you miss the visual separation that the old block provided, then you could add a single line of dashes before the first comment. However, this should not be necessary; after all, there should not be many entities in the same file in the first place.

Use Blank Space   Summary comments usually indicate a new step in a train of thought. Therefore, it makes sense to separate them from the preceding code with a blank line. A blank line after the comment is unnecessary, and it would weaken the connection between the comment and the code that follows. A blank line before the summary comment is also not strictly necessary when there is a change in indentation before and after the comment because the different indent level already provides enough visual separation.

It is also a good idea to use one whitespace between a delimiter symbol (--, /*, or */) and the comment text. The lack of surrounding whitespace makes the comment look crowded, and we want the beginning of the text to be clearly identifiable. Moreover, in block comments, the whitespace creates an alley between the comment symbols and the comment text. When we move our eyes to the next line, they will be automatically drawn to beginning of the word instead of the delimiter symbols, improving readability.

17.5.2   Language and Content

Make Comments Clear, Correct, and Concise   When skimming through code, we often use comments to decide whether to read the corresponding statements. Unnecessary words and convoluted language slow down our reading, so avoid them both. Following this advice, instead of:

```
-- The purpose of this function is to find the minimum value in the input array.
```

or:

```
-- Function that finds the minimum value in the input array.
```

Prefer the shorter and imperative version of the sentence:

```
-- Find minimum value in the array.
function minimum(values: integer_array) return integer;
```

The imperative mood works especially well for short or summary comments. In these cases, write as if you were giving an order to the code that the comment describes.

You do not need to write complete and grammatically correct sentences in every comment, but the comment needs to be precise and readable. Try not to be cryptic, and avoid unnecessary abbreviations. What is the point of adding a comment if it is harder to read than the code?

```
return_value := minimum(valid_values);  -- init ret val w/ min valid val
```

The previous comment reminds us of why we should avoid endline comments. It is hard to write a clear, complete, and precise sentence with the leftover space from a line of code.

Record Your Thoughts, Then Iterate Writer William Zinsser reminds us that "a clear sentence is no accident. Very few sentences come out right the first time, or even the third time."[4] This is also true of a good comment. To overcome writer's block when creating a comment, start by writing whatever crosses your mind, without worrying about form. Then iterate a few times, making the comment more precise and compact.

This advice is intended for those who struggle to start writing, but it also serves for those who write comments in a hurry and then move on. Remember to read what you wrote, making sure that it will be understood correctly. Iterate as needed.

Consider Your Audience When writing a comment, imagine you are writing for a programmer who is familiar with the language but not as intimate with the project or code as you are. Try to anticipate any questions this programmer might have. However, do not use comments to explain language features. Leave that for the books instead.

17.5.3 General Commenting Guidelines

Do Not Use Comments to Make Up for Bad Names This is a corollary to the principle of expressing ourselves in the code. Sometimes you will find a comment that seems useful, is written clearly, and provides relevant information. Nevertheless, it is only necessary because an object or routine was poorly named. In the following example, the comment helps the reader figure out the role of signal wr:

```
signal wr: boolean;   -- true when a write has been requested
```

If we move the relevant information into the code, then the comment becomes irrelevant:

```
variable write_requested: boolean;
```

The main benefit of using meaningful names rather than comments is that the reader does not need to refer back to the declaration to recall an object's meaning. Considering the large number of names in a typical design, relying on names rather than comments can make a big difference in the time and effort it takes to understand the code.

Consider Using a Variable or Routine Instead of a Comment   This is another corollary to the principle of expressing ourselves in the code. Sometimes you will find an explanatory comment describing an operation. The operation may be a boolean test, as in the following example, or a sequence of statements:

```
-- if pixel is visible
if pixel.x >= VIEWPORT_LEFT and pixel.x <= VIEWPORT_RIGHT and
    pixel.y >= VIEWPORT_TOP and pixel.y <= VIEPORT_BOTTOM
then
    ...
end;
```

Granted, the comment is useful. It tells you something that could take a while to figure out from the code. There are, however, at least two other ways to achieve the same result without comments. The first is to use an explanatory variable:

```
pixel_is_visible :=
    pixel.x >= VIEWPORT_LEFT and pixel.x <= VIEWPORT_RIGHT and
    pixel.y >= VIEWPORT_TOP and pixel.y <= VIEPORT_BOTTOM;

if pixel_is_visible then
    ...
end if;
```

The use of an explanatory variable moves knowledge from the comment into the code proper. A side benefit is that it makes it easier to debug expressions: when we split a complex expression into several explaining variables, we can examine and debug each part individually.

The second way to move knowledge from comments to code is to move the test to a function:

```
function pixel_is_visible(pixel: pixel_type) return boolean is
begin
    return (
        pixel.x >= VIEWPORT_LEFT and pixel.x <= VIEWPORT_RIGHT and
```

```
        pixel.y >= VIEWPORT_TOP and pixel.y <= VIEPORT_BOTTOM
    );
end;

if pixel_is_visible(pixel) then
    ...
end;
```

An advantage of this approach is that functions are easier to test, compared with variables embedded in the code. For instance, we can write a simple testbench that calls the function with known parameters and checks the expected return values. Logic that is buried inside the code, as in the original version of this example, can be hard to test.

Document Any Surprises, Workarounds, or Limitations   Sometimes we need to write code that is less than optimal or do things in an unusual way to circumvent a problem with a compiler or another tool. At other times, the behavior of a certain piece of code may be surprising at first glance. In those cases, we can save the reader some time by using a comment. It can also prevent other developers from "optimizing" or "fixing" the unusual code.

Document the Present, Not the Past   Do not keep comments that say how things used to be done in the past. The purpose of comments is to explain why the code works now. Leave the changes history for your revision control tool.

Write Code That Does Not Require End-of-Block Comments   Some programmers like to add endline comments after the closing keywords of a block, as in the following example. The reasoning is that this would make the code clearer because it tells the reader to which statement or declaration the keywords belong.

```
dff: process (clock) begin
    if rising_edge(clock) then
        if reset = '1' then
            q <= '0';
        else -- if reset = '0'
            q <= d;
        end if; -- if reset
    end if; -- if rising_edge(clock)
end process; -- dff
```

Although this could make sense for long blocks of code with deeply nested control structures, in short and straightforward code, this practice clutters the statements and causes more harm than good. Instead of using comments to cope with the complicated logic, use proper indentation and simpler control structures to make the code easier to understand. Consider moving the statements inside a loop or conditional statement to a separate function, thus reducing the number of indentation levels.

Avoid Nonlocal Information   When commenting a declaration, avoid writing about how and where it will be used in the larger context of the system. It may seem like useful information, but the truth is that the declared entity has no control over where it will be used. At its declaration, write from a neutral point of view, without assuming where it will be instantiated. Try to leave this kind of information for the places where the entity is used instead.

Comment as You Go   Do not wait until you have finished implementing a functionality to go back to the code and comment it. The best time to comment is while the code is still fresh in your head. With time, you will learn what deserves to be commented. As a side benefit, the process of writing will help you organize your thoughts and may lead to improvements in the chosen solution.

17.5.4   Guidelines for Specific Kinds of Code

Commenting Files   Many project standards require a block comment at the top of every source file. This kind of comment is called a *file header*. Having a file header in each file is good advice, provided that you follow two basic principles. First, you should minimize the amount of duplicate information between the code and header. Second, you should put in the header only information pertaining to the file as a whole. Many real-life examples violate these two principles. To illustrate what belongs in a file header, let us look at an example containing many typical problems (listing 17.7).

Listing 17.7 A file header with many typical problems

```
 1   -- Copyright (C) 2014 FlitMining Inc.
 2   --
 3   -- License: This is free and unencumbered software released into the
 4   --          public domain. Anyone is free to copy, modify, publish, use,
 5   --          compile, sell, or distribute this software, for any purpose.
 6   --          The author or authors dedicate any and all copyright interest
 7   --          in the software to the public domain. The software is provided
 8   --          "as is", without warranty of any kind. In no event shall the
 9   --          authors be liable for any claim, damages or other liability.
10   --
11   -- Project:      Flitcoin Miner
12   -- File name:    flitcoin_miner_top.vhd
```

```
13    -- Module name:      flitcoin_miner_top
14    --
15    -- Purpose:          This file contains the top module for the project.
16    --                   It instantiates the flitcoin_miner entity and connects
17    --                   its ports to the board peripherals.
18    --
19    -- Dependencies:     numeric_std, std_logic_1164, text_io, flit_components
20    -- Target device:    Speedster X-7 FPGA
21    -- Tool version:     MegaSynthesis Lite v10.1
22    --
23    -- Creation date:    07/01/2014
24    -- Last modified:    11/06/14
25    -- Author:           A. Prentice (a.prentice@flitmining.com)
26    -- Maintainer:       Jim Hacker (j.hacker@flitmining.com)
27    --
28    -- Revision History
29    --
30    -- Revision  Date       Author       Change
31    -- --------  --------   ----------   -----------
32    --    1.0    10/08/14   Jim Hacker   Initial revision
33    --    1.1    11/05/14   Jim Hacker   Added initial values for simulation
34    --    1.2    11/06/14   Jim Hacker   Fixed typos
```

The top of the header contains legal and licensing information. This kind of information is useful in case you stumble on the file and need to know whether you can use it. It is also a mandatory policy in many companies. However, you do not need to clutter your code with the legal mumbo-jumbo of a full software license. Instead, put the license in a text file in the project folder and refer to it in the file header:

```
-- Copyright (C) 2014 FlitMining Inc.
-- Released under the XYZ license. See license.txt for copying permission.
```

Lines 11 to 13 are debatable. Adding the project name is probably safe: it does not change often, and this name is not repeated in the code. The file name, however, is redundant with the file system. Repeating the name in the header only makes it harder to rename the file when you find a better name for it. The module name (line 13) also duplicates information defined in the code. Using names inside comments poses an additional problem: names in the source code can be changed easily with refactoring tools, whereas names within comments are much harder to change. Automated renaming tools can never be sure about the relationship between a string in a comment and an object in the code.

Lines 15 to 17 contain a brief description of the file's purpose. Or is it the module's? It is hard to tell because the description mixes up the roles of the file and the module it contains (in this case, a design entity). If you want to keep this section in the file header, then try not to trump the module's responsibility. Information such as the design approach used in the construction of the module is important, but it is more suitable for the module header. A better approach would be to say that the file contains the module's implementation and then describe the module in a separate header comment associated with the design entity.

The list of dependencies in line 19 is totally redundant with the code. It makes no sense to keep such a list by hand when you can look it up directly from the code. Besides, if you fail to satisfy a dependency, your compiler will tell you right away.

The target device and tool version in lines 20 and 21 can be useful information, but do they belong in the file header? If your code uses no device-specific features, then do you really need to mention a target device? Furthermore, is the tool version a property of a single file, or does it pertain to the project as a whole? This is probably a project-wide information, so it belongs more accurately in an external *readme* file.

Lines 23 to 26 contain the authors' names and the file creation date. The problem with these fields is not that they are useless, but that there are much better tools for this job. Putting in the author's name and contact information is generally considered a good practice. However, in a team project, it may be hard to tell on a file-by-file basis who is an author, who is a maintainer, and who is a developer who made a small change or fix to the code. Leave this kind of detail to your version control tool; it will tell you the complete lineage of the file if you want. For the same reason, the full changes history (lines 28–34) does not belong in a file header. See the discussion about journal comments in section 17.3.3 for more details.

To summarize, keep your file headers clean, containing just the information that is essential and guaranteed to be useful. File headers occupy a prominent place in the file; any noise or disinformation here is highly visible. Furthermore, only add information that pertains to the file as a whole. If you have something to say about a module that is implemented in the file, then go to the module's declaration and add a comment there. Finally, avoid project-wide information and any content that makes your files harder to change. Listing 17.8 shows the file header from listing 17.7 after removing most of the redundant and nonlocal information.

Listing 17.8 A file header after removing most of the redundant and nonlocal information

```
1    -- Copyright (C) 2014 FlitMining Inc.
2    -- Released under the XYZ license. See license.txt for copying permission.
3    -- --
4    -- Project    Flitcoin Miner
5    -- Purpose    This file contains the top module for the project.
6    -- Author     A. Prentice (a.prentice@flitmining.com)
```

One could think that the header in listing 17.8 looks too simple and that much of the important information is missing. Note, however, that we did not remove the important information; we simply moved it to where it properly belongs. Revision history belongs in the revision control tool, a description about the operation of a module belongs in the module header, and project-wide information belongs in a project *readme* file.

Commenting Design Entities   Entities are the primary hardware abstraction in VHDL and the building blocks of structural design. To use an entity effectively, a developer must be able to treat it as a black box without peeking at its implementation. Ideally, all the information required to use an entity should be communicated through its interface. In practice, if you need to disclose additional information to make an entity more usable, then an entity header is the right place for it.

Listing 17.9 shows an entity header for a module that calculates the square root of unsigned numbers, using Dijkstra's algorithm.[5]

Listing 17.9 An entity header documenting an entity declaration

```
 1 -- Compute the square root of an unsigned value using Dijkstra's algorithm.
 2 --
 3 -- The computation is iterative and calculates the root of an N-bit wide unsigned number
 4 -- in log2(W)+1 clock cycles (time between 'load' is asserted externally and 'done' gets
 5 -- asserted by the entity).
 6 --
 7 -- The algorithm is described in the paper "Area-efficient implementation of a fast
 8 -- square root algorithm" (DOI: 10.1109/ICCDCS.2000.869869), and uses only additions,
 9 -- subtractions, and logic shifts.
10 --
11 -- Limitations: works only for even values of W. Throws an error otherwise.
12 --
13 entity dijkstra_square_root is
14    port (
15        -- Iteration clock (rising edge)
16        clock: in std_logic;
17        -- Synchronous reset
18        reset: in std_logic;
19
20        -- Assert high to load a new radicand and start the computation
21        start: in std_logic;
22        -- Asserted high when root has been calculated
23        done: out std_logic;
24
```

```
25      radicand: in unsigned;
26      root: out unsigned
27   );
28 end;
```

Using the code in listing 17.9 as an example, let us discuss what makes a good entity header. The trick about commenting design units effectively is to give enough information about their usage without duplicating information or exposing unnecessary implementation details. Some developers have the misconception that their designs will be easier to use if they mention all sorts of implementation details in an entity header. Actually, a good entity has an interface that is easy to use without any knowledge of its implementation details.

So what should be included in an entity header? To start off, remember that an entity is a module, and a good module should support a central purpose. Start the comment with one or two sentences describing this purpose. Next, if the interface alone is not clear enough to figure out how to use the module, then add any instructions you find necessary. For instance, describe any limitations on the input values or values that would cause the module to malfunction. If relevant, you can also describe the overall design approach.

Note that the header in listing 17.9 does not include a description of the entity's ports. Following the principle of locality, it is better to describe the ports where they are declared—in the entity's port list (lines 15, 17, 20, and 22). Note that the ports radicand and root (lines 25–26) do not have any comments associated with them; it would be hard to write a description that is not redundant with the port names. Descriptions such as "the output value" or "the calculated square root" do not add any information. In such cases, not writing a comment is better than adding a redundant description.

Finally, avoid saying anything about where the entity will be used and for what purpose. Remember that the entity has no control over where it will be instantiated. Describe the interface clearly, but leave the question of where it will be used for the higher level entities that instantiate the module.

Commenting Packages   Packages are a nice way to group behavior that supports a central goal and must be reused across a design. A well-documented package needs two types of comments: a package header, containing a high-level description of the package's functionality; and a description of each member in the public interface whose name is not self-explanatory. Listing 17.10 shows a package interface documented using comments.

Listing 17.10 Documenting a package interface with comments

```
1 -- Data types and routines for graphics operations. Supports both 24-bit
2 -- truecolor and 5-bit paletted colors.
```

```
 3 package graphics_pkg is
 4
 5     constant SCREEN_WIDTH_IN_PIXELS: integer := 1920;
 6     constant SCREEN_HEIGHT_IN_PIXELS: integer := 1080;
 7
 8     -- A point defined by its x and y coordinates.
 9     type point_type is record
10         x, y: integer;
11     end record;
12
13     -- A rectangle defined by its edge coordinates.
14     type rectangle_type is record
15         left, top, right, bottom: integer;
16     end record;
17
18     -- True if the point is inside the rectangle (including its edges).
19     function rectange_contains_point(rectangle: rectangle_type; point: point_type)
20         return boolean;
21
22     -- The minimum bounding box of two rectangles.
23     function bounding_box(rectangle_1, rectangle_2: rectangle_type)
24         return rectangle_type;
25
26     -- True if the two rectangles have at least one point in common.
27     function rectangles_intersect(rectangle_1, rectangle_2: rectangle_type)
28         return boolean;
29
30     ...
31 end;
```

A good way to start the header is with one or two sentences describing the package's purpose. It is not a problem if this description is short; sometimes it is hard to write a long description without going into details about the package contents. Also, do not replicate information that can be read directly from the code (such as the module name) or that does not pertain to the package proper (such as the file name). Read the guidelines about commenting files to know what should go in the file header and what belongs in the package header. For example, if you need to provide additional information about the package interface, such as how the types, routines, and global objects are used together, then the package header is the right place.

The usual members of a package are constants, types, and subprograms. For each member whose name is not self-explanatory, add a descriptive comment. Where the name alone is enough (as for the constant SCREEN_WIDTH_IN_PIXELS), there is no need for a comment.

A common doubt when documenting packages is where to put the routine headers: should they go in the package declaration or in the package body? Because the public interface is the only part of the package that external users should know, it makes more sense to place the comments in the public region—the package declaration. Also, human readers, documentation generators, and IDEs all look for function descriptions in the public region. The only exception is when you need to describe how a function is implemented for maintenance purposes; in this case, the comment should go in the package body.

Commenting Routines

Routines vary greatly in complexity and scope. At one end of the spectrum, there are simple subprograms that take no arguments and have self-explanatory names. At the other end, there are procedures with several input and output parameters for which no perfect name exists. For this reason, guidelines that require us to use the same header template for every routine are probably a bad idea. They lead to aberrations such as the header shown in listing 17.11.

Listing 17.11 An inadequate function header

```
 1 ---------------------------------------------------------------------------
 2 -- Function name:    rectangle_contains_point
 3 --
 4 -- Purpose:          Check if the given point coordinates are within the limits
 5 --                   determined by the given rectangle.
 6 --
 7 -- Algorithm:        If the point's x coordinate is within the rectangle's
 8 --                   left and right edges, and the point's y coordinate is
 9 --                   between the top and bottom edges, than the point is
10 --                   contained within the rectangle.
11 --
12 -- Inputs:           rectangle     The rectangle
13 --                   point         The point
14 --
15 -- Return value:     true if the point is contained within the rectangle,
16 --                   false if the point is outside the rectangle.
17 --
18 -- Author:           Jim Hacker
19 --
20 -- Creation date:    06/01/2015
21 --
22 -- Modifications:    none
23 ---------------------------------------------------------------------------
```

```
24 function rectangle_contains_point(rectangle: rectangle_type; point: point_type)
25     return boolean is
26 begin
27     return (
28         (point.x >= rectangle.left) and (point.x <= rectangle.right) and
29         (point.y >= rectangle.top) and (point.y <= rectangle.bottom)
30     );
31 end;
```

You should be able to point out the problems with the header in listing 17.11. To name a few, it duplicates information, such as the function name; it contains information that should be handled by the version control system; and it adds redundant descriptions for names that were already clear enough. For such a simple function, the header in listing 17.12 is much more adequate.

Listing 17.12 A better function header

```
1 -- True if the point coordinates are within the given rectangle, including
2 -- the rectangle's edges.
3 function rectangle_contains_point(rectangle: rectangle_type; point: point_type)
4     return boolean is
5 begin
6     ...
7 end;
```

Because a routine is a procedural abstraction with clear inputs and outputs, it can also be seen as a module. Therefore, the same advice that applies to other kinds of modules is also valid here: start the header with one or two sentences describing the routine's purpose. Ideally, the name should tell as much as possible about what the routine does, but some functions may perform a sequence of actions for which finding a fully descriptive name is hard. A good header helps mitigate this problem, as in the following example:

```
-- Move sprite using its internal movement vector; the animation
-- frame is also updated according to the sprite's FSM.
procedure update_sprite(signal sprite: inout sprite_type) is
begin
    ...
end;
```

When you need to describe the input and output parameters, it is better to do so right next to their declarations and not in the function header. The example in listing

17.13 uses endline comments to document function parameters. Note that this is not necessarily a good choice because they leave little room for the comment text. Generally, it is better to use a single-line comment before each parameter, as shown in the interface list of listing 17.9.

Listing 17.13 Using endline comments to document parameters (not always a good choice)

```
1 -- Generic function for subtracting two floating point numbers. The function is
2 -- flexible enough to be used with any rounding style, a configurable number of
3 -- guard bits, and with or without support for subnormal numbers.
4 function subtract(
5     minuend: unresolved_float;                    -- Left operand
6     subtrahend: unresolved_float;                 -- Right operand
7     constant ROUND_STYLE: round_type := round_nearest;  -- Rounding style
8     constant GUARD: natural := 3;                 -- Number of guard bits
9     constant DENORMALIZE: boolean := true         -- Turn on subnormal number
10                                                   --    processing
11 ) return unresolved_float;
```

Use your judgment to decide whether you need to provide any additional information to the routine's users. Depending on the situation, you can provide this information in the routine header or parameter list. If a routine raises an error under a certain combination of inputs, then document it in the header (listing 17.14, lines 2–3). If the routine has any limitation on the input ranges, then use ranges and assertions when possible; in some cases, however, it may be useful to add a note in the header or next to the parameter declaration.

Listing 17.14 Using comments to document error conditions and limitations on the input values

```
1 -- Return x to the power of y (x**y).
2 -- For x = 0, y must be > 0.
3 -- For x < 0, y must be a round number (an integer multiple of 1.0).
4 function "**"(x, y: real) return real;
```

Commenting Statements   We close this section with a few recommendations for commenting at a finer level of granularity. The first thing to remember is that commenting individual statements should rarely be necessary. This kind of comment is usually redundant or a shallow explanation of the code.

Instead, a better approach is to comment paragraphs of code. A paragraph is a short sequence of related statements that accomplish a single subtask (see section 18.4). Most paragraphs deserve a summary comment (one or two sentences describing the section of code). The reader may use such comments to get an overview of the code or decide whether to read the actual statements. Other statements that are good candidates for a summary comment include control structures (such as *if*, *for*, and *case* statements) and *process* statements.

In any case, remember to keep all comments at the level of intent, and avoid describing *how* the code works. Focus instead on *why* the statements or paragraphs exist in the code.

17.6   Self-Documenting Code

We close this chapter with a discussion of the ultimate goal of code readability: *self-documenting code*. Self-documenting code is code that can explain itself. This kind of code has two important characteristics: it is less dependent on external documentation, and it relies less on comments to describe how it works.

The documentation of a software project consists of two parts: *internal* documentation written in the source code itself and *external* documentation kept outside the code. The external documentation of a project starts well before the first line of code is written. Requirements, specifications, and design documents are typical examples. Although external documentation plays a role that code alone cannot, it comes with its own problems. To name a few, the documents must be kept up-to-date with the source code, and there are no guarantees that the code really does what is written in a separate document.

The key motivation for self-documenting code is to obviate the need for external documentation. Contrary to what one might think, the best way to achieve self-documenting code is not by adding comments to the source code. The best way to write code that can stand on its own is to use good design principles, avoid clever solutions, and keep the code simple and well factored. Here are some techniques to help make our code more self-documenting:

- **Use good design principles.** Hierarchy and modularity make the parts of a design easier to locate. Abstraction hides implementation details and reveals the big picture of a design. Any design principle aiming at simplicity, understandability, and maintainability helps reduce the need for external documents.
- **Create clear interfaces that are obvious to use.** If you create interfaces that cannot be misconstrued, then they will not require much in terms of documentation.
- **Avoid clever solutions.** They usually need to be explained. Prefer simple and straightforward logic that is self-evident to the reader.
- **Use the code layout to evidence its logical structure.** Visual presentation is an effective way to communicate the code's hierarchy and control flow logic.
- **Minimize nesting.** Deeply nested structures are harder to understand and usually call for explanatory comments.

- **Choose meaningful names.** If we choose names that fully and accurately describe our objects and routines, then we do not need to provide any further explanation.
- **Never use a comment when you can use an object or a routine.** In many cases, a comment can be avoided by introducing an explanatory step in a computation. The new steps serve as signposts, helping guide the reader through the logic of the computation.
- **Use named constants instead of magic numbers.** This makes expressions and conditions much easier to understand. It also makes the code easier to modify.
- **Use language features to communicate your intent.** Instead of using comments to describe an object, look for an alternative to express the same information in the code. If a variable can assume only a limited set of values, then constrain it with a range. If an object can assume only a number of states, then use an enumerated type.
- **Use coding conventions and standards.** If you do things consistently, then there is less need to document individual decisions. Readers can reuse their knowledge to understand other parts of the system.
- When there is no other way to make your code clearer, **use a comment.**

18 Excellent Style

Coding style or *programming style* is a set of conventions, rules, and guidelines used when writing source code. It is a cross-cutting theme involving topics as varied as how to choose names, use comments, and arrange the source code on the screen. The first two topics—names and comments—were the subject of the previous chapters. Code layout, the most visible aspect of coding style, is discussed in this chapter.

18.1 Programming Style

Programming style is the sum of all the decisions about the organization and presentation of the source code, including its layout, structure, choice of names, and preferred coding approaches. Most stylistic choices affect only nonfunctional characteristics of the code and have no influence in the system behavior. Nevertheless, programming style can greatly impact how easy the code is to read, understand, and maintain. Here are some examples of style choices:

- A programmer chooses to limit the width of all lines of code to 80 characters or less to make the entire source visible without horizontal scrolling.
- A programmer chooses to indent the code using spaces rather than tabs to make it look the same in all text editors and tools.
- A programmer chooses to name constants in `ALL_CAPS` to make them visually distinct from signals and variables.
- A programmer decides to limit subprograms to 40 lines of code or less to ensure that all the routines are focused and easy to understand.

Programming style is a broad subject, encompassing both high-level principles and nit-picky details. Although some recommendations may look arbitrary, the good ones are based on fundamental design principles, empirical evidence, or common sense. Whatever the case, we can always use a set of design qualities such as consistency, clarity, simplicity, and conciseness to evaluate any style choice.

EXAMPLES OF PROGRAMMING STYLE CHOICES		
FORMATTING AND LAYOUT	**STRUCTURE AND GROUPING**	**NAMING CONVENTIONS**
• Indentation style • Indentation size • Indentation character (tabs × spaces) • Letter case • Use of whitespace • Use of optional parentheses	• Choice of control structures • Grouping of statements • Ordering of statements • Statements per line • Maximum nesting • Maximum file length • Maximum routine length	• Naming guidelines • Suffixes and prefixes • Use of abbreviations
		COMMENTS • What to comment • What not to comment • How to comment

Figure 18.1
Examples of programming style choices.

Figure 18.1 shows many examples of style choices organized by category. In this book, recommendations about the choice and use of statements were presented in the chapters about sequential, concurrent, and assignment statements (chapters 9–11); naming recommendations were presented in chapter 16; and commenting guidelines were presented in chapter 17. This chapter focuses on the style choices related to source code formatting and layout.

18.1.1 Programming Style Matters

Code layout is often overlooked because it does not change what a program does, only how it looks. However, code is meant for more than interpretation by a computer. Before a product is ready to ship, its source code must be read, understood, and manipulated by humans countless times.

Is it worth putting effort into aspects of the code that do not affect its behavior? Many different studies indicate that it does.[1,2,3] Consistent indentation, communicative names, and good comments have been correlated with better program comprehension, improved readability, and lower incidence of bugs. More important, code that cannot be easily understood, corrected, and extended has no practical value in real-life projects. If the code is hard to understand and cannot be safely modified, then it may be easier (or more cost-effective) to rewrite it from scratch rather than try to fix it.

To illustrate what programing style means and its effects on source code, let us look at an example. Take a look at the following *process* statement and try to understand what it does and how it works:

```
proc: process (all) begin
cy <= (others =>'0');
case o is when A => for i in rp' range loop
r (i) <= oa(i) xor ob(i) xor cy(i); cy(i+1)
<= (oa(i) and ob(i)) or (oa(i) and cy(i)) or
(ob(i) and cy(i)); end loop;
when B => r <= oa and ob; when C => r <= oa or ob;
when others => r <= (others => '0'); end case;
end process;
```

This example is a little contrived because it packs a lot of bad practices into a few lines of code, but many of its problems are common in real projects. Instead of meaningful names, the code uses short and cryptic abbreviations. Because there is no indentation, the code layout does not represent its logical structure. Can you tell quickly how many choices there are in the *case* statement? Even if you can find out what this code does, how confident can you be that you understood it correctly?

The code can be made much easier to understand by applying a few simple techniques. If we use good indentation, communicative names, and a single statement per line, then it becomes much more readable:

```
alu_operation: process (all) begin
    carry <= (others => '0');
    case operation is
        when ALU_ADD =>
            result <= operand_a xor operand_b xor carry;
            for i in result_port'range loop
                carry(i+1) <=
                    (operand_a(i) and operand_b(i)) or
                    (operand_a(i) and carry(i)) or
                    (operand_b(i) and carry(i));
            end loop;
        when ALU_AND =>
            result <= operand_a and operand_b;
        when ALU_OR =>
            result <= operand_a or operand_b;
        when others =>
            result <= (others => '0');
    end case;
end process;
```

Granted, the formatted version is longer than the original one. But how do you feel about telling what it does? With proper indentation, we can see the conditions that control each operation. With meaningful names, we can have a good grasp of the code's intent and function. There is no point in making the code shorter if it is also harder to understand.

In the formatted version, the overall program flow is much clearer. Indentation alone can show the flow of control; there is no need to read the individual statements to know whether they are subordinate to one another. This hierarchical layout saves precious time when skimming through large sections of code. Moreover, because the objects are named precisely after what they represent, there is no need to decipher their names or memorize their meaning. If we put the code aside for a few weeks, then it should be fairly easy to understand when we read it again. Finally, note that the revised code has the same amount of comments than the original: none. A less seasoned programmer could have thought that the best way to make the code easier to understand would be to add an explanatory comment. As discussed extensively in chapter 17, comments should never be used to make up for bad code. When a section of code is unclear, the right thing to do is to rewrite it.

This example showed some of the benefits of a good programming style. It also demonstrated that we can greatly improve readability and understandability by following a few simple rules. In larger projects and organizations, such rules are explicitly documented and named a *coding standard*. This standard offers two important benefits. First, it makes the entire codebase in the organization more uniform, eliminating irregularities and other distractions. Second, it helps disseminate good programming practices. A good set of guidelines allows less experienced developers to produce code with a quality level that meets the company standards.

18.2   Code Layout

Code layout is the visual appearance of source code. It is also the most visible aspect of programming style. On a purely textual level, all layout techniques consist of adding spaces between the functional elements of the code. Therefore, layout does not affect any aspect of the compiled application, such as size, performance, or correctness. However, it can have a great impact on how easy the code is to read, understand, and maintain.

We can deliberately choose the position of statements and other constructs to make our code easier to work with. For that purpose, we use the fundamental concepts of whitespace, grouping, and separation, much like a graphic designer. Unlike in graphic design, however, aesthetics comes second to more objective quality measures. But do not be disappointed if you take pride in the appearance of your source code: in practice, when we follow a basic set of principles aimed at improving readability and presenting information in a logical way, the result is also aesthetically pleasing. In any case, we must remember that readability and maintainability come first, or we may develop habits that undermine the qualities we would like to maximize.

The importance of a good layout is that it augments the information provided by the source text. A compiler uses only the textual part of statements and declarations, but humans infer a lot from the relative positioning of source code elements. Therefore, as we lay out the code on the screen, the most important consideration is to make the visual presentation of the source text match the structure seen by the compiler.[4] A good layout allows us to see scopes, emphasizes the subordination between statements, and highlights the differences and similarities between lines of code. This effectively increases the information bandwidth between the code on the screen and our brains. Conversely, a bad layout makes the code harder to read and conveys misleading information.

Because layout does not change what the code does, it is hard to measure the effectiveness of any layout technique. Consequently, it is common to find conflicting recommendations aiming at the same goal, especially when it involves a subjective quality, such as readability. To prevent personal preferences and aesthetics from interfering with this discussion, the goals of a layout scheme must be stated clearly. This is the subject of the next section.

18.2.1 Layout Goals

Any layout technique should support the high-level goals of improving readability, understandability, and maintainability. Here we state our goals more objectively so that they can be used to evaluate any proposed layout technique.

Accurately Match the Logical Structure of the Code A good layout scheme should make the control structures (*if-else*, *case-when*, *while-loop*, *for-loop*, etc.) clearly visible to the programmer. It should be possible to visualize the hierarchy of the code (subordination between declarations or statements) without reading the individual statements. All scopes and nested statements should be clearly identifiable.

Improve Code Readability The ultimate measure of utility for a layout technique is how much it makes the code easier to read and understand. The only problem is that it is not possible to measure readability objectively, so we need to use heuristics to determine what makes the code more readable. For example, we can use the results of program comprehension studies or well-established visual information principles such as grouping, proximity, and regularity.

Because we are also readers, a technique that is always available is to make our own assessment of a layout style. However, we must be careful when subjecting any method to our personal evaluation because we may be strongly biased by aesthetic preferences and habit. Once we build familiarity with a certain style, anything that looks different might feel strange at first. Moreover, what pleases our eyes is not always more effective. A famous study by Miara et al[5] evaluated the effect of the amount of indentation in program comprehension and found significant differences in understandability.

Programmers were given four different versions of the same code: unindented or indented with two, four, or six spaces. The researchers found a significant effect: the programmers performed better when two or four spaces were used. However, when asked for a subjective rating of the difficulty to read each version, the programmers rated the version with six spaces as the least difficult, when in fact this version yielded results that were almost as bad as the one with no indentation at all. The researchers' conclusion was that the programmers found the deeply indented program more visually pleasing because it spreads out neatly the constructs of the language, but this did not in fact improve program comprehension.

The key takeaway is that we must remember that "more readable" does not necessarily equal "more aesthetically pleasing." Too many pseudo-guidelines exist on the grounds of making the code more readable whose secret goal is only to make it look nice on the screen.

Afford Changes   One of the most important characteristics of a design is its ability to tolerate changes. We put a lot of effort into making a design more modular, free of duplicate information, and loosely coupled, just to make it more amenable to changes and improvements. We should aim for the same with any layout technique. A layout scheme that makes it hard to add, remove, or change a line of code has less value than a scheme that affords changes. If a layout technique forces you to modify several lines of code to make a single change, then look for a better technique. When changes are harder than necessary, they will be done less often, and many improvements may be left out.

Be Simple (Have Few Rules)   A layout scheme should be composed of a few simple rules that are easy to follow. Remember that it may be used by many programmers with different skill levels. The resulting layout should also look simple, with no more complexity than necessary.

Be Consistent (Have Few Exceptions)   Coding conventions and formatting rules may have to be interpreted by many developers in a team. Ideally, two programmers following the same set of rules should arrive at the same (or nearly the same) layout. This is only possible if the rules are unambiguous and do not conflict with each other. When evaluating a technique, give preference to rules that have no exceptions.

Be Easy to Use   Besides being simple, the rules should require little effort to implement and maintain.

Be Economic   Avoid layout schemes that take up more screen real estate than necessary. However, do not sacrifice any of the other goals in the name of a more economic layout. There is no point in making the code shorter if it is also harder to read.

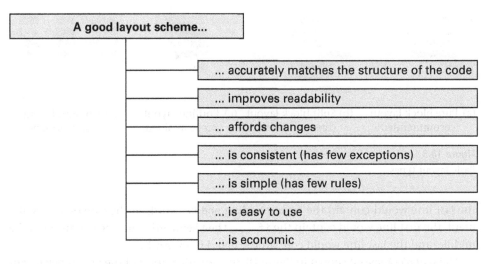

Figure 18.2
Characteristics of a good layout scheme.

Summary of Layout Goals   Figure 18.2 summarizes the goals that a good layout scheme should aim for. The goals are listed in order of roughly decreasing priority. When you consider any layout technique, rule, or guideline, check it against the characteristics shown in the figure to assess its practical value.

As a final remark, keep in mind that the end use of the code may affect the relative importance of the layout goals. When the code is intended for print, for example, the author may care little about maintainability and changeability. However, space might be at a premium. The goals in figure 18.2 were ordered according to the priorities of production code, in real-life projects, where teamwork and changeability are highly important. Again, remember that "readable" is different from aesthetically pleasing. If a layout technique makes the code look neat in detriment of the other goals, then it loses points in the utility scale.

18.2.2   Basic Layout Types
Most code layouts fall into one of the basic types shown in figure 18.3, in which each line of code is represented by one or more rectangles. This abstract representation allows us to discuss the advantages and disadvantages of each layout type. In this discussion, a *block* is a group of statements, declarations, or values enclosed between delimiters such as *begin-end* keywords or opening and closing parentheses.

The *pure block* layout shown in figure 18.3a is the most basic layout structure. Examples of where it can be used in VHDL include entity declarations and *if-then* statements. In the case of an entity, the first line would contain the keywords entity ... is, and

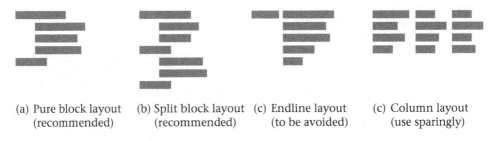

(a) Pure block layout   (b) Split block layout   (c) Endline layout   (c) Column layout
   (recommended)   (recommended)   (to be avoided)   (use sparingly)

Figure 18.3
Basic layout types.

the last line would contain the corresponding end keyword. In the case of an *if* statement, the first line would contain the if ... then keywords with the condition in the middle, and the last line would contain the end if keywords.

The *split block* layout shown in figure 18.3b is an adaptation of the pure block layout for constructs that have more than one inner region, such as an architecture body or *if-then-else* statement.

The *endline layout* shown in figure 18.3c is a general template for several layout techniques that violate indentation rules and align elements at the middle or end of a line. An example would be a procedure call whose arguments span multiple lines and are aligned to the right of the procedure name. In the figure, the left-side rectangle would be the procedure name, and all other rectangles would be the arguments. In this case, the distance from the left margin would be determined by the length of the routine name, violating the standard indentation amount.

Finally, the *column layout* shown in figure 18.3d is used when groups of similar statements or declarations are written in a tabular format. An example could be the port list of an entity, if we aligned the port names, modes, and types into three columns of text.

As with any other layout convention, we can use the goals of figure 18.2 to evaluate the merits of the four basic layout types. The pure block layout accurately matches the structure of the code. It clearly shows the limits of a construct and the subordination between statements. It makes the scope of a declaration highly visible. Each pure block can contain other blocks, mirroring the hierarchy that exists in the code when statements are nested. This layout also makes it easier to skim through the code: we can easily skip through entire blocks and scopes without needing to read the inner statements. Compared to the styles of figure 18.3c and d, pure blocks are more amenable to changes because changing one line never affects the others. It is also easy to use and lends itself naturally to indentation: we only need to indent the inner region by the standard indentation amount. The pure block is the fundamental layout type, and it is highly recommended for all statements and declarations that have a single inner region.

The split block layout is a generalization of the pure block layout. We should choose between them based on the statement or declaration we want to write. If the construct has more than one inner region, then we should use the split block layout. This layout type has the same advantages as the pure block layout, and it is also highly recommended.

Endline layout is a different story. It can be used with sequences of statements, but it is more common in routine calls or association lists. When used with statements, it may look like this:

```vhdl
calc_preamble: process (clock, reset)
              begin
                  if reset then enabled <= '0';
                                preamble <= '0';
                  elsif rising_edge(clock) then case s is
                                               when 1|2 => enabled <= '1';
                                               when 3|4 => preamble <= '1';
                                               when others => enabled <= '0';
                                                             preamble <= '0';
                                               end case;
                  end if;
              end process;
```

When used with association lists (for instance, in a routine call), it looks like this:

```vhdl
test_result := test_alu_operation( alu_op => op_mul,
                                   op_a => x"0000_0001",
                                   op_b => x"0000_0000",
                                   expected_result => x"0000_0000" );
```

Let us evaluate this kind of layout against the objective goals of figure 18.2. In the first example, the two inner regions of the *if-elsif* statement are not at the same distance from the left margin, so they do not accurately represent the code hierarchy. As for readability, the code is much wider than it needs to be, so it forces us to read sideways as well as from top to bottom. As for affording changes, if we rename the function test_alu_operation or the variable test_result in the second example, then we are forced to realign all the other lines as well. Moreover, the indentation level is not consistent across different statements in the code: instead of using a standard amount of indentation, the distance from the left margin is determined by the names of functions and objects used in the code, causing each group of statements to have its own indentation level. It is definitely not easy to use: it takes a significant amount of effort to format the code as shown in the examples. It is also not economic: in the two previous examples, the area nearly doubles when compared with the same code formatted using the pure block style. Endline layout scores negative points on all items of the scale. Overall, this style is best avoided.

If endline layout is not recommended, then what are the alternatives? In the first example (process `calc_preamble`), the alternative is to use pure blocks, split blocks, and standard indentation techniques (described in section 18.3). In the function call example, the alternative is to emulate a pure block by breaking the line after the opening parenthesis:

```
test_result := test_alu_operation(
    alu_op => op_mul,
    op_a => x"0000_0001",
    op_b => x"0000_0000",
    expected_result => x"0000_0000"
);
```

In this way, the code is just as readable, and we get all the benefits from the pure block layout.

The last kind of layout, shown in figure 18.3d, is column layout. Column layout usually appears where similar structures are repeated across several lines. Two common examples are the port list of an entity and association lists such as port maps. It can also appear in a sequence of similar declarations or statements, as show in the following object declarations and signal assignments:

```
architecture example of column_layout is
    signal    pc:                        std_ulogic_vector(31 downto 0) := x"0000_0000";
    signal    next_pc:                   std_ulogic_vector(31 downto 0) := x"0000_0000";
    constant INTERRUPT_VECTOR_ADDRESS: std_ulogic_vector(31 downto 0) := x"FFFF_FF80";
    signal    read:                      std_ulogic                     :=            '0';
    ...
begin
    read            <= mem_read or io_read;
    pc              <= next_pc;
    interrupt_address <= INTERRUPT_VECTOR_ADDRESS or interrupt;
    ...
end;
```

First of all, we have to admit that this layout is aesthetically pleasing. The shapes are geometric and beautifully aligned. This layout was created by someone who really cares about the visual presentation of source code.

Now let us evaluate this layout according to the goals of figure 18.2. As for readability, it has advantages and disadvantages. In the vertical dimension, it is easy to scan the names, types, and initial values in the declarations. However, it is hard to match the values with their names horizontally because they are too far apart. The same is true for the signal assignment statements: the assignment source and target, which are conceptually closely related, are separated by a distance determined by the length of the longest name. As for changeability, this layout has the same problems as endline layout: if we rename the longest declaration, then we need to fix all the others. Keep in mind that the same name can be used in several different files, so we may have to hunt for occurrences in the entire project to fix their layout.

So where is column layout really useful? Its main advantage is that the column edges provide "visual handrails" that make it easy to scan vertically any of the columns.[6] Therefore, column layout is recommended only where it is important to scan the information by any of the columns, the horizontal lines are conceptually related (as in the elements of an array), and the lines are unlikely to change (consider object names and expressions in assignments as likely to change). If the structure is table-like but there is no real need to scan it by any column other than the first, then column layout is not necessary. The typical use case for column layout is to provide values for 2D or nested arrays. Between unrelated statements, it is not recommended.

18.3 Indentation

In the previous sections, we have discussed how a good layout can highlight the structure of the code. In this section, we present the chief technique to make the visual presentation of source code reflect the structure seen by the compiler. In programming, *indentation* is the addition of leading blank space before each line of code to convey information about scopes, nesting, and subordination between statements.

Why Use Indentation?   Indentation is a powerful technique; it allows us to infer a lot of information from the code without reading its statements. For example, we can get a feel for the code complexity by looking at the indentation outline: the more ragged it is, the more complex the code is likely to be. We can also estimate the level of nesting used in the code: the farther apart from the left margin, the more deeply nested the code is.

Indentation is also helpful when skimming through the code. Because it makes it easy to see the beginning and end of blocks of code, we can hop over an entire section when it is not related to our task at hand. It also helps us see when a given line of code is subordinate to a statement or declaration. For example, when a line is too long, we can break it at a sensible place and indent the continuation line to indicate that it is subordinate to the line above it.

Good indentation leads to better code comprehension and a lower incidence of bugs. To illustrate some of the benefits provided by indentation, let us compare two

```
alu: process (all) begin          alu: process (all) begin
carry <= (others =>'0');             carry <= (others =>'0');
case operation is                    case operation is
when ALU_ADD =>                         when ALU_ADD =>
result <= a xor b xor carry;               result <= a xor b xor carry;
for i in result_port'range loop            for i in result_port'range loop
carry(i+1) <=                                 carry(i+1) <=
(a(i) and b(i)) or                               (a(i) and b(i)) or
(a(i) and carry(i)) or                           (a(i) and carry(i)) or
(b(i) and carry(i));                             (b(i) and carry(i));
end loop;                                   end loop;
when ALU_AND =>                         when ALU_AND =>
result <= a and b;                         result <= a and b;
when ALU_OR =>                          when ALU_OR =>
result <= a or b;                          result <= a or b;
when others =>                          when others =>
result <= (others => '0');                 result <= (others => '0');
end case;                             end case;
end process;                       end process;
```

(a) Without indentation.   (b) Using indentation.

Figure 18.4
Two versions of the same process (a simplified ALU).

versions of the same code, first with no indentation (figure 18.4a) and then using a proper indentation scheme (figure 18.4b).

In the nonindented version, even simple questions require careful examination of the code. How many case alternatives are there? Where does the loop start and where does it end? In the indented version, it is easy to see that the code contains a single process, and the bulk of it is a case statement. It is easy to identify each alternative in the case statement, and it is easy to skip over the irrelevant parts until we find the specific statements that we want to debug.

Because indentation is such a useful technique, in practice all programs use some sort of indentation. However, programmers are rarely taught how to indent their code properly. Most learn by example, which may cause them to pick an inconsistent style because the samples differ widely from one another. The next topics introduce the basic indentation principles and provide examples of consistent indentation.

Basic Indentation Principles   In the source code, certain statements and declarations have inner regions that can contain other statements or declarations. Such constructs are called *compound* statements or declarations. Many compound structures can be nested, creating a hierarchy in the source file. At the top of the hierarchy is the file level, where we find, for instance, entities, architectures, and packages. Immediately inside an entity declaration, there are optional generic and port clauses, which may

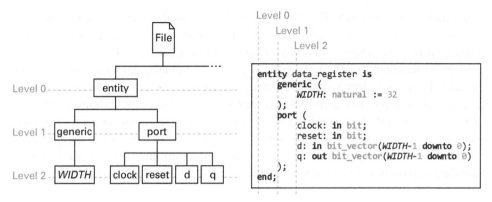

(a) Source file hierarchy.   (b) Indented source code.

Figure 18.5
The basic indentation principle: the amount of leading whitespace in each line is proportional to
its level in the source file hierarchy.

also contain declarations. If we look at the source file hierarchy (figure 18.5a), starting
from the bottom, we can see that each port or generic (at level 2) is subordinate to the
respective *port* or *generic* clause (level 1), which is in turn subordinate to the entity declaration (level 0). The basic principle behind indentation is to add leading whitespace
to each line of code in proportion to its level in the source file hierarchy. Figure 18.5b
demonstrates how this applies to the hierarchy of figure 18.5a.

An implication of this principle is that the lines should always be indented by multiples of a standard amount (e.g., two or four spaces). When we allow arbitrary indentation levels, this undermines the strength of the indentation scheme.

An Example The basic indentation algorithm is best explained by example. Figure
18.6 shows a file containing an architecture body. We start from line 1 at indent level
0 (the level corresponding to the file scope). For each *simple* statement, declaration,
or clause, we maintain the current indent level. When a *compound* construct is found,
we write the lines that mark the beginning and end of the construct (in this case,
architecture and end architecture) at the current indent level and then increase
the indent level between those lines by one. In the case of an architecture body, the
keyword begin (line 7) divides the declaration in two regions of equal importance so
it gets indented at the same level as the limiting keywords (it uses a split-block layout).
Inside the declarative part (lines 5–6) and the statement part (starting at line 8), the
current indent level is increased to 1. We proceed at this level until the next compound
statement: the process statement in line 11. Again, we write the limiting keywords at
the current level (lines 11 and 30) and increase the indent level for the inner region.

Note that a comment (line 12) should always be indented at the current indent level, not at the left margin of the file. Also note that in this example one indentation level corresponds to the width of four space characters in the source text.

For the remaining code, the algorithm proceeds similarly, but two points warrant further clarification. Inside the `case` statement, the inner region is indented one level, but after each alternative (keyword when) the following line is also indented one level. This is done to indicate the subordination between the statements and the case alternative; however, this extra indentation level is optional and not used in many coding standards. The other point to consider is what to do with lines that do not fit within the right margin (set at 50 characters in this example). In this case, we break the line at

Indent Level Line #

```
0    1   library ieee;
0    2   use ieee.numeric_std.all;
     3
0    4   architecture rtl of alu is
1    5       signal a, b: unsigned(DATA_WIDTH downto 0);
1    6       signal carry: unsigned(DATA_WIDTH downto 0);
0    7   begin
1    8       a <= '0' & unsigned(operand_a);
1    9       b <= '0' & unsigned(operand_b);
     10
1    11      alu: process (all) begin
2    12          -- Set default value for carry signal
2    13          carry <= (others => '0');
2    14          case operation is
3    15              when ALU_ADD =>
4    16                  result <= a xor b xor carry;
4    17                  for i in result_port'range loop
5    18                      carry(i+1) <=
6    19                          (a(i) and b(i)) or
6    20                          (a(i) and carry(i)) or
6    21                          (b(i) and carry(i));
4    22                  end loop;
3    23              when ALU_AND =>
4    24                  result <= a and b;
3    25              when ALU_OR =>
4    26                  result <= a or b;
3    27              when others =>
4    28                  result <= (others => '0');
2    29          end case;
1    30      end process;
     31
1    32      carry_out <= carry(carry'left);
1    33      ...
0    34   end architecture;
```

Figure 18.6

Example of the basic indentation algorithm showing the indent levels.

a logical place and add one indent level to all the continuation lines (lines 19–21). The rest of the code is indented following the same rules. As a final remark, note that we must decrement the indent level whenever the closing keyword of a compound statement is found (lines 22, 29, 30, and 34).

Typical Indentation of VHDL Statements   Based on the simple layout algorithm introduced in the previous section, figure 18.7 shows typical indentation templates for some VHDL statements and declarations. Naturally, other styles are possible, but before you adopt them, check that they do not violate the basic indentation principles introduced in the previous topic.

The *if* statement is noncontroversial—virtually every programmer uses the layout scheme shown in (a). The *case* statement appears in two flavors: a more compact one (c), and one that makes clear the subordination between the *case* and *when* clauses (b). Each style has its supporters, and either one works in practice. No matter which one you pick, the important thing is to choose a style and apply it consistently.

The *process* statement is also shown in two versions: one with a declarative part (d), and one with only a statement part (e). In the version without a declarative part (e), the keyword begin was put in the same line as the process keyword, using a pure block layout. Alternatively, the keyword begin could also appear on a line of its own, indented at the same level as the process keyword; in this case, the statement would be using a split block layout similar to figure 18.3b but with an empty upper half. Both styles work in practice. The conditional (f) and selected (g) signal assignments are shown in the figure with a line break after the assignment symbol, so that all the alternatives align vertically under the signal name or expression. Another alternative is to write as much of the statement as possible in the first line and then break the line as explained in section 18.5.1.

Figure 18.8 shows typical indentation templates for entities, architectures, and subprograms. Figures (a) and (b) show an entity declaration and an architecture body. Figures (c) and (d) show two alternative ways to indent a subprogram. The first one (c) groups all the parameters in a single line, which works well if the subprogram has few parameters. Figures (d) and (e) show how the parameters can be arranged one per line, which is a good recommendation when the subprogram has more than a few parameters.

Indentation Amount and Character   An indentation convention involves at least two choices: which character to use as whitespace (space or tab), and the size of an indentation step (number of equivalent fixed-width characters). As in any other choice involving multiple advantages, disadvantages, and computer scientists, this tends to spark heated debates. As in most layout choices, picking a reasonable style and using it

```
if condition_1 then
    statement_1;
    statement_2;
    ...
elsif condition_2 then
    statement_3;
    statement_4;
    ...
else
    ...
end if;
```

(a) if-then-elsif-else

```
case expression is
    when choice_1 =>
        statement_1;
        statement_2;
        ...
    when choice_2 =>
        statement_3;
        statement_4;
        ...
    when others =>
        ...
end case;
```

(b) case-when
(2-level version)

```
case expression is
    when choice_1 =>
        statement_1;
        statement_2;
        ...
    when choice_2 =>
        statement_3;
        statement_4;
        ...
    when others =>
        ...
end case;
```

(c) case-when
(compact version)

```
process (sensitivity_list)
    declaration_1;
    declaration_2;
begin
    statement_1;
    statement_2;
end process;
```

(d) Process with declarative part

```
process (sensitivity_list) begin
    statement_1;
    statement_2;
end process;
```

(e) Process without declarative part

```
signal_name <=
    value_1 when condition_1,
    value_2 when condition_2,
    value_3 when others;
```

(f) Conditional signal assignment

```
with expression select signal_name <=
    value_1 when choice_1,
    value_2 when choice_2,
    value_3 when others;
```

(g) Selected signal assignment

Figure 18.7
Indentation templates for VHDL statements.

consistently is more important than the actual choice. In any case, keep in mind that any good source code editor makes the differences negligible, and it is possible to convert between files written in different styles.

As for the choice of whitespace character (spaces or tabs), any source code editor can be configured to do one of two things when you hit the tab key. One is to insert a special tab character (ASCII code 0x09) in the source text, which is interpreted as "advance to the next fixed-width column." The other is to insert as many space characters as needed (ASCII code 0x20) to advance to the next fixed-width column. In both cases, the standard column width is configurable in the editor.

Indenting using tabs offers several benefits. The indentation width can be set according to the preference of the reader (independently of what the original author used). Because each tab character occupies several columns of text, they are faster to navigate through using the cursor keys; they are also easier to delete. Indentation tends to be more regular because only multiples of the column width are allowed. However, the author must be careful and use tabs only to indent the beginning of a line (up to the current indent level) and then use spaces after that. Otherwise the code will look broken when read in an editor configured with a different tab width.

```
entity ent_name is                          architecture arch_name of ent_name is
    generic (                                   declaration_1;
        generic_declaration_1;                  declaration_2;
        generic_declaration_2               begin
    );                                          statement_1;
    port (                                      statement_2;
        port_declaration_1;                 end;
        port_declaration_2
    );
end;
```

(a) Entity declaration.   (b) Architecture body.

```
function function_name(arg_1: type_1; arg_2: type_2) return type_3 is
    declaration_1;
    declaration_2;
begin
    statement_1;
    statement_2;
end;
```

(c) Subprogram with parameter list in a single line.

```
function function_name(                     procedure procedure_name(
    arg_1: type_1;                              class_1 arg_1: mode_1 type_1;
    arg_2: type_2                               class_2 arg_2: mode_2 type_2
) return type is                            ) is
    declaration_1;                              declaration_1;
    declaration_2;                              declaration_2;
begin                                       begin
    statement_1;                                statement_1;
    statement_2;                                statement_2;
end;                                        end;
```

(d) Subprogram with one parameter   (e) Subprogram with one parameter
    per line (function).   per line (procedure).

Figure 18.8
Indentation templates for VHDL entities, architectures, and subprograms.

Indenting using spaces offers several benefits. The author can guarantee that the code will look the same in every editor or tool. This is an advantage when you need to open the same source file in several different tools, and you want the code to look the same in all of them, but you do not want to spend time configuring each tool. Spaces are simple to use—there is nothing to learn, and there is no way to misuse them. Spaces play nice with endline comments and with any ASCII art you draw in the code.

In practice, authors who find more value in the *configurability* of the presentation use tabs. Authors who value more the *stability* of the presentation use spaces. This decision is also subject to existing coding standards. For instance, the European Space Agency (ESA) VHDL guidelines[7] mandate the use of spaces, explicitly forbidding the use of the tab character.

The other choice is the indentation size: the width of each tab or the equivalent number of spaces. This choice has a component of taste, and different languages have different recommendations. Ruby code uses two, Python code uses four, and the Linux kernel conventions for C use eight.

Most users tend to gravitate between two and four, and either value will work well in practice. You may notice that with two spaces, the indentation outline is not so apparent. However, with four spaces, we run out of horizontal space earlier as the indent level gets deeper. If you find this a compelling reason to use a narrower indent width, remember that excessive indentation is a sign of deep nesting and overly complex code. In fact, some authors defend that one of the advantages of a wider indentation width is to prevent the programmer from using too many levels of nesting.

Between two or four spaces, which are the most used values, there are not strong arguments for choosing one over the other. If you are not inclined either way, then the clearer outline and the pressure against deep indentation may be a tiebreaker in favor of four spaces. The examples in this book were all written using spaces as the indentation character and with an indentation width of four.

18.4   Paragraphs of Code

In English or in code, any complex subject must be divided into chunks so that we can absorb it one piece at a time. Because code is normally read from top to bottom, we need to give the reader vertical cues when a new step is reached, a subtask is completed, or the main subject changes. This is not different from the use of paragraphs in written prose. A good technique to tell the reader that a series of statements belong together is to group them into paragraphs, with the beginning of each paragraph indicated by a blank line and possibly a summary comment.

The following example shows a monolithic block of code without any clear divisions:

```
text_ram_x := pixel_x / FONT_WIDTH;
test_ram_y := pixel_y / FONT_HEIGHT;
display_char := text_ram(text_ram_x, text_ram_y);
ascii_code := character'pos(display_char);
char_bitmap := FONT_ROM(ascii_code);
x_offset := pixel_x mod FONT_WIDTH;
y_offset := pixel_y mod FONT_HEIGHT;
pixel := char_bitmap(x_offset)(y_offset);
```

In the way it is written, it may give the impression that the block needs to be read and understood all at once. However, it is actually composed of three subtasks, clearly divided by the outputs they produce: every few lines of code produce a value that is used by the next subtask. We can make the code easier to digest by grouping together the statements that are closely related and separating the subtasks with a blank line. For extra readability, we can add a summary comment before each subtask:

```
-- Find character in text RAM corresponding to x, y coordinates
text_ram_x := pixel_x / FONT_WIDTH;
test_ram_y := pixel_y / FONT_HEIGHT;
display_char := text_ram(text_ram_x, text_ram_y);

-- Get character bitmap from font ROM
ascii_code := character'pos(display_char);
char_bitmap := FONT_ROM(ascii_code);

-- Get pixel value from character bitmap
x_offset := pixel_x mod FONT_WIDTH;
y_offset := pixel_y mod FONT_HEIGHT;
pixel := char_bitmap(x_offset)(y_offset);
```

18.5 Line Length

VHDL does not specify a maximum length for lines in a source file. However, most programmers find it useful to set a personal limit and break any line that would extend beyond that. Because the compiler does not make a distinction between tabs, spaces, and line breaks, we can break a line at any place where a space is allowed.

How long should we allow a line to be? Examples of values recommended in various coding standards include 72, 80, 100, and 120 characters per line, with a general preference toward the shorter values. If your IDE has a default value for this setting, then it is likely configured to 80 columns. The reasons behind this number are historical; they can be traced to old text-only monitors that had a fixed width of 80 columns and to 80-column punch cards used by the programmers of old. However, although today's hardware can be pushed well beyond this number, there are real advantages to choosing a limit that is much less than your monitor can afford:

- It allows us to work on two or more source files side by side. This is useful to compare different files or different versions of the same file, a common task in software projects. An 80-column limit gives you enough room to have two files open side by side, plus any annotations and user interface panels from your IDE.
- Navigating and skimming through the source code is much easier when there is no need to scroll horizontally or read the code from left to right as well as from top to bottom. Although word wrapping could be used, it has its own disadvantages: it messes with indentation, it makes it hard to select rectangular regions of text, and it makes it difficult to match line numbers when two files are open side by side.
- We can use the central region of the screen for code and the edges for other useful information, such as a tree view for the project files on the left and a design hierarchy or source code outline on the right.

- Long lines are often a sign of complicated logic, deep nesting, excessive indentation, or routine calls with too many parameters. Limiting the line length also limits the amount of complexity we can put in any single line.

As a rule of thumb, the 80-column limit is still widely recommended and in use; 100 or 120 columns are also acceptable, but you may run out of space when you need to show two windows side by side in an IDE.

Most source code editors offer a feature called "print margin" or "vertical ruler," which shows a vertical line on the screen at a given column. This limit is not strictly enforced, but it works as a reminder to keep our lines below the configured width. In any case, remember that the ultimate goal is readability. If running a little over the limit makes a line more readable than wrapping it, then it is acceptable to do so.

18.5.1 Wrapping Long Lines

When a line is too long to fit into the limit set by our standards, the common practice is to break the line at a proper place and indent its continuation on the next line. This should be done consistently and in a way that supports the layout goals of figure 18.2 and does not violate the indentation rules of section 18.3. Here we will see a set of recommendations that support all the layout goals, keep the code easily modifiable, and are consistent, simple, and easy to use. This list is based on the work of Steve McConnell in *Code Complete* 2nd Ed., one of the most successful and influential books on software development. It contains hundreds of pages filled with the nitty-gritty details about the practice of writing software.

An important point to note is that we may be highly biased by the code we are used to seeing and working with. When we see code that is formatted differently, it may look strange only because we are not used to its style. It is important to train ourselves to keep an open mind and give other promising practices a try before ruling them out.

Break the Line at a Point That Clearly Shows It Is Incomplete When we need to break an expression or statement, it is a good idea to clarify that it continues on the next line. Although the lack of a semicolon is enough for the compiler, we can provide more visual clues by breaking the line at a point that is obviously incorrect:

```
hash_key := ip_addr_1 & ip_addr_2 &   -- Concatenation symbol & makes incompleteness obvious
    port_number_1 & port_number_2;

sprites_intersect := not (             -- Opening parenthesis makes incompleteness obvious
    (sprite_1.y + HEIGHT < sprite_2.y) or    -- or operator makes incompleteness obvious
    (sprite_1.y > sprite_2.y + HEIGHT) or    -- or operator makes incompleteness obvious
    (sprite_1.x > sprite_2.x + WIDTH) or     -- or operator makes incompleteness obvious
    (sprite_1.x + WIDTH < sprite_2.x)
);
```

Keep Related Pieces Together   We would like the reader to finish a complete thought before moving to the next line. In the following example, the comma at the end of the line makes it clear that it has a continuation; however, it breaks the line in the middle of a function call:

```
bounding_box := rectangle'(minimum(rect_1.top, rect_2.top), minimum(rect_1.left,
    rect_2.left), maximum(rect_1.bottom, rect_2. bottom), maximum(rect_1.right,
    rect_2. right)
);
```

By breaking each line only after a complete function call, we make the arguments easily identifiable:

```
bounding_box := rectangle'(
    minimum(rectangle_1.top, rectangle_2.top),
    minimum(rectangle_1.left, rectangle_2.left),
    maximum(rectangle_1.bottom, rectangle_2. bottom),
    maximum(rectangle_1.right, rectangle_2. right)
);
```

This second version is two lines longer than the original example, but the improved clarity more than compensates for the extra lines. Remember there is no point in making the code shorter if it is also harder to understand.

Consider Breaking down a List to One Item per Line   For lists that appear in statements, such as the argument list in a routine call or a port map in a component instantiation, consider breaking the list so that each item has an entire line for itself. This is especially important if the items are complex expressions. It also makes it easier to use named association or occasional endline comments. Instead of:

```
paddle <= update_sprite(paddle, paddle_position.x + paddle_increment.x,
    paddle_position.y + paddle_increment.y, vga_raster_x, vga_raster_y, true);
```

Consider writing:

```
paddle <= update_sprite(
    sprite => paddle,
    sprite_x => paddle_position.x + paddle_increment.x,
    sprite_y => paddle_position.y + paddle_increment.y,
    raster_x => vga_raster_x,
    raster_y => vga_raster_y,
    sprite_enabled => true -- paddle must always be visible
);
```

Using one item per line, it is easier to discern the number of arguments and the position of each one in the list.

Make the End of a Multiline Statement or Declaration Obvious   In the previous examples, the end delimiter of each list (the closing parenthesis and semicolon) was placed on its own line and vertically aligned with the corresponding statement. This makes it clear that the inner elements are enclosed between the delimiters. Another benefit of this approach is that it facilitates shifting the items up and down, adding new items, or removing them from the list.

Indent the Next Line by the Standard Amount   The last guideline is to always use the standard indentation amount for continuation lines. Consider the following indentation style, which is a form of endline layout (figure 18.3c):

```
paddle <= update_sprite(paddle, paddle_position.x,
                        paddle_position.y, vga_raster_x,
                        vga_raster_y, true);
```

This is a popular style, especially among beginners. The technique shown in the example consists in indenting continuation lines at the same level of the first argument in a function call. This is an example of a layout style in which aesthetics conflicts with more objective layout goals—it may be aesthetically appealing, but it is a maintenance headache. Consider what happens if we rename the signal called `paddle` to `player_sprite`: we break the indentation of all subsequent lines, which will have to be manually realigned. The same happens if we rename the function `update_sprite`. Remember that the same name could be used in several different files, so we may have to search for and fix all the occurrences. This may look like a minor issue, but renaming objects and routines is a highly common development task, and it contributes a lot to improving code readability. A layout scheme should not stand between us and any practice that improves the quality of our code.

Another problem with this layout is that it undermines the strength of a consistent indentation scheme. Indentation works best when the columns and steps are clearly identifiable. If instead of using a standard amount we let the names of functions and objects determine the distance from the left margin, we weaken the visual handrails that evidence the code structure. Finally, this layout wastes a lot of space (all the area underneath the signal name and the function name, in the example) and leaves little room for actual code in each continuation line.

If you feel tempted to use this style or any form of endline layout, then try using the technique of emulating pure blocks described in section 18.2.2.

18.6 Spaces and Punctuation

Good horizontal spacing helps make individual statements more readable. We have seen how blank lines can form groups of related statements by separating distinct steps in the code. Spaces can do the same in the horizontal dimension: we can use them to keep together things that are closely related and separate things that are disconnected. Achieving this effect requires a judicious use of whitespace: we will not improve readability if we add gratuitous whitespace around every word, symbol, and operator in the code.

Punctuation marks are important in VHDL because the language uses them as delimiters. Because whitespace around delimiters is optional and can be used in any amount, different styles exist, with varying degrees of effectiveness. To evaluate the possible ways of laying out punctuation symbols, we can use the objective layout goals from section 18.2.1, such as consistency and readability. As with any other style issue, we should aim at a rule that works well and has as few exceptions as possible. It should give the right visual cues about the code and help our readers to find the relevant information effortlessly.

The Ada Quality and Style Guide (AQ&S), the recommended style guide for all DoD programs,[8] has good advice on this. The recommendation is simple: use the same spacing around punctuation as you would in regular prose. The rationale behind this rule is that many of the delimiters (commas, semicolons, parentheses, etc.) are familiar as normal punctuation marks. Therefore, it is distracting to see them in code used differently from normal text. This rationale leads to the following rule: never add a space before a comma, colon, or semicolon, and always add a space after any of these symbols.

As for parentheses, the same rationale applies. There are, however, two justifiable exceptions. First, we should never add a space between the name of an array and the parentheses used for accessing its elements. This is justified because the two form a single construct and are undisputedly closely related. Second, and for the same reason, we should never add a space between a routine name and the parentheses around its parameter list. These two exceptions are both justifiable and traditional in virtually any programming language.

These rules are summarized in table 18.1. They are simple to follow, easy to remember, economic, and consistent. Note, however, that the use of parentheses in expressions follows different principles, described in the next topic.

Here is an example of a declaration that does not follow the rules of table 18.1. In the following function declaration, the extraneous whitespace makes it look like there is more to the declaration than there really is. It calls too much attention to accessory symbols that do not have any relevant meaning and exist only to satisfy the language syntax:

```
function add ( addend : signed ; augend : signed ) return signed ;
```

The following example uses only the necessary amount of whitespace to provide visual separation between the important elements. The punctuation looks more natural because it is used as in regular prose:

```
function add(addend: signed; augend: signed) return signed;
```

Some coding conventions include a large number of rules and exceptions, such as "a colon must be preceded by a space except where it appears after a label" or "a semicolon should be preceded by a space except at the end of a statement or declaration." It is debatable whether such rules actually improve or worsen readability. In the interest of keeping the rules simple, easy to remember, and easy to apply, it is recommended that no such exceptions be made.

Finally, an important point to note about the space before an opening parenthesis is that language keywords are not function names. Therefore, the rule to keep them together does not apply. Instead, it is recommended to add spaces before and after every keyword, as shown in the following examples:

```
if (cond)      process (cond)     port map (...)     -- OK, space used after keyword
if(cond)       process(cond)      port map(...)      -- NOT OK, missing space after keyword

rising_edge(clk)      to_slv(bv)     -- OK, no space after function name
rising_edge (clk)     to_slv (bv)    -- NOT OK, extraneous space after function name
```

Parentheses and Spaces in Expressions In expressions, we should use more parentheses than strictly necessary. Suppose you come across the following expression:

```
a + b mod c sll d;
```

Table 18.1 Recommended spacing and punctuation rules for statements and declarations

Punctuation symbols (comma, colon, semicolon)	General rule	Use the same space around punctuation as you would in regular prose. Never add a space before a symbol. Always add a space after a symbol.
	Exceptions	None.
Parentheses	General rule	Add a space before an opening parenthesis. Add a space or a punctuation symbol after a closing parenthesis.
	Exceptions	Array indices and routine parameters; expressions.

Can you say with confidence which operators are evaluated first? Now suppose you find this other version of the same expression:

```
(a + (b mod c)) sll d;
```

Now it is clear that the mod operator is applied first, followed by the addition and shift operators. To understand the first version, the reader must know the rules that govern the precedence of operators. By adding extra parenthesis, we make the expression accessible to any reader.

As for the use of spaces in expressions, it is difficult to provide hard-and-fast rules. Consider the following expression:

```
(-b+sqrt(b**2-4*a*c))/2*a;
```

The terms and factors look a bit too crowded. The expression can be improved somewhat by adding spaces between them to show the logical chunks more clearly:

```
(-b + sqrt(b**2 - 4*a*c)) / 2*a;
```

However, when we blindly add spaces around all operators and operands, the situation is no better than where we began:

```
( - b + sqrt( b ** 2 - 4 * a * c ) ) / 2 * a;
```

Better than trying to come up with an arbitrary rule, use your judgment to decide when and where to add spaces in an expression. Adding spaces around operators usually helps, but if you want to remove them to accentuate the precedence of operators, it is reasonable to do so. Adding spaces in the inner side of parentheses is usually unnecessary, but if you want to make the code inside the parentheses stand out, it is reasonable to do so.

18.7 Letter Case

When we choose a naming convention for a project, it is a good idea to establish some general formatting rules for the names of objects, routines, and other named entities. Besides giving the code a more consistent presentation, we can use the visual appearance of a name to convey additional information about the entity it represents.

In a project, many names will be like simple nouns, composed of a single word. Other names will be like compound nouns, made of multiple words. If we just concatenated the names end to end, as in instructionvector or numshipsinview, then our names would be hard to read. There are two general solutions to this problem. One is to embed an underscore between consecutive words, as in `instruction_vector` or `num_ships_in_view`. The other is to capitalize the beginning of each word, as in `InstructionVector` or `numShipsInView`. These techniques correspond to two widely used capitalization conventions:

- *CamelCase* is the practice of composing names by joining consecutive words without a separating character, marking the beginning of each word with an uppercase letter. The term CamelCase comes from the bumps that form in the middle of a name and resemble the humps of a camel. There are two variations: `UpperCamelCase` (in which the first letter is capitalized) and `lowerCamelCase` (in which the first letter is in lowercase).
- *snake_case*, in contrast, uses only lowercase letters and separates words with underscores. It is also called `lowercase_with_underscores`. A common variation is `SCREAMING_SNAKE_CASE`, or `ALL_CAPS`, in which all letters are capitalized.

When we define a coding standard, we can associate different case conventions with different kinds of named entities in the code. For example, we could determine that all constants and generics must be written in `ALL_CAPS` and all other names in `snake_case`. In this way, when we see this statement in the code:

```
if count_value < COUNT_MAX then ...
```

we will know immediately that the value on the left is an object that can change, and the one on the right is a constant. This helps us concentrate our debugging efforts on the parts of an expression that are allowed to change.

Before we choose to use `ALL_CAPS` for certain names in our code, we should be aware of some inconveniences associated with this style: it is attention-grabbing, looks a little rough on the eyes, and is less convenient to type. Therefore, we should avoid using all caps for names that appear frequently in the code, such as VHDL keywords. A better approach is to use a source code editor with syntax highlighting and display the language keywords in boldface or with a different color.

An important thing to note is that while it is recommended to use different conventions for different kinds of names in a project, it is best to avoid mixing two different conventions within a single name, as in `convert_ToSynthComplexType` or `Names_Like_This`. This undermines the coding standard, making it necessary for us to remember which part of a name uses which convention. It also makes the rules more complex to apply and memorize. Finally, it makes the code look less uniform. For each kind of name, it is better to choose one of the widely adopted conventions and stick with it.

A particular case worth mentioning is the use of acronyms in names. When a name uses an acronym, a common and simple convention is to treat it as any regular word.

This prevents any inconsistencies and avoids clashes with conventions that require the first letter to be in lowercase. What should we do if our standard requires us to write variable names in lowercase but the first part of a name is an acronym? The best way to solve this is to drop the problem altogether by treating all acronyms as regular words. Thus, instead of `UDPHdrFromIPPacket`, we would write `UdpHdrFromIpPacket` (or `udp_hdr_from_ip_packet`, in snake case). Microsoft .NET naming conventions[9] recommend treating all acronyms longer than two letters as regular words. For the sake of simplicity and to avoid arbitrary exceptions, it is probably better to use the same rule for all acronyms, regardless of their length.

Even if we do not use case conventions to distinguish among different kinds of names, it is still worthwhile to choose one convention and use it consistently. Between `CamelCase` and `snake_case`, there is little reason to choose one over the other. Either will work in a project if used consistently. `CamelCase` names will be a little shorter on average because they do not use word separators. However, the standard VHDL packages use `snake_case` and `ALL_CAPS` extensively (although not consistently), so this may be a good choice if you want to maintain a uniform look between your code and the standard libraries. This book uses `snake_case` for all names except constants and generics, which are written in `ALL_CAPS`.

18.8   Other Layout and Style Recommendations

On a Team Project, Have Everyone Agree on a Style and Use It from the Beginning
Before starting a project, a team should spend some time deciding which set of conventions to use and then register it in a project document maintained together with the source code. Because everyone has their own favorite rules, if you start a project without an explicit coding standard, then coding inconsistencies will build up fast. Later, trying to convince someone to change styles might start a minor war.

If you want to save some time and avoid fruitless discussions, then you are invited to use the *VHDL Coding Style Guide* provided in the companion website for this book.

Learn to Adapt to Different Coding Conventions   Everyone has his or her own favorite coding conventions, but when we are working on a team, we must play by the team rules. For this reason, the ability to adapt to and work with different coding standards is a valuable skill for a programmer. You will not be able to change every line of code you see to suit your preferences, so the best approach is to learn how to be productive in different environments.

Choose a Low-Maintenance Style   One of the main goals of this chapter has been to sell the idea that the visual presentation of source code is important. However, we must keep in mind that the time hitting the space bar or the tab and delete keys is time during which we could be working on the next feature or functionality. We need to strike a balance between a nice-looking layout and the amount of typing and maintenance effort. Automatic formatting tools are a nice promise, but they tend to lose all the fine details that we would like to keep in the formatting. So far the best approach is to choose a formatting style that is visually appealing, easy to use, and amenable to changes.

Part VI   Synthesis and Testbenches

19 Synthesis

This chapter presents the key characteristics of VHDL code that is intended for synthesis and the main differences from code that is intended for simulation only. It also provides a brief overview of register-transfer level (RTL) design and the synchronous design methodology. Together, RTL and synchronous design form a powerful toolset that helps developers of any skill level to create systems that are simpler, more reliable, and easier to design.

19.1 Introduction

HDL synthesis is the process of translating a description in a hardware design language into a circuit that performs the described function. The main input to the synthesis process is HDL code, but the tool also uses device-specific information, design constraints, and synthesis libraries for the target technology. The output of the synthesis process is a structural description composed of basic hardware elements such as registers, gates, and other structures available in the target device or library.

There is a fundamental difference in how the VHDL code is used by the simulator and the synthesis tool. In simulation, after the design has been reduced to a series of processes communicating via signals, the code can be executed directly by the simulation kernel. The transformations performed during elaboration are relatively straightforward and follow a precise set of rules specified in the language manual.

A synthesis tool, in contrast, must make an *interpretation* of the VHDL code. Certain statements and constructs map more directly to hardware structures. For example, a concurrent signal assignment whose value is given by a boolean expression maps directly to a block of combinational logic. However, other hardware structures must be described with a large number of statements, which may be intertwined with other constructs pertaining to unrelated parts of the circuit. The synthesis tool must examine the source code and infer hardware elements from the constructs used in the code. This process is less straightforward and involves more decisions than the elaboration process used in simulations.

This need for interpreting the meaning of a description leads to a high variability in synthesis results: different tools may accept or reject different constructs, and some will infer different circuits from the same source code. Traditionally, each tool required its users to write VHDL code according to a set of code *templates* to ensure the correct recognition of hardware structures. These templates were defined and documented by the tool vendor and had to be learned by the designers. A developer needed to adapt to what was accepted by each tool or find a subset accepted by most vendors.

To minimize the discrepancy between synthesis tools, a new standard was created. The *IEEE Standard 1076.6 for VHDL RTL Synthesis* aims to provide a precise syntax and semantics for all compliant synthesis tools, much like the standard that defines the VHDL language provides precise information for simulators. This synthesis standard is relevant for both users and vendors. For a model designer, it describes which constructs are available for use in synthesizable code. For a tool designer, it specifies which constructs should be supported, not supported or ignored. The set of constructs supported for synthesis is called the *synthesizable subset* of VHDL.

Currently, there is still significant variability across tools, but the basic principles of hardware inference are relatively consistent. Because tools are allowed to support other constructs beyond those required by the standard, vendors have used this to differentiate from each other and obtain a competitive advantage. The result is that much of what was not synthesizable some years ago is now largely supported. Another practical consequence is that the definitive source about what is supported or not by a tool is still the tool manual. The observations in this chapter are based on what is supported according to the synthesis standard and what works in practice with most tools, with additional remarks where actual tool support is significantly different from the standard.

19.2 What Is the Difference?

There are two main differences between code that is intended for synthesis and code that is intended for simulation only. The first one is that synthesizable code *must imply a hardware architecture*. In other words, it must have a meaning in hardware, and this should correspond to the designer's intent. The practical consequence is that, in general, it is not possible to write a model using any construct that works in the simulator and later turn it into a synthesizable model only by changing a few statements. Code that is intended for synthesis should be written as such from the start. In other words, we should not start coding before we have a clear picture of the intended hardware.

The second difference is about the constructs available for use in the code. While simulation code can use all the language features available in VHDL, synthesizable code is subject to two kinds of restrictions. First, there are *syntactic* restrictions: certain statements, clauses, and types are available only in simulation code. Second, there are *semantic* restrictions: the statements must be used in a proper way to imply specific hardware elements. These two limitations are elaborated in the following topics.

19.2.1 Syntactic Restrictions

Syntactic restrictions define which statements or clauses are allowed in synthesizable code. Syntax differences are not the biggest distinction between synthesizable and non-synthesizable code. They are not hazardous either because syntax errors are detected immediately by the compiler. Consequently, it is not particularly helpful to distinguish the two types of code in terms of synthesizable or nonsynthesizable statements. Nevertheless, there are some syntax restrictions to be aware of.

The first syntactic restriction is about statements that control the passage of time, which are obviously nonsynthesizable because no hardware technology supports arbitrary delays. However, the timing clauses are not disallowed in synthesizable code—they are simply ignored by the synthesis tool. In other words, they will not cause the synthesis process to fail. In the following assignment, the *after* clause is ignored when the statement is synthesized:

```
q <= d after 10 ns;   -- 'after' clause is ignored during synthesis
```

The *for* clause in a *wait* statement is also ignored. *Wait* statements are allowed in synthesizable code only to imply a signal edge. In the following *wait* statement, the *for* clause is ignored:

```
wait until rising_edge(clock) for 10 ns;   -- 'for' clause ignored in synthesis
```

Other syntactic constructs simply do not make sense in a synthesizable design and are therefore ignored or not supported. An example of an ignored construct is a file *declaration*, which can exist in the code but is disregarded during synthesis. A file *operation*, in contrast, is not supported: if a model includes a statement that opens or reads a file, then it will be in error. Similarly, declarations of access types are simply ignored, but the use of access types or allocators is an error.

Certain constructs are supposed to be ignored or not supported as per the synthesis standard, but in practice they are supported by many tools. An example is the use of initial values for signals and variables, which should be ignored according to the synthesis standard but are supported in many FPGAs (see section 19.5 about the use of initial values). Another example is the *assert* statement, which should be ignored but is used by some tools to stop the synthesis process if the assertion is violated. Yet another example is the *while-loop* statement, which is listed as not supported in the standard but is accepted by some tools. This is also true for shared variables, deferred constants, and physical types.

Finally, a few constructs are mistakenly thought to be nonsynthesizable but are actually supported as per the synthesis standard. For example, aliases are fully supported. All operators are supported, although a tool is not required to implement the exponentiation operator for bases other than two. Multidimensional arrays are also supported.

19.2.2 Semantic Restrictions

Semantic restrictions are those that cannot be expressed using the language grammar alone. For instance, if a statement accepts only certain values of a type, this is a semantic restriction. The templates that must be observed to imply certain hardware structures are also a form of semantic restriction. In this topic, we touch briefly on the semantic restrictions of some specific constructs.

Many of the semantic restrictions about the use of individual statements can be explained by intuition if we think about the meaning of the statement in hardware. For example, *for* loops with dynamic ranges (ranges that may change during execution time) are not synthesizable. This makes sense if we recall that loops are implemented by replicating hardware, and it is not possible to change the number of hardware elements dynamically. (A common workaround is to wrap the loop body in an *if* statement that checks the value of the current iteration, as shown in the example of figure 10.12.)

A similar limitation is the use of recursion in subprograms. Recursion is allowed, as long as the number of recursions is given by a constant value. Likewise, array slices must be static because it does not make sense to change the size of a bus or the number of storage elements dynamically.

The *case* statement also has a number of restrictions. Choices that include 'Z', 'U', 'X', 'W', or '-' should be ignored. The synthesis tool should compile the circuit as if the choice did not exist. For the proper handling of *don't care* inputs, we should use the matching case statement (case?) as shown in the example of section 10.3.1.

Process statements are required to include in their sensitivity lists all the signals that are read inside the process, except those read only on a clock edge. This is important to prevent simulation mismatches. In simulation code, we can specify arbitrarily complex behavior using the sensitivity list of a process. In synthesizable code, most of this behavior would not find a correspondence in hardware. In practice, if faced with code that violates this restriction, most synthesizers will issue a warning and ignore the sensitivity list.

19.2.3 Synthesizable Types

Few types are absolutely nonsynthesizable. Objects of type *time* and *real* cannot be synthesized; however, they can be used in expressions as long the resulting value is of a synthesizable type. Access and file types are never synthesizable. Physical types are listed as nonsupported in the synthesis standard, but some tools will accept them.

Table 19.1 Synthesizable VHDL types (available by default or in a standard library)

Scalar type	1D array, unresolved	1D array, resolved	Package (library)
boolean	boolean_vector	—	standard (std)
bit	bit_vector	—	standard (std)
integer	integer_vector	—	standard (std)
character	string	—	standard (std)
std_ulogic	std_ulogic_vector	—	std_logic_1164 (ieee)
std_logic	—	std_logic_vector	std_logic_1164 (ieee)
—	u_unsigned	unsigned	numeric_std* (ieee)
—	u_signed	signed	numeric_std* (ieee)
—	u_float	float	float_pkg (ieee)
—	u_ufixed	ufixed	fixed_pkg (ieee)
—	u_sfixed	sfixed	fixed_pkg (ieee)

*Also available in package numeric_bit.

The synthesizable types specified in the synthesis standard and in the VHDL Language Reference Manual are listed in table 19.1. The type names are shown in three different columns to emphasize their relationships; types in the same row are related either because one is an array type of the other or because one is resolved and the other is unresolved. User-defined types based on the types shown in the table are generally synthesizable.

As for enumeration types, boolean, bit, std_ulogic, and std_logic are synthesized as single-bit values. For type boolean, the value true is synthesized as a logic 1 and false as a logic 0. For type bit, '1' is a logic 1 and '0' is a logic 0. For std_ulogic and std_logic, '1' and 'H' synthesize as a logic 1 and '0' and 'L' as a logic 0. In practice, values other than '0', '1', 'Z', and '-' should not be used. *Don't cares* can be used in assignments to allow the tool to perform logic optimizations; the tool may replace a *don't care* with a logic 0 or 1, whichever results in a smaller circuit.

User-defined enumeration types are encoded in a tool-dependent fashion. If we want to control how each value is represented in terms of bits, then we can use the synthesis attribute enum_encoding or create a custom conversion function to a logic vector type (e.g., std_ulogic_vector). An example of how to use the enum_encoding attribute is shown in section 19.6.1.

Type integer and its subtypes are synthesized as vectors of bits, using the minimum size required to represent the specified range. However, if a range does not contain the value zero, then it will be extended to include it. For example, if the range is declared as 14 to 15, then values of this subtype will be encoded as if the range were from 0 to 15 and will require four bits per value.

If the range does not include negative values, then the objects are encoded in unsigned binary form. If the range includes negative values, then objects are encoded in two's complement form. If a range is not specified, then the integer will be encoded using 32-bit two's complement representation (from -2,147,483,648 to +2,147,483,647 inclusive). This highlights the importance of always constraining the ranges of integer objects in synthesizable code, or we may waste precious device resources.

The way integer objects are encoded has an important ramification in the sizing of integers in expressions. The rules of the language state that operations with subtypes are performed as if the objects were of the base type; a range check is performed only after the expression has been evaluated, at the moment of an assignment. Consider the case where a subtraction is performed between two integers with a range from 0 to 15. The result can be any value from -15 to +15. This means that even though the operands could be represented with unsigned 4-bit values, the set of possible results requires a signed 5-bit value. Hence, the compiler will expand the operands as necessary in the middle of an expression and truncate the results when an assignments is made.

The relevance of all this to the circuit designer is that these transformations can cause a potential discrepancy between simulations and the generated hardware. In the simulator, some of the input combinations may be invalid and would cause errors upon assignment, thus interrupting the simulation. However, this possible discrepancy will not prevent the description from being synthesized. In the resulting circuit, however, a condition that would cause an error in the simulator will silently cause the result to be truncated or rolled over to accommodate it in the allocated number of bits. The upshot is that the result of the synthesized circuit will be correct for all input values that simulate without error. For other values, the results may have been truncated.

As for array types, the synthesis standard imposes a number of restrictions, but in practice the tools are much more flexible in what they accept. The standard specifies that the indices must be integer values, and elements must be integers, enumerations, or 1D arrays of bits. In practice, most tools accept arrays whose elements are records or other user-defined types. Indices of enumeration types are also generally supported.

Abstract floats (objects of type real) are not directly synthesizable as constants, variables, or signals. However, most tools accept them in expressions if the result type is synthesizable. This can be used to write formulas in the code using real numbers as operands, avoiding the inclusion of literal numbers calculated outside the source code. The result can be assigned to objects of integer, fixed-point, or synthesizable floating-point types using the appropriate type conversions. For an example, see listing 13.1.

A final point to note about types is that after a circuit has been synthesized, there will be no more error checking in range limits, array indices, or overflow. In simulation, ranges and indices are constantly checked and can abort the simulation or print warning messages. In a synthesizable model, some kinds of errors are checked when the design is synthesized, but once the model is converted to a circuit, there is no built-in error checking. For example, an integer counter that would overflow and generate an error in a simulation will quietly roll over and restart from zero in a synthesized circuit.

This kind of error may be hard to catch because it can manifest itself far from the point of origin. This is one more reason to perform simulations at the RTL level before synthesizing a model.

19.3 Creating Synthesizable Designs

VHDL allows us to model at various levels of abstraction, from purely behavioral down to gate level. However, synthesizable designs are usually synchronous systems written at the register-transfer level (RTL). Especially for the less experienced designers, it is a good idea to stick with synchronous designs at all times. For complex or algorithm-oriented systems, the RTL design methodology provides an organized approach to guide the design process. Together, these two techniques greatly improve the odds that a VHDL design will be translated to a circuit that works as intended by the designer.

19.3.1 RTL Design Methodology

Register-transfer level (RTL) design is a digital design methodology in which a circuit is described as a series of registers and transfer functions that define the data flow between the registers. The registers are synchronous storage elements (flip-flops), and the transfer functions are blocks of combinational logic.[1] The RTL methodology consists of a variety of design techniques and approaches, but it can be roughly summarized in three steps: a designer specifies the registers in a system, defines the transfer functions between them, and creates the control logic to orchestrate the sequence of operations and data transfers.[2]

The term RTL can refer to the level of abstraction of a design or to this design methodology. Usually, there is no need to differentiate between the two because one implies the other; the RTL methodology produces designs at the RTL level of abstraction. Figure 19.1 shows a generic diagram of a design at the RTL abstraction level.

RTL is a medium-level methodology; it is not too high that there is no obvious connection with the hardware and not too low that every element is a physical hardware component. RTL allows the designer to specify a hardware architecture without

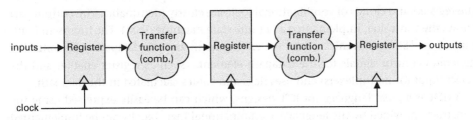

Figure 19.1
Abstract view of a design at the RTL level of abstraction.

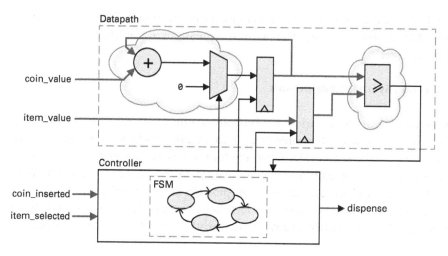

Figure 19.2

Example of a design using the RTL methodology—vending machine.

thinking about every gate, leaving to the tool the job of performing logic optimization. The process of automatically converting an RTL description to a gate-level implementation is called *logic synthesis*. Writing at the RTL level results in a design that is easily understood by synthesis tools and minimizes the potential hazards of a circuit created following an unstructured approach.

RTL design is great for modeling circuits that perform computations in steps or when a pure dataflow implementation would be infeasible or too expensive. More specifically, it provides a systematic approach to transforming an algorithm into a hardware circuit. In a typical RTL design, the circuit is divided into two parts: a *datapath* that performs the data transformations and keeps calculation results, and a *controller* that orchestrates the entire process. Figure 19.2 shows an example of a design at the RTL level. The design is a vending machine with four inputs, shown on the left, and one output, shown on the right. The inputs interface with a currency detector and a user keyboard, and the output controls the dispenser. The upper part shows the datapath; the registers and clouds of combinational logic are clearly identifiable. The bottom part shows the controller, implemented as a finite state machine (FSM). The inputs and outputs of the FSM communicate with the environment and the datapath. The controller generates control signals for the datapath elements (such as register enables and the select input for multiplexers) and uses flags and values calculated in the datapath.

VHDL is a great language for RTL designs, which can be easily expressed using the constructs provided by the language. Combinational logic blocks can be implemented as concurrent statements or combinational processes. Registers can be inferred from clocked processes that assign values to signals of any synthesizable data type. The controller FSM is easily implemented with one or more processes in one of the several coding styles available for implementing FSMs. Section 19.4 shows how these different kinds of hardware elements can be inferred from VHDL code.

19.3.2 Synchronous Circuits

In a *synchronous circuit*, all flip-flops are clocked by the same clock signal. Synchronous circuits are made of two general components: combinational logic blocks and edge-sensitive storage elements. The circuit shown in figure 19.1 is an example of a synchronous circuit; RTL is an inherently synchronous design methodology.

In a synchronous circuit, the storage elements are allowed to change state only at discrete moments, dictated by the rising or falling edge of a free-running clock signal. Combinational logic is inserted between the output of a register and the input to the next. Because combinational logic implies some amount of propagation delay, the interval between successive clock edges must be long enough to allow the combinational logic to stabilize at the input of the next flip-flop. The longest combinational delay and the registers setup time determine the maximum operating frequency of a synchronous circuit.

Synchronous systems impose a strict discipline on the designer, but they offer many benefits in return. The circuit is easier to design because all that matters for its functional correctness is the data transformations at the RTL level. The circuit operation is more predictable because it has a higher immunity to hazards and variations of temperature, voltage, and fabrication process. The circuit is easier to test and verify because of its deterministic history-dependent behavior. In fact, most established circuit testing methods presume synchronous operation.[3] In summary, synchronous systems make the job simpler for both the designer and the tools. Most digital devices today are synchronous, and most of the tools are aimed at designing, optimizing, and verifying synchronous circuits.

Another advantage is that the synchronous design paradigm can be expressed in a small number of guidelines that are easy to follow. Adhering to these guidelines makes the circuit more dependable and ensures correct timing under all valid operating conditions. The recommendations are especially important for less experienced designers, who are more likely to resort to unconventional clocking approaches and asynchronous controls. Later, as we gain experience, we may choose to deviate from specific recommendations as long we know their implications.

The following is a list of guidelines and recommendations for the design of synchronous circuits. These rules are given in the *Reuse Methodology Manual (RMM) for System-on-a-Chip Designs, 3rd Ed.*, a book that resulted from a joint effort sponsored by Synopsys and Mentor Graphics and is used as a development standard by many IC designers. For the full set of best practices, check the RMM.[4]

- **Use a single clock for the entire design.** Although this is not always possible in more complex designs, it is a goal worth pursuing.

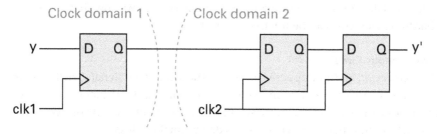

Figure 19.3
Use of a synchronizer circuit to transfer a single bit between clock domains.

- **Avoid manually generated clocks.** Do not generate clock signals using combinational logic or clock dividers. Always use clocks that are properly generated and distributed.
- **Use only edge-based storage elements (flip-flops).** Do not use latches. Check the warnings provided by your tool to ensure that latches were not inferred unintentionally.
- **Clock all flip-flops at the same edge.** Clocking at both clock edges complicates the design, making it necessary to consider the duty-cycle in timing analysis.
- **Provide a reset value for each flip-flop.** The ability to put the entire circuit into a known state is invaluable for testing and debugging.
- **Avoid complicated and conditional resets.** Use a simple reset signal for all registers in the design. If an asynchronous reset is used, then it must be deasserted synchronously.
- **Do not use the flip-flop reset and preset signals except for initialization.** Reset and preset are asynchronous control signals and complicate the design.
- **Avoid combinational feedback.** If feedback must be used, then make sure the signal passes through a register.
- **Synchronize all signals that cross clock domains.** This includes inputs generated by external systems, natural or manual external inputs, and signals from circuit blocks that use different clocks. Use the circuit of figure 19.3 to transfer single-bit values and avoid combinational logic between the clock domains. This is sufficient to prevent metastability in most synchronous designs.

19.4 Inferring Hardware Elements

This section describes how to write VHDL code that implies specific hardware elements or structures.

19.4.1 Combinational Logic
Combinational circuits are those whose outputs depend only on the current input values and not on any internal state or previous input history. This kind of circuit has

no memory or feedback loops, and the output values are given as a logic function of the present inputs. In a combinational circuit, when a new input is applied, the output eventually settles to a value that depends only on the presented inputs (after a certain delay).

There are two ways to model combinational logic in VHDL: using concurrent signal assignments (presented in section 11.1.3) or using *process* statements (covered in section 9.2). In a sense, any concurrent signal assignment is effectively a process, as was described in section 9.2.3. Nevertheless, it is useful to make a distinction between concurrent assignments and processes because as we write the code, we must choose between the two kinds of constructs.

A concurrent signal assignment is appropriate for modeling combinational logic because it specifies the signal value as an expression that is reevaluated whenever one of the signals used in the expression changes. In other words, the concurrent signal assignment is implicitly sensitive to all the signals read in the expression. This is precisely the behavior expected from a block of combinational logic. The actual hardware structures inferred for the block of combinational logic are given by the operators used in the expression, which can include all the operators seen in chapter 8.

We can use simple, selected, and conditional signal assignments depending on the kind of structure that we want to model. In selected and conditional assignments, besides the logic implied by the expressions, the synthesized circuit will include the control logic inferred from the conditions, as well as the control structure that chooses between the possible expressions (typically, a multiplexer).

```
-- Simple signal assignments can implement boolean or arithmetic functions:
mul <= op_a * op_b;
start <= en and cs and req and not busy;

-- Selected signal assignments are useful when the conditions
-- are mutually exclusive:
with operation select alu_result <=
    op_a + op_b when op_add,
    op_a - op_b when op_sub,
    op_a * op_b when op_mul,
    x"0000" when others;

-- Conditional signal assignments can be used when the conditions
-- are not mutually exclusive:
encoded_data <=
    "11" when sel(3) = '1' else
    "10" when sel(2) = '1' else
    "01" when sel(1) = '1' else
    "00";
```

The other way to model combinational logic is with a nonclocked process. This is a good alternative if the logic is complex or if the computation is best described in an algorithmic fashion (e.g., when it includes a loop statement). It is also a good choice if the computation involves several auxiliary or intermediate values.

When a signal or variable is assigned inside a process, it will model combinational logic if all the following conditions are met:

- All signals read in the process are in the sensitivity list (or the sensitivity list is the keyword all);
- In the case of a signal, the object is assigned a value in every possible execution path inside the process;
- In the case of a variable, the object is always assigned a value before it is read;
- The assignment is not under control of a clock edge.

The first condition ensures that the synthesized circuit will match the simulation results. The three other conditions ensure that no storage elements are inferred (we will get back to this subject when we discuss unintentional storage elements in the next section). The following example shows a process that observes these conditions and therefore models combinational logic:

```
-- Combinational process that counts the number of bits '1' in signal
-- input_vector and outputs the count in signal num_bits_one. Signal
-- input_vector is the only signal read inside the process.
count_bits_one: process (input_vector)
    -- Auxiliary variable to accumulate the partial sum inside the loop.
    variable bit_count: integer range 0 to 31;
begin
    -- Unconditional assignment to a known value ensures that no storage
    -- elements are inferred and the variable models combinational logic.
    bit_count := 0;

    for i in input_vector'range loop
        if input_vector(i) then
            bit_count := bit_count + 1;
        end if;
    end loop;

    -- The signal is always assigned a value in the process,
    -- so it models combinational logic.
    num_bits_one <= bit_count;
end process;
```

One final point to note about expressions is that type conversion functions that do not change the bit-level representation of an object do not incur any hardware overhead. For example, calling the function to_unsigned to convert from a natural to an unsigned value does not cost any hardware because the natural was already represented using the same format. By the same reasoning, a shift operation with a constant shift amount does not imply any hardware.

19.4.2 Sequential Logic

In a sequential circuit, the outputs depend not only on the current inputs but also on previous system states. Sequential circuits maintain their internal state in storage elements. There are two general kinds of storage elements based on the conditions under which their values are updated. Edge-sensitive storage elements are called *flip-flops* and are updated on the transition of a clock signal. Level-sensitive storage elements are called *latches*. Latches are controlled by an enable input; when enabled, the latch is *transparent* and propagates the input value to the output. When disabled, the latch holds its current value. In synchronous circuits, we use only registers, and latches should be avoided. We will see how to model each kind of storage element in turn.

Edge-Sensitive Storage (Registers) A *register* is composed of one or more flip-flops that share the same control logic and store a single- or multibit signal. Edge-sensitive storage elements are updated on the transition of a clock signal. In VHDL code, clock signals are no different than any other signal; their only requirement is to be of type bit, std_ulogic, or one of their subtypes (such as std_logic). We imply that a signal is a clock when certain statements are executed on the transition of that signal.

A register is inferred when a signal or variable is updated on a clock edge—in other words, when we do an assignment inside a region of a process that is subject to a clock edge. The edge condition can be modeled in two basic forms. The first is to use a process with a sensitivity list and then test for the clock edge using an *if* statement (figure 19.4a). The second is to use a *wait* statement that suspends until a clock edge occurs; in this case, the other statements inside the process are implicitly subject to this condition (figure 19.4b). The *wait* statement should be the first or last statement in the process.

We can test for a clock edge by using the predefined functions rising_edge or falling_edge. We could also use an expression combining a clock level and a test for an event (as in the expression clock='1' and clock'event). However, for clarity and better handling of metalogical values, it is recommended to use the predefined functions when we test for a clock edge.

The sensitivity list of a clocked process should include only the clock signal and any asynchronous control signals required by the storage element, such as a reset or preset. The asynchronous signals must be checked outside the test for the clock edge.

```
process (clock) begin                      process begin
    if rising_edge(clock) then                 wait until rising_edge(clock);
        q <= d;                                q <= d;
    end if;                                end process;
end process;
```

(a) Using a sensitivity list.   (b) Using a *wait* statement.

```
process (clock, reset) begin               process (clock) begin
    if reset_n = '0' then                      if rising_edge(clock) then
        q <= '0';                                  if reset_n = '0' then
    elsif rising_edge(clock) then                      q <= '0';
        q <= d;                                    else
    end if;                                            q <= d;
end process;                                       end if;
                                               end if;
                                           end process;
```

(c) With asynchronous reset.   (d) With synchronous reset.

Figure 19.4
Examples of clocked processes that imply edge-sensitive storage (registers).

Figure 19.4c shows an example of a D-type flip-flop with an asynchronous reset. The reset signal is in the sensitivity list because the process must be executed every time reset changes and not only on clock edges.

If a synchronous reset is used, then in practice it is no different than a logic gate in series with the flip-flop data input, and the reset signal should not be in the sensitivity list. In the example of figure 19.4d, the original data input would be combined with the active-low reset signal using an *and* gate.

Although the examples in figure 19.4 show a single-bit register, signals d and q could be of any synthesizable type, and the source value in the assignments could be any boolean or arithmetic expression. In this case, the expression would be synthesized as a block of combinational logic feeding the register input.

It is not only signal assignments that can imply a register. A variable in a clocked process can also imply a register if it is read before being assigned a value. To understand this, recall that variables are not reinitialized between successive runs of a process. Therefore, when a process resumes, its variables still hold the same values from the previous run. When we read the value of a variable before assigning to it in a process, the synthesis tool will understand that we are referring to its previous value and infer a storage element.

In the following example, there is an execution path (when condition is false) in which the variable auxiliary_value is read before it gets assigned a value. Therefore, a register is inferred for the variable:

```
process (clock)
    variable auxiliary_value: std_ulogic := '0';
begin
    if rising_edge(clock) then
        if condition then
            auxiliary_value := a xor b;
        end if;

        -- A register is inferred for the variable 'auxiliary_value'
        -- because it is read before being assigned a value when
        -- 'condition' is false.
        y <= auxiliary_value;
    end if;
end process;
```

The upshot is that if a variable is not intended to model a register, then it should always be assigned a value before it is read. We will get back to this subject when we discuss unintentional storage elements in the next sections.

Level-Sensitive Storage (Latches)   In a synchronous design, only edge-sensitive storage elements should be used. However, level-sensitive storage may happen unintentionally, sometimes because of coding styles that lead to the inference of latches. If for no other reason, it is important to know how to model latches to avoid them in our synchronous designs.

Latches can be inferred for signals or variables, but the conditions are different in each case. Before we delve into the details of latch inference, it should be clear that latches can only be inferred in assignments that are not under control of a clock edge—otherwise a register would be inferred instead. Furthermore, it will be implicit in our discussion that all signals read in the process are in the sensitivity list.

For a signal, a latch is inferred if a nonclocked process assigns a value to the signal in one execution path but does not assign a value in another path. This can be explained as follows. Whenever a process has one or more assignments to a signal, a single driver is created for that signal. This driver is like a virtual hardware device that is always sourcing a value to the signal.[5] When an assignment happens, it changes the value sourced by the driver. If no assignment happens, the driver keeps sourcing the same value indefinitely.

Now consider the case of a process with a single *if-then-else* statement. Suppose this process assigns a value to the signal when the *if* condition is true but does not make an assignment to the signal otherwise. Then a latch will be implied because the signal value must be preserved when the condition is false. We can generalize this reasoning by stating that a latch is inferred whenever a signal is assigned in one execution path and is not assigned in some other path.

For a variable, the condition is somewhat different. A latch is inferred when a variable is read in a nonclocked process before being assigned a value. The explanation is similar as for the inference of registers from a variable in a clocked process. Variables in a process are initialized only once and keep their values between successive runs of the process. When a process is reactivated, if we read the value of a variable, then it will have the value from the previous run of the process. This implies storage, which means a latch if the process is nonclocked. However, if we always assign a value to the variable before reading it, there is no need for storage.

We can summarize the rules for the inference of latches in a *combinational* process as:

• For a signal, a latch is inferred if the signal is assigned in the process but not in all possible execution paths.
• For a variable, a latch is inferred if the variable is read before it is assigned a value.

In both cases, the assignments must not be under control of a clock edge, and the sensitivity list should contain all signals read in the process. A process with an incomplete sensitivity list does not imply a latch; it is a modeling error.

In *Circuit Design and Simulation with VHDL*, Pedroni offers the following alternative explanation. A combinational circuit can be represented by a truth table. If the compiler is not able to completely assemble that table, then latches will be inferred.[6] This is a clear and simple way to summarize the conditions for the inference of latches.

On a final note, if a concurrent statement has an equivalent process that falls under the above conditions, then it will also imply a latch. This could happen, for instance, in an incomplete signal assignment. In the following concurrent statements, the first assignment is incomplete because q1 receives a new value only when enable is high; at other times, it will hold its value. In contrast, q2 always receives a value because of the *else* clause. Finally, q3 does not imply a latch because it models a clocked process (and therefore it implies a register).

```
q1 <= d when enable;              -- implies a latch
q2 <= d when enable else '0';     -- implies a combinational function (d and enable)
q3 <= d when rising_edge(clock);  -- imples a register
```

Unintentional Storage Elements (Latches or Registers)
Now that we have covered the basics of when edge- and level-sensitive storage elements are inferred, we can summarize the conditions under which they are implied and declare a few rules to avoid the unintentional inference of storage elements (for both latches and registers). These are the general rules for inference of storage elements:

- For a signal, a *register* is inferred whenever the signal has an assignment under control of a clock edge.
- For a variable, a *register* is inferred whenever the variable has an assignment under control of a clock edge and is read before it is assigned a value.
- For a signal, a *latch* is inferred whenever the signal is assigned in a combinational process but not in all possible execution paths.
- For a variable, a *latch* is inferred whenever the variable is read before it is assigned a value and the assignment is not under control of a clock edge.

Table 19.2 summarizes the rules for the inference of storage elements in combinational and clocked processes. When we look at these conditions the other way round, we arrive at the rules to avoid unintentional latches in a combinational process described in table 19.3.

While coding, we can use two techniques to ensure that these rules are observed:

- Assign a default value to an object at the beginning of a process, outside any conditional statement.
- Use only complete assignments—assignments that are unconditional or where a value is assigned to an object in all execution paths.

Figure 19.5 demonstrates the use of the two techniques with an *if* statement. Figure 19.6 demonstrates the use of default unconditional assignments in a *case* statement.

Table 19.2 Rules for the inference of storage elements (latches and registers)

Assignment type	Object class	Situation	Inferred element
Nonclocked (combinational)	Signal	The signal has an assignment, but not in all execution paths.	Latch
	Variable	The variable is read before it is assigned a value.	Latch
Clocked (under a clock edge)	Signal	The signal has an assignment.	Register
	Variable	The variable is read before it is assigned a value.	Register

Table 19.3 Rules for avoiding unintentional latches in combinational processes

Object class	Rule
Signal	For every signal that is assigned in a combinational process, make sure that the signal is assigned a value in every possible execution path of the process.
Variable	For every variable that is assigned in a combinational process, make sure that the variable is always assigned a value before it is read.

```
process (all) begin          process (all) begin          process (all) begin
   if condition then            if condition then            q3 <= '0';
      q1 <= d;                     q2 <= d;                   if condition then
   end if;                     else                             q3 <= d;
end process;                      q2 <= '0';                 end if;
                              end if;                     end process;
                           end process;
```

(a) Latch inferred due to   (b) Latch prevented with   (c) Latch prevented with a
    incomplete assignment.   an *else* clause.   default, unconditional
                                                                  assignment.

Figure 19.5
Unintentional latch inference and its prevention in an *if* statement.

```
process (current_state) begin          process (current_state) begin
   case current_state is                  -- Default assignments prevent
      when red =>                         -- the inference of latches
         red_light <= '1';               red_light <= '0';
         green_light <= '0';             green_light <= '0';
         yellow_light <= '0';            yellow_light <= '0';
      when green =>
         red_light <= '0';
         green_light <= '1';             case current_state is
         -- Oops, missing assignment        when red =>
         -- to signal yellow_light             red_light <= '1';
      when yellow =>                        when green =>
         red_light <= '0';                     green_light <= '1';
         green_light <= '0';                when yellow =>
         yellow_light <= '1';                  yellow_light <= '1';
   end case;                            end case;
end process;                         end process;
```

(a) A latch is inferred because of a   (b) Latch prevented with default
    missing assignment.   initial assignments.

Figure 19.6
Unintentional latch inference and its prevention in a *case* statement.

Finite State Machines   A finite state machine (FSM) is an abstraction of a sequential circuit whose possible internal states are enumerated by the designer. The FSM is always in exactly one state (called its current state) and can transition to a different state in response to an event. To design an FSM, we must specify its states, input signals, output signals, next-state function, and output function. The outputs of an FSM may depend on its current state and current inputs.

An FSM can be implemented as a synchronous digital circuit using a register to keep its current state, and combinational logic to implement the next-state and output

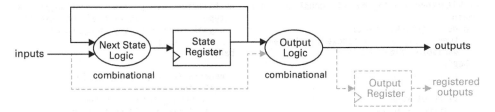

Figure 19.7
Finite state machine (FSM) model.

functions. FSMs are great for managing complexity because they allow us to concentrate on one part of the problem at a time, corresponding to one of the machine states.

At the circuit level, an FSM can be broken down into the parts shown in figure 19.7: a combinational block that determines the next state based on the current state and inputs (the *next state logic*), a register that stores the current state (*state register*), and a block of logic that produces the output values, which can be combinational or registered. The optional output register may be included depending on the application.

The choices for implementing an FSM in VHDL refer to where to place the statements corresponding to each of the blocks. In theory, each block could be coded independently; the combinational blocks could even be coded with concurrent statements. In practice, designers tend to choose one of two main styles or some of their variations. The two styles are commonly known as *single process* or *two process*. Both styles are common, and there is a big component of personal taste when choosing between them. In the single-process FSM, all the blocks are implemented in the same process statement. In a two-process FSM, the most common division is to code the state register as a clocked process and the next state and output logic blocks as a combinational process. Because the main characteristic of this approach is the separation between the state register and any combinational logic, other variants where the combinational logic is split into more processes can be considered equivalent to the two-process style for the purposes of this discussion.

Figure 19.8 shows the two implementation approaches for the same FSM. The circuit is a turnstile controller with two inputs and one output. The FSM has two states: locked and unlocked. In the locked state, the turnstile waits until the input ticket_accepted is true. In the unlocked state, the turnstile waits until the input pushed_through is true. The single output, named lock, is true when the FSM is in the locked state.

Each style has advantages and disadvantages. We can start our discussion with the advantages of the single-process style (figure 19.8a). First, we can use a variable for the state, thus keeping the implementation details of the FSM entirely encapsulated within the process. Note that the relevant declarations (type state_type and variable state) are made at the process level, leaving the architecture declarative part uncluttered.

```
architecture single_proc of turnstile_fsm is    architecture two_proc of turnstile_fsm is
begin                                             type state_type is (locked, unlocked);
  process (clock, reset)                          signal current_state, next_state:
    type state_type is (locked, unlocked);          state_type;
    variable state: state_type;                  begin
  begin                                             -- State register:
    -- Single process for state register,         process (clock, reset) begin
    -- next-state logic, and output logic:          if reset then
    if reset then                                     current_state <= locked;
      state := locked;                              elsif rising_edge(clock) then
      lock <= '1';                                    current_state <= next_state;
    elsif rising_edge(clock) then                   end if;
      case state is                               end process;
        when locked =>
          if ticket_accepted then                   -- Next-state and output logic:
            lock <= '0';                            process (all) begin
            state := unlocked;                        case current_state is
          else                                          when locked =>
            lock <= '1';                                  if ticket_accepted then
            state := locked;                                lock <= '0';
          end if;                                          next_state <= unlocked;
        when unlocked =>                                 else
          if pushed_through then                           lock <= '1';
            lock <= '1';                                    next_state <= locked;
            state := locked;                              end if;
          else                                          when unlocked =>
            lock <= '0';                                  if pushed_through then
            state := unlocked;                              lock <= '1';
          end if;                                          next_state <= locked;
      end case;                                         else
    end if;                                               lock <= '0';
  end process;                                            next_state <= unlocked;
end;                                                    end if;
                                                      end case;
                                                    end process;
                                                  end;
```

(a) Single-process style.   (b) Two-process style.

Figure 19.8
Two ways of implementing an FSM in VHDL.

Second, it requires a single object to keep the state (instead of one object for the current and another for the next state). Third, because the entire process is clocked, there is no risk of accidental latches. However, all outputs are registered by default, which is not always necessary or wanted. In the diagram of figure 19.7, this corresponds to including the optional output register by default. Furthermore, if the FSM includes more complex structures and operations, such as timers, local counters, and additional registers, then a single-process solution may become too complicated.

The two-process style also offers several advantages. First, it is easier to choose whether to register the outputs: we have the choice of putting the output logic in the combinational process, in the clocked process, or in a third process. Second, it resembles more closely the parts of the block diagram and gives each process a clear responsibility, which is favored by many designers. However, because the next-state logic is a combinational process, it is possible to imply unwanted latches unless all outputs and the next state signal are assigned in all possible execution paths. Another disadvantage in VHDL versions prior to 2008 is that the designer must manually update the sensitivity list.

The inference of latches in the two-process style deserves further discussion. Latches will be inferred only in the cases where the FSM code incompletely specifies the assignments for the next state or outputs. Therefore, the fact that latches were inferred is actually a warning that part of the FSM specifications may have been overlooked. In this case, we can use the warnings to our advantage—a completely specified FSM should never imply any latches. Therefore, if our main reason for choosing the single-process style is to prevent latches, then we may be hiding some deeper design problems.

Regardless of the chosen style, in the two examples of figure 19.8, the code is more verbose than it needs to be. We can clean it up with two simple changes.

First, it is not necessary to write redundant assignments that keep the machine in the same state it already is. In the present example, this corresponds to all the assignments to state or next_state made in the *else* branch of the *if* statements. In the single-process style, we can simply remove the redundant assignments. In the two-process FSM, we also need to add a default assignment (next_state <= current_state) before the *case* statement to prevent unwanted latches.

Second, when the outputs have sensible "default" values or the same value for most states, we can add default assignments to the outputs before the *case* statement. In this way, inside the *case* statement, we only need to assign to the outputs when the intended value is different from the default. When we add default assignments for all outputs as well as for the next state signal, this eliminates the possibility of unwanted latches in the two-process approach. However, it spreads the statements that define the behavior of a state, making it harder to analyze any state in isolation. In any case, because this approach can provide a substantial reduction in the amount of code, it is recommended where it makes the code significantly shorter.

19.4.3 Three-State Logic

To model a three-state bus, we assign the high-impedance value 'Z' to a signal. There are two typical situations where we may use three-state buses. In the first situation, there is a single-sourced signal (which may be a scalar or an array) that must be put in high-impedance state under a certain condition (figure 19.9a). This is the case of a signal that is connected to a three-state bus outside the chip or outside the VHDL model. The second case is when there are multiple signals (scalars or arrays) wired together into a single signal (figure 19.9b). In this case, the bus is inside the VHDL model.

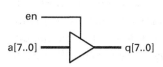
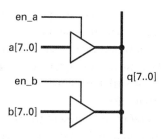

(a) A single signal (q) that can assume
   a high-impedance value. The three-state
   bus is outside the VHDL model or chip.

(b) A three-state bus that is
   part of a VHDL model.

Figure 19.9
Uses of the high-impedance value 'Z' in a VHDL model.

In the case of figure 19.9b, signal q should be of a resolved type, such as std_logic, std_logic_vector, or one of the resolved types in table 19.1. This is necessary because there are multiple drivers for the same signal (q), so their values must pass through a resolution function. In the case of figure 19.9a, an unresolved type such as std_ulogic and std_ulogic_vector would also work because no resolution function is necessary.

Here is the code for the example of figure 19.9a:

```
library ieee;
use ieee.std_logic_1164.all;

entity three_state_single_driver is
    port (
        en: in std_logic;
        a: in std_ulogic_vector(7 downto 0);
        q: out std_ulogic_vector(7 downto 0)
    );
end;

architecture rtl of three_state_single_driver is
begin
    q <= a when en else "ZZZZZZZZ";
end;
```

Here is the code for the example of figure 19.9b:

```
library ieee;
use ieee.std_logic_1164.all;
```

```
entity three_state_multiple_drivers is
    port (
        en_a, en_b: in std_logic;
        a, b: in std_logic_vector(7 downto 0);
        q: out std_logic_vector(7 downto 0)
    );
end;

architecture rtl of three_state_multiple_drivers is
begin
    q <= a when en_a else "ZZZZZZZZ";
    q <= b when en_b else "ZZZZZZZZ";
end;
```

In the second example, each concurrent signal assignment is equivalent to a process that assigns a value to q when the corresponding enable signal is active. Therefore, each assignment has its own driver for signal q.

19.4.4 ROM Memories

The synthesis standard describes two ways to model a ROM: using a constant array or using a *case* statement. Listing 19.1 shows how a ROM can be modeled using an array.

Listing 19.1 Modeling a ROM using a constant array

```
 1 library ieee;
 2 use ieee.std_logic_1164.all;
 3 use ieee.numeric_std.all;
 4
 5 entity sync_rom_array is
 6     port (
 7         clock: in std_ulogic;
 8         address: in unsigned(3 downto 0);
 9         data_out: out unsigned(7 downto 0)
10     );
11 end;
12
13 architecture rtl of sync_rom_array is
14     signal data_out_int: natural range 0 to 255;
15     type rom_contents_type is array (natural range <>) of integer range 0 to 255;
16     constant rom_contents: rom_contents_type := (
17          13,  37,  62,  86, 109, 131, 152, 171,
18         189, 205, 219, 231, 240, 247, 252, 255
19     );
```

```
20 begin
21     data_out <= to_unsigned(data_out_int, data_out'length);
22
23     process (clock) begin
24         if rising_edge(clock) then
25             data_out_int <= rom_contents(to_integer(address));
26         end if;
27     end process;
28 end;
```

One advantage of this approach is that we can move the constant to a package so that the memory contents do not interfere with the rest of the code. This is especially important for large memories.

The second way to model a ROM is using a *case* statement. In this style, all the assignments to the ROM object should be done from within a single *case* statement (listing 19.2).

Listing 19.2 Modeling a ROM using a *case* statement

```
1 library ieee;
2 use ieee.std_logic_1164.all;
3 use ieee.numeric_std.all;
4
5 entity sync_rom_case is
6     port (
7         clock: in std_ulogic;
8         address: in unsigned(3 downto 0);
9         data_out: out unsigned(7 downto 0)
10    );
11 end;
12
13 architecture rtl of sync_rom_case is
14     signal address_int: natural range 0 to 15;
15     signal data_out_int: natural range 0 to 255;
16 begin
17
18     address_int <= to_integer(unsigned(address));
19     data_out <= unsigned(to_unsigned(data_out_int,8));
20
21     process (clock) begin
22         if rising_edge(clock) then
23             case address_int is
24                 when 0 =>  data_out_int <=  13;
```

```
25                  when 1 =>  data_out_int <=   37;
26                  when 2 =>  data_out_int <=   62;
27                  when 3 =>  data_out_int <=   86;
28                  when 4 =>  data_out_int <=  109;
29                  when 5 =>  data_out_int <=  131;
30                  when 6 =>  data_out_int <=  152;
31                  when 7 =>  data_out_int <=  171;
32                  when 8 =>  data_out_int <=  189;
33                  when 9 =>  data_out_int <=  205;
34                  when 10 => data_out_int <=  219;
35                  when 11 => data_out_int <=  231;
36                  when 12 => data_out_int <=  240;
37                  when 13 => data_out_int <=  247;
38                  when 14 => data_out_int <=  252;
39                  when 15 => data_out_int <=  255;
40             end case;
41         end if;
42     end process;
43 end;
```

The synthesis tools can usually infer a ROM automatically from these templates. Sometimes, however, it may be necessary to give the synthesis tool a hint by using synthesis attributes (presented in section 19.6), such as the rom_block attribute or another attribute specified by the tool vendor.

19.4.5 RAM Memories

RAM memories can be easily described using array signals or variables. Generally, a synthesis tool can recognize when an array is used as a RAM and infer the correct hardware structure. Later, it will choose the appropriate hardware elements available in the target technology to implement each memory. Before writing a model, however, we should check the kinds of memory structures available in the target technology. If we model some behavior not supported by the dedicated hardware structures (e.g., asynchronous memory reads in some FPGAs), then the tool will be forced to implement the memory using logic resources.

The code in listing 19.3 is an example of a synchronous read, synchronous write RAM.

Listing 19.3 A RAM with synchronous read and write

```
1 library ieee;
2 use ieee.std_logic_1164.all;
3 use ieee.numeric_std.all;
```

```
4
5 entity sync_ram is
6     generic (
7         ADDRESS_WIDTH: natural := 8;
8         DATA_WIDTH: natural := 8
9     );
10     port (
11         clock: in std_ulogic;
12         address: in std_ulogic_vector(ADDRESS_WIDTH-1 downto 0);
13         write_enable: std_ulogic;
14         data_in: in std_ulogic_vector(DATA_WIDTH-1 downto 0);
15         data_out: out std_ulogic_vector(DATA_WIDTH-1 downto 0)
16     );
17 end;
18
19 architecture rtl of sync_ram is
20     constant RAM_DEPTH: natural := 2**ADDRESS_WIDTH;
21     type ram_contents_type is array (natural range <>) of std_ulogic_vector;
22     signal ram_contents: ram_contents_type
23         (0 to RAM_DEPTH-1)(DATA_WIDTH-1 downto 0);
24     signal address_int: natural range 0 to RAM_DEPTH-1;
25 begin
26     address_int <= to_integer(unsigned(address));
27
28     process (clock) begin
29         if rising_edge(clock) then
30             data_out <= ram_contents(address_int);
31             if write_enable then
32                 ram_contents(address_int) <= data_in;
33             end if;
34         end if;
35     end process;
36 end;
```

19.5   Initialization and Reset

The design of the reset mechanism is an important step in the RTL methodology. The designer must ensure that the circuit starts in a known and well-defined state to guarantee that the system will work correctly. Here we cover two typical design choices in the design of a circuit reset mechanism.

Initialization with Default and Initial Values versus Explicit Reset Signal   The first choice is whether to use default and initial values in signal and variable declarations.

Recall from the discussion about initial values in simulation (section 14.7) that a signal declaration may include an optional *default expression*, which provides a *default value* for the signal:

```
-- The decimal value 100 is the 'default value' for signal 'countdown':
signal countdown: unsigned(6 downto 0) := d"100";
```

Similarly, a variable declaration includes an optional *initial value expression*, which defines an *initial value* for the variable.

```
-- The enumeration literal 'unlocked' is the 'initial value' for variable 'state':
variable state: state_type := unlocked;
```

The VHDL synthesis standard states that default and initial expressions should be ignored for signals and variables. This is a reasonable constraint because many implementation technologies do not support arbitrary initialization of storage elements. In an ASIC, for instance, we need to assume that the flip-flops will be initialized to random values unless we provide an explicit initialization mechanism.

FPGAs, in contrast, always initialize to a known state, and some technologies support the use of initial values specified in the VHDL code to provide register power-up values. In such technologies, both signals and variables can use initial values as long as they are synthesized as registers. Initializing a signal that is driven from combinational logic would have no effect.

Compared with the creation of an explicit reset signal, the use of initial values has advantages and disadvantages. The main advantages of initial values are that they cost no routing or logic resources and provide a safe initialization mechanism that is immune to routing delays. They can also define the initial contents of block memories in some FPGAs. In contrast, the main advantages of explicit reset signals are that they are more portable and allow us to reinitialize our circuit at any time. They also make it possible to reset only the application circuit rather than the entire FPGA chip. If necessary, it is possible to create multiple reset signals and reinitialize different parts of a circuit independently. Finally, in a simulation, we can put the design into a known state at any time.

If you do not have a strong opinion one way or the other, then you may want to follow the recommendations in section 19.3.2 about synchronous designs and use an explicit reset signal for all registers in your design. For code that is intended for simulation only, initial values can be used without restriction.

Synchronous versus Asynchronous Reset When you choose to use an explicit reset signal, there is still another choice to make: should the reset be synchronous or asynchronous?

An asynchronous reset has two main advantages. First, it does not add another level of logic in series with the register data input because it uses dedicated control pins. Second, it works even while the clock signal is not running. However, an asynchronous reset introduces a potential hazard: if the clock signal is deasserted at the same time as a clock transition happens, the circuit could go into a metastable and unknown state. A synchronous reset avoids this problem but requires extra logic at the data input and does not work unless the clock is running. The recommended solution is to combine the best of the two approaches and use an asynchronous reset with synchronous release. This requires a circuit similar to the one shown in figure 19.3. A good reference for safe reset circuits is *The Art of Hardware Architecture* by Mohit Arora. The reset synchronizer circuit is shown by Arora in figure 2.40.[7]

19.6 Pragmas

Pragmas are source code constructs that provide additional information to a compiler about how to process the source text. Pragmas are not considered an official part of a programming language and do not appear in the language grammar. Consequently, they tend to vary widely across compilers. The VHDL synthesis standard defines two kinds of pragmas: synthesis attributes and metacomments. We will see each one in turn.

19.6.1 Synthesis Attributes

Synthesis attributes belong to the category of user-defined attributes introduced in section 8.3. *User-defined* attributes are constants that can be associated with named entities to provide additional information about them. *Synthesis attributes* are user-defined attributes whose names and interpretation are specified in the IEEE 1076.6 synthesis standard. These attributes can be associated with individual named entities in a design, such as signals, entities, or types, providing a finer control over the synthesis of a model.

Before the standardization of synthesis attributes, vendors defined their own attributes and metacomments, which varied widely across tools. The synthesis standard provides an initial standardization effort by defining a common set of attributes that should be supported by all tools. These attributes are declared in a standard package named *rtl_attributes*. Because VHDL allows us to redeclare a user-defined attribute that is already declared, there are two ways to use the synthesis attributes in our models. One is to include the package *rtl_attributes* in the design with a *use* clause. The other is to redeclare manually the attributes we want to use, using the syntax for user-defined attributes introduced in section 8.3.

Table 19.4 lists all the synthesis attributes specified in the IEEE 1076.6 synthesis standard. It is worth skimming through the name and meaning of each attribute to

Table 19.4 Synthesis attributes specified in IEEE Std 1076.6

Attribute	Applies to	Type	Meaning
keep	signal, variable, entity, component declaration/ instantiation	boolean	Item should not be deleted or replicated during synthesis. Entity or component should not be optimized internally.
create_ hierarchy	entity, block, subprogram, process	boolean	The design boundary around the item should be preserved.
dissolve_ hierarchy	entity, component declaration/ instantiation	boolean	Entity should be deleted from the hierarchy and its content instantiated in the parent item.
sync_set_reset	signal, process, block, entity	boolean	Tool should use set/reset pins in a flip-flop, instead of implementing control logic with its data input.
async_set_ reset	signal, process, block, entity	boolean	Tool should use set/reset pins in a latch, instead of implementing control logic with its data input.
one_hot, one_cold	signal	boolean	Identifies a collection of signals in which only one is active at a time. The synthesis tool should not implement priority logic among them.*
infer_mux	label	boolean	*Case* or selected signal assignment should be implemented as a mux rather than using random logic.
implementation, return_port_ name	procedure, function, label	string	Ignore the body of a subprogram; use given entity or library cell instead.
rom_block	constant, variable, signal	string	Item should be implemented as a ROM.
ram_block	constant, variable, signal	string	Item should be implemented as a RAM.
logic_block	constant, variable, signal	string	Item should be implemented as logic (not as a RAM or ROM).
combinational	process, conditional/ selected assignment	boolean	Item implies only combinational logic. Feedback loops should be used instead of latches.
gated_clock	signal, process	boolean	Register clock and enable should be combined logically to create a gated clock.
enum_encoding	type, subtype	string	Specifies a custom bit-level encoding for enumeration literals.*
fsm_state	signal, variable, type, subtype	string	Specifies the state vector encoding for FSMs: "BINARY", "GRAY", "ONE_HOT", "ONE_COLD", or user-defined.*
fsm_complete	signal, variable, type, subtype	boolean	Specifies that unmapped states should transition automatically to the default assignment (others clause).
buffered	signal	string	Specifies that the signal requires a high-drive buffer.

*Use of this attribute may cause simulation mismatches.

know what is available or will be available soon. Keep in mind that many synthesis tools may not support a specific attribute yet, but they might provide an equivalent one. The definitive source of information for attributes and metacomments is still the tool manual.

An important point is that some of the attributes (the ones marked with an asterisk in the table) may cause mismatches between simulation results and the synthesized circuit. To understand why this happens, consider the enum_encoding attribute, which allows us to define a custom bit-level encoding for enumeration literals. The VHDL language specifies that values in an enumeration have an ascending order from left to right. Therefore, if we declare the following type:

```
type alphanum_digits is (zero, one, two);
```

then the expressions (zero < one), (one <= two), and (two > zero) are all true. If we use this kind of comparison in our code, then the simulator will use the ordering implied by the type definition. However, by using the enum_encoding synthesis attribute, we may specify a different behavior for the synthesized circuit. Some synthesis tools disallow certain operations and operators when the attribute is used. Others issue a warning and implement the comparator using the user-specified encoding. In any case, we have created a mismatch between simulation and synthesis.

Nevertheless, apart from these possible mismatches, synthesis attributes can be useful. An example of a commonly used attribute is the fsm_state synthesis attribute, which specifies the encoding for an FSM state at the bit level. Suppose we have the following type and signal declarations for the state register of an FSM:

```
type state_type is (s0, s1, s2, s3, s4, s5);
signal current_state: state_type;
```

There are two ways to specify an encoding using the fsm_state attribute: we can apply the attribute to the state *type* or the state *signal*. The latter allows us to specify different encodings for each FSM that uses the type. The following example illustrates the first alternative, applying the attribute directly to the state type:

```
attribute fsm_state: string;

-- The designer can choose one of the attribute specifications below:
attribute fsm_state of state_type: type is "GRAY";
attribute fsm_state of state_type: type is "ONE_HOT";
attribute fsm_state of state_type: type is "111 000 101 010 110 011";
```

Many synthesis tools provide some version of this attribute, although not always with the name fsm_state or with the attribute values shown in table 19.4. Remember to check your tool manual for the supported attributes.

A final point that deserves clarification is the difference between the enum_encoding and fsm_state attributes. Here is a list of their main differences:

- fsm_state is intended for use with FSMs only; enum_encoding can be used with any application of an enumeration type.
- enum_encoding can only be applied to types or subtypes; therefore, it is "global" to all objects of that type. fsm_state can be used with a type or with a specific signal or variable, thus allowing us to choose different encodings for different FSMs.
- enum_encoding allows only user-defined encoding vectors (e.g., "111 000 101"); fsm_state allows user-defined encoding, plus the predefined values "BINARY", "GRAY", "ONE_HOT", and "ONE_COLD".
- Finally, if both attributes are specified for the same named entity in a design, then fsm_state takes precedence over enum_encoding.

19.6.2 Metacomments

Besides synthesis attributes, the synthesis standard defines a second way for a VHDL model to communicate with the synthesis tool. *Metacomments* are comments that provide additional information to a tool about how to process the source text.

The synthesis standard defines only two metacomments, which are used together to indicate that a piece of code should be ignored by the synthesis tool. Everything between the metacomments

```
-- rtl_synthesis off
```

and

```
-- rtl_synthesis on
```

should not be included in the synthesized model. Nevertheless, the source text between the metacomments must be valid VHDL code. A typical use is to ignore file and I/O operations in the middle of synthesizable code:

```
architecture rtl of multiplier is
begin
    y <= a * b;
```

```
-- rtl_synthesis off
process
    -- Code using file and I/O operations
    ...
end process;
-- rtl_synthesis on
end;
```

Just like with attributes, before the synthesis standard was published, vendors defined their own metacomments, which varied widely. We should check the tool manual for the list of metacomments recognized by any specific tool.

20 Testbenches

A testbench is simulation code that applies some stimulus to a design, observes its response, and verifies that it behaves correctly. In real-life projects, verification tasks consume more time and effort than RTL design. Verification engineers outnumber logic designers with ratios of up to four to one,[1] and the verification effort is already between 60% and 80% of the total design effort.[2]

Making sure that a design works correctly is obviously an important thing for a company. What may be less obvious is that testbenches are just as important for our day-to-day work as developers. The introduction of testbenches in a VHDL design has a huge impact on our ability to produce high-quality code. When we write a testbench for a model, our code becomes self-checking; it can tell us immediately when we make a change that would cause the design to stop working correctly. Working with this kind of code offers tremendous benefits. The most obvious advantage is that our code will have fewer bugs. The not so obvious effect is that we will be more inclined to improve the code because we are not afraid to break it. Because we will do more of these little improvements, our code will be cleaner, better organized, and easier to work with. The net effect is that, for all but the most trivial designs, writing good testbenches helps us develop better and faster.

This chapter takes a practical approach to teaching the design and implementation of testbenches. It is structured as a series of examples for verifying different kinds of circuits, including combinational, sequential, and FSM code. Each example introduces new concepts and techniques in the design and construction of testbenches. The last examples introduce the topics of functional coverage and random-based verification, two major trends in modern functional verification. The concepts and techniques presented in this chapter will help you write better testbenches in VHDL so that you can apply all the practices taught in this book without worrying about breaking your code.

20.1 Introduction

A *testbench* is a VHDL entity that instantiates a model, applies a set of inputs, examines the outputs, and produces a pass/fail result. The testbench code is simulation code; the instantiated model is synthesizable code. The main goal of a testbench is to verify the device functionality and ensure that it matches its specifications and the designer's intent.

Figure 20.1 shows how a testbench is used with a simulator. The testbench reproduces the real operating conditions of the design under verification (DUV). It performs many individual checks, each one aimed at demonstrating that part of the specifications are met by the design. If the DUV passes all tests and the tests cover all the specifications, then the design can be declared functionally correct. A testbench run fails when any of the individual checks produces an output that is different from the expected.

Testbenches can be developed before, after, or concurrently with the RTL design. From a product development perspective, testbenches are important because they let the development team know when the design is finished and whether it is correct. From the designer's perspective, testbenches offer many benefits during the development of RTL code. Besides confirming that any new features we implement are correct, testbenches give us confidence to change a design knowing that we will not introduce bugs that will go unnoticed. This has a tremendous impact on the quality of our code. In a design that is not covered by tests, the first solution that seems to work will tend to be used. The problem with this approach is that the first solution is rarely the simplest or the best; if we do not spend some time cleaning up the code after we have made it work, the quality of the source code will deteriorate over time. A good set of tests helps you avoid this fate.

An important characteristic of a VHDL testbench is that it can be written using only VHDL code. With other hardware languages, this is not always the case. VHDL allows us to reuse our knowledge and tools for both design entry and verification.

Compared with code that is written for synthesis, the main difference of testbench code is that it can use the full expressive power of the language. Some of the constructs that are especially useful in simulations include shared variables, file input and output, and all forms of the *wait* statement.

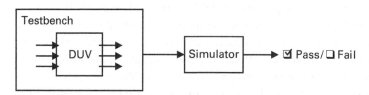

Figure 20.1
Using a testbench to verify a VHDL design.

Specification   RTL or   Silicon
                                       Netlist

Figure 20.2
Difference between the test and verification processes.

20.1.1   Test or Verification?

In hardware development, the term *testing* is usually associated with manufacturing tests. Thus, the purpose of testing is to verify that a *device* was manufactured correctly. In contrast, *verification* means demonstrating that a design correctly implements its specifications. Thus, the purpose of verification is to ensure that a *design* was created correctly (figure 20.2).[3] In other words, verification demonstrates that the intent of a design was preserved in its implementation.[4]

Writing testbenches is a verification task. Nevertheless, the word *test* is occasionally used in the context of verification. For example, each of the checks performed on the DUV is also called a *test* or a *testcase* in some contexts.

20.1.2   Functional Verification

Functional verification is the part of the verification process that considers only the logical behavior of a design. It does not include, for instance, timing verification. It is also not concerned with ensuring that an RTL description was translated correctly to a netlist or circuit. When a design is implemented as an RTL description, functional verification is the process of demonstrating that the RTL code matches its specifications. Most testbenches are used for functional verification and so are the examples presented in this chapter.

Functional verification can be described as the task of answering two questions about the functionality of a design: "Does it work?" and "Are we done?" Does-it-work questions address the *functional correctness* of the design. Are-we-done questions address the *functional coverage* provided by our verification suite.[5] In this chapter, we will see testbenches that address both kinds of questions.

20.1.3   What Should We Test?

When we are new to verification, one of the greatest difficulties is knowing *what* to test (another is knowing *how* to test it). In a formal design process, after the specifications are done, much effort is put into enumerating all the features and deciding how each of them will be verified. The result is called a *verification plan*. It specifies what testbenches will be created, which features will be exercised under which conditions, and the expected responses of the DUV. Once this is done, the verification task becomes a matter of implementing the verification plan.

For our testbenches, we will use a simple set of "Does it work?" and "Are we done?" questions, created from the short design specifications provided in the examples. Then we will implement testcases for one or more features by writing code that exercises the device in the conditions required to observe the feature in operation.

To illustrate this process, consider the case of a BCD to 7-segment decoder—a device that receives a single decimal digit in binary form and shows it in a 7-segment display. What kind of questions do we need to answer to ensure that our design works? Because this design has a simple transfer function, we could write our verification questions by enumerating all of its inputs and outputs:

- Does the display show the digit 0 when the input is the binary value "0000"?
- Does the display show the digit 1 when the input is the binary value "0001"?
- Does the display show the letter F when the input is the binary value "1111"?

There are two things to notice about these questions. First, it would be more practical to group them into a single question such as, "Does the display show the correct digit for all possible input values?" Second, we still need to rephrase them using terms that relate to the interface presented by the device under verification—its input and output ports. For a combinational circuit, we could use a truth table and then ask a single question: "Does the device present the correct output values for all the input values presented in the truth table?"

In any case, our goal is to write a set of yes/no questions corresponding to the features described in the design specifications. Then for each question, we will write code that applies stimuli to our design and checks whether its responses correspond to the expected values.

The questions above are all "does-it-work" questions, aimed at checking the *functional correctness* of the design. Because this design has a small number of inputs, we can enumerate and test all of them. However, if a design has more than a moderate number of inputs, outputs, or internal registers, the number of combinations quickly approaches the limits of what is viable to simulate. In such cases, we need another set of questions aimed at answering, "Which of all the possible conditions do we need to simulate to be confident that the design is correct?" These are *functional coverage* questions. For example, in the design of a CPU, we could ask:

- Has every instruction been tested with every register it can use as an input?
- Has every legal transition been tested in the instruction execution FSM?

Once we have asked the right questions, creating a testbench is a matter of writing the code to check that they are true. In a real design, there could be hundreds of such questions. Therefore, it is important to organize our testbenches so that a good part of the verification infrastructure can be reused across different tests.

20.1.4 Self-Checking Testbenches

A testbench that checks the outputs of the design against a set of expected responses is called a *self-checking testbench*. The term *self-checking* is used to distinguish it from other kinds of testbenches that do not check the outputs of the DUV, leaving this work for the developer. In practice, a testbench that does not check the DUV outputs is useless for verification. It also does not have the desirable characteristics that motivate us to write tests—it does not tell us when we break something, it does not produce a pass/fail result, and it does not save us time because we need to inspect the results manually. Non-self-checking testbenches may be useful during the initial development and implementation of a model when we are still debugging the code. However, if you have to visually inspect dozens of waveforms to know whether a design works or a change broke anything, then you will have a strong incentive *not* to change anything. It is strongly recommended that all testbenches be self-checking. This is the only kind of testbench we will see in this chapter.

20.1.5 Basic Parts of a Testbench

Because testbenches are just VHDL code and can use all the statements available in the language, their implementation details will vary. However, certain basic elements are present in every testbench. Figure 20.3 shows the major parts of a self-checking testbench. Although the stimulus generator and response checker appear as single blocks, in practice they may be composed of several smaller blocks with more specific purposes. Depending on the size and complexity of the testbench, each block can be represented by different VHDL constructs. For example, in a complex and reusable testbench, each part could be an instantiated component. In a simple testbench, each block could be a process, procedure, or series of statements.

The DUV is present in every testbench. It is often a component or entity that is instantiated in the testbench, although it could be a package whose subprograms are tested directly in the testbench code. The DUV is also known by other acronyms, created from the possible combinations of the words design, circuit, module, unit, verification, and test. Some of the most common include DUV, DUT, UUT, and MUT.

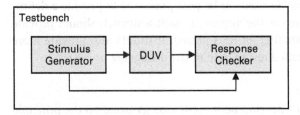

Figure 20.3
Basic parts of a testbench.

The stimulus generator produces the necessary signals to exercise the DUV. These signals are called *stimuli* (plural) or *stimulus* (singular). There are several ways to generate stimuli. One is to write them directly in the VHDL code. Another method is to read the input vectors from a file. A third alternative is to use random values. When the values are chosen explicitly, the performed checks are called *directed tests*. When the input values are chosen randomly, the checks are called *random tests*. Random tests must specify a series of limits and constraints to restrict the applied values to a valid range. This approach is called *constrained random verification*. If the stimulus generator uses coverage information to select random values that have not yet been tested, then the approach is called *coverage-driven randomization* or *intelligent coverage*.

The last major part of a testbench is the response checker. Its main goal is to compare the results produced by the DUV with values that are known to be correct, which may come from a number of possible sources. For example, we could write the expected responses directly in the testbench code using assertions. We could also use a list of input values and the corresponding expected outputs; such values are often called *golden vectors*. Another possibility is to use a second model that is known to have the same functionality expected from the DUV, called a *golden reference model*. Finally, we could create another implementation of the DUV (or part of it) at a higher level of abstraction. Any alternative implementation of an operation performed by the device is called a *transfer function* of the functionality under test.

Depending on the size and complexity of the testbench, the stimulus generator and response checker could be broken down into several smaller modules. Many of these modules have specific roles and appear regularly in testbenches. Such modules are called *verification components*. We will introduce those elements as they appear in our examples.

20.1.6   What is the Output of a Testbench?

As shown in figures 20.1 and 20.3, the testbench is a self-contained system and a closed environment. Consequently, as a design entity, it does not have any input or output ports. The outputs of a testbench are the results of each individual test that it performs on the DUV, which are output as text messages on the system console and usually saved to a file. In many cases, this file needs to be post-processed to provide a definitive answer about the tests. Whatever the approach, each testbench should provide a definitive message indicating whether all tests passed. It should also provide more detailed information about each failed test to help in the debugging process.

20.1.7   Test-Driven Development

Test-driven development (TDD) is a development methodology based on the principle of using tests to guide the design and implementation processes. TDD turns the conventional development approach on its head, putting the task of writing tests before the task of writing application code.[6] Among other benefits, TDD encourages simple design, helps ensure that all features are tested, and fosters good design principles.

TDD is characterized by the repetition of a short development cycle (figure 20.4) consisting of the following steps[7]:

- Write a simple test for a new or improved functionality.
- Run all tests and see the new test fail (this is the "red" stage).
- Write the minimum amount of code to make the test pass. At this stage, we can commit any sins to get to green fast.[8]
- Run all tests; they should all pass (this is the "green" stage).
- Refactor the code to remove duplication and bring it to an acceptable quality standard.

This cycle is summarized by the "TDD Mantra": "Red, Green, Refactor." This cycle allows us to focus on a single, clear goal at a time. For TDD to work, the increments must be small and simple, and the entire loop should only take minutes. Each test that works brings us one step closer to the finished solution.

TDD has some implications for the way we conduct our development tasks.[9] First, we must design iteratively and incrementally, preferably in small steps. This allows us to learn from the code already implemented and use it as feedback for our design decisions. Second, each developer must write his or her own tests. Note that this does not preclude acceptance tests or other system tests required in later verification phases. Third, our tests must provide rapid feedback from any code changes and should let us know immediately when we break something. Fourth, to make such frequent testing possible, our design must be structured as many loosely coupled and strongly cohesive modules (which is a good recommendation in itself).

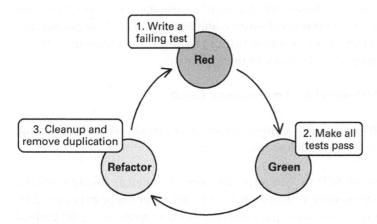

Figure 20.4
The TDD cycle: Red, Green, Refactor.

Depending on your development style, it may seem counterintuitive to write tests first. However, this approach offers many benefits:

- It tends to produce simpler designs by encouraging us to write just enough code to make the current test pass. This prevents gold-plating and premature generalization.
- It encourages us to think about the requirements before writing the code. We can use the tests to clarify our design ideas and imagine the best interface for each task.
- It helps us know when we are done implementing a feature: as soon as the test passes.
- It helps ensure that every feature gets tested. The tests then become an executable specification and a regression suite to prevent errors from creeping into the system.
- It helps ensure that the tests really work. Seeing the test fail proves that it can detect a feature that is not working. Seeing it pass later proves that it verifies the expected behavior.
- It leads to better code—code that is more modular, less coupled, and more reusable.
- It leads to code that is easier to use and has a better interface because it encourages the designer to think from the client side.

The basic TDD steps are mostly tool-independent, so we can follow them with any development process already in place. In practice, however, the tests are often written with the help of a unit testing framework. For VHDL, the most advanced framework is VUnit, based on the popular xUnit architecture that is available for several other languages. VUnit provides the infrastructure to run unit tests with the help of Python scripts, for advanced features, or with pure VHDL code for better compatibility. It offers improved logging support, file dependency scanners, and automatic compilation and execution of test suites. The project is open sourced and can be downloaded from the project page on GitHub.[10]

Because it can represent a significant change in development practices, TDD may not be the best approach for everyone. However, unit testing and TDD are gaining traction in many VHDL circles. Because unit testing can be used with or without TDD, you are highly encouraged to try at least one of these techniques.

20.2   Example #1: Testbench for a Combinational Circuit

We will start our walkthrough with a simple testbench for a combinational circuit. Suppose we are given the following design specifications.

Design Specifications—A BCD to 7-Segment Decoder   The design should convert a 4-bit binary input representing a decimal digit to a 7-bit output appropriate for driving a 7-segment display. When the input value does not correspond to a valid decimal digit, the display should be off. The truth table for the circuit is shown in figure 20.5. The value should be presented at the output in at most 15 ns.

Decimal	BCD	a b c d e f g
0	0000	0 1 1 1 1 1 1
1	0001	0 0 0 0 1 1 0
2	0010	1 0 1 1 0 1 1
3	0011	1 0 0 1 1 1 1
4	0100	1 1 0 0 1 1 0
5	0101	1 1 0 1 1 0 1
6	0110	1 1 1 1 1 0 1
7	0111	0 0 0 0 1 1 1
8	1000	1 1 1 1 1 1 1
9	1001	1 1 0 1 1 1 1
10–15	----	0 0 0 0 0 0 0

Figure 20.5
Truth table for the BCD to 7-segment decoder.

The first step in planning the testbench is to write down our does-it-work and are-we-done questions. The does-it-work questions can be read directly from the textual specifications or from the truth table. As for the are-we-done questions, because every input is associated with an arbitrary output value, we need to test them exhaustively. This is not a problem because the number of inputs is small. The timing requirement will be addressed in a later iteration of our testbench.

Based on the described features, we could write down the following questions in our verification plan:

- Does the display show the correct digit for all input values from 0 to 9 as given in the truth table?
- Is the display off for all input values from 10 to 15?
- Have all the rows in the truth table been tested, and did they produce the correct response?

Depending on how we organize our testbench, we could combine the questions or write a separate test for each one of them. Because the three questions use the same table, it is probably easier to combine them into a single test.

First Iteration—A Linear Testbench At this point, we are ready to start designing our testbench using the concepts seen in the previous sections. We will not discuss the implementation of the DUV because it is not relevant to our verification task. This makes our job one of *black-box* testing: the DUV is tested without any knowledge of its internal implementation. In case you want to experiment with the source code, all the code used in the examples can be downloaded from this book's companion website.

To ease our way into writing testbenches, we will start with the simplest solution that works. For a testbench, this means instantiating the DUV in a top-level design entity and then generating the stimuli and checking the responses in a single *process*

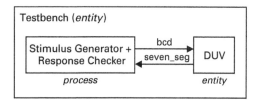

Figure 20.6
Block diagram of a linear testbench for the BCD to 7-segment decoder.

statement. This kind of testbench is called a *linear testbench* because everything happens in a single thread of execution. Figure 20.6 shows a block diagram for this testbench.

With the testbench design done, we can start writing its code. Listing 20.1 shows the skeleton code for our linear testbench.

Listing 20.1 Skeleton code for the linear testbench of figure 20.6

```
1    entity bcd_to_7seg_tb is
2    end;
3
4    architecture testbench of bcd_to_7seg_tb is
5        signal bcd: std_logic_vector(3 downto 0);
6        signal seven_seg: std_logic_vector(6 downto 0);
7    begin
8        duv: entity work.bcd_to_7seg
9            port map(bcd => bcd, seven_seg => seven_seg);
10
11       stimuli_and_checker: process
12       begin
13           report "Testing entity bcd_to_7seg.";
14
15           -- TODO: Add testbench code here...
16
17           report "End of testbench. All tests passed.";
18           std.env.finish;
19       end process;
20   end;
```

There are some points to note about this initial code:

- The entity declaration (lines 1–2) has no input or output ports because the testbench is a closed system and a self-contained environment.

- The testbench entity is named after the DUV by adding the suffix _tb (line 1). This convention facilitates distinguishing between synthesis code and testbench code; it also makes it easy to match the RTL code with the corresponding testbench.
- The architecture declares all the signals necessary to communicate with the DUV (lines 5–6). To prevent any confusion, the signal names match the DUV ports.
- The DUV is instantiated directly from library work, using a component instantiation statement (lines 8–9).
- Because this is a simple testbench, a single process controls the entire test. It is also responsible for finishing the simulation once all tests have completed. The procedure finish (line 18) is declared in package env and terminates a simulation.

Although incomplete, this skeleton code can be compiled and run. As you write testbenches (or any code for that matter), it is a good idea to run frequent smoke tests after any modification. If we run the code from listing 20.1, then we should see the following message in the simulation console:

```
# ** Note: End of testbench. All tests passed.
```

The next step is to flesh out the main process, starting with the declarations that will be needed in our test. The best way to specify a small truth table is with a constant array. For the element type, we can use a record whose fields correspond to the DUV inputs and outputs. It is not a good idea to use a single std_logic_vector for holding both the stimuli and response because it would not carry any information about the meaning of the individual values. Remember the recommendation to keep as much information as possible in the code and as little as possible in our heads. The declarations for the stimuli and responses are:

```
type stimulus_response_type is record
    bcd: std_logic_vector(3 downto 0);
    seven_segment: std_logic_vector(6 downto 0);
end record;

type stimulus_response_array_type is array (natural range <>) of
    stimulus_response_type;

constant TEST_DATA: stimulus_response_array_type := (
    (x"0", "0111111"),
    (x"1", "0000110"),
    (x"2", "1011011"),
    (x"3", "1001111"),
    (x"4", "1100110"),
    (x"5", "1101101"),
```

```
    (x"6", "1111101"),
    (x"7", "0000111"),
    (x"8", "1111111"),
    (x"9", "1101111"),
    (x"a", "0000000"),
    (x"b", "0000000"),
    (x"c", "0000000"),
    (x"d", "0000000"),
    (x"e", "0000000"),
    (x"f", "0000000")
);
```

Because the object and types are used only in the main test process, we can declare them locally in the process declarative part.

The next step is to apply the stimuli and check the DUV outputs. Because all the test data is provided in an array, it is easy to read in a loop. Each iteration applies one test vector to the DUV input, waits for some amount of time for the signals to propagate, and then checks whether the DUV output presents the expected value. This can be done with an assertion whose condition is true when the actual value matches the expected value:

```
for i in TEST_DATA'range loop
    bcd <= TEST_DATA(i).bcd;
    wait for TEST_PERIOD;

    assert seven_seg = TEST_DATA(i).seven_segment
        report "Error: DUV.seven_seg /= TEST_DATA.seven_seg"
        severity failure;
end loop;
```

In principle, for a functional verification, the constant TEST_PERIOD could be a very short delay, because a combinational RTL model only needs delta time to update its outputs.[11] However, using time delays that are compatible with the circuit specifications has two advantages. First, the simulation waveforms will correspond to the normal operating conditions of the finished design, making them easier to interpret and correlate with the original specifications. Second, we can reuse the same testbench for the synthesized gate-level netlist produced by the synthesis tool, allowing us to perform dynamic timing checks besides functional verification. In this version of the testbench, we chose a period of 20 ns; in the upcoming iterations, we will verify that the correct value is presented in at most 15 ns and that it remains stable for the last 5 ns of the test period.

Listing 20.2 shows the finished code for the linear testbench. The code can still be improved, but it does demonstrate that the implemented design matches its specifications.

Listing 20.2 Finished testbench code for the BCD to 7-segment decoder

```
 1   architecture testbench of bcd_to_7seg_tb is
 2       signal bcd: std_logic_vector(3 downto 0);
 3       signal seven_seg: std_logic_vector(6 downto 0);
 4   begin
 5       duv: entity work.bcd_to_7seg
 6           port map(bcd => bcd, seven_seg => seven_seg);
 7
 8       stimuli_and_checker: process
 9           report "Testing entity bcd_to_7seg.";
10
11           type stimulus_response_type is record
12               bcd: std_logic_vector(3 downto 0);
13               seven_segment: std_logic_vector(6 downto 0);
14           end record;
15
16           type stimulus_response_array_type is array (natural range <>) of
17               stimulus_response_type;
18
19           constant TEST_DATA: stimulus_response_array_type := (
20               (x"0", "0111111"),
21               ...
22               (x"f", "0000000")
23           );
24
25           constant TEST_PERIOD: time := 20 ns;
26       begin
27           for i in TEST_DATA'range loop
28               report "Testing DUV with input " & to_string(TEST_DATA(i).bcd);
29               bcd <= TEST_DATA(i).bcd;
30               wait for TEST_PERIOD;
31
32               assert seven_seg = TEST_DATA(i).seven_segment
33                   report "Error: DUV.seven_seg /= TEST_DATA.seven_seg"
34                   severity failure;
35           end loop;
36
37           report "End of testbench. All tests passed.";
```

```
38          std.env.finish;
39      end process;
40  end;
```

Second Iteration—A Modular Testbench   The testbench from the previous example could benefit from several improvements. For example, it would be a good idea to separate the responsibilities of the `stimuli_and_checker` process into a stimulus generator process and a response checker process. Although the benefits are not so apparent in a small testbench, there are several advantages to keeping the two tasks in separate threads. First, each process becomes shorter, more focused, and easier to understand. Second, as the testbench grows, it is easier to move each process to a proper entity. Third, depending on the DUV behavior, it is not always possible to apply one input, wait for an output, and then check the response. When the DUV is pipelined, for example, we would need to account for the latency between the input and output values, complicating the test procedure. This approach would also prevent the DUV from operating at its maximum throughput rate.

Moving the response checking to a separate process is easy. The only problem is that the checker needs to correlate the output read from the DUV with the corresponding golden output vector. We could assume that the stimuli will always be presented in the order they appear in the table and keep an incrementing counter in the checker process; however, it is bad form for one module to make assumptions about the behavior of another module. Alternatively, we could have the generator send the expected response to the checker, but then it would be using information that is not related with its central goal of generating the stimuli. For now, we will compromise a little and have the generator send to the checker the index of the current stimulus in the test data table. This can be done easily with a signal. Ideally, we would like to make the two processes even less coupled, but we can use this opportunity to demonstrate how two processes in a testbench can use a signal to coordinate their operation.

In a simulation, a signal can carry both a data value and an activation request: if a process is waiting for a transaction on that signal, then when an assignment is made the process will resume execution and can read the signal value. In figure 20.7, the signal `test_vector_index` informs the checker process that a new input has been provided to the DUV; it also provides the index of the current stimulus in the test data table. The checker can access the same data table used by the generator and look up the expected response using the index value.

We could also make the testbench less verbose and more informative by declaring a few auxiliary subprograms. It is generally a good idea to create one or more packages with routines commonly used in our testbenches. Tests need to be easy and fast to write, or we will write fewer of them. The testbenches in this chapter will assume the existence of a procedure called `print` that outputs a string or the value of an

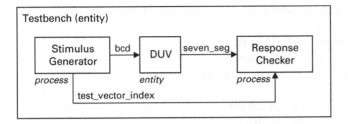

Figure 20.7
Separating stimulus generation from response checking.

object of any type. We will also use a procedure named `assert_eq` (assert equal), which tests whether two values are equal and causes an assertion violation otherwise. It also prints a more informative message, including the actual and expected values:

```
procedure assert_eq(message: string; actual, expected: std_ulogic) is
begin
    if actual = expected then
        print("- " & message);
    else
        report "Error in test '" & message & "'" & LF & "       " &
        "actual: " & to_string(actual) & ", " &
        "expected: " & to_string(expected)
        severity failure;
    end if;
end procedure;
```

To test this procedure, we could manually insert an error in the test data. For example, if we changed the third row of the test data table, then we could see the following simulation output:

```
- DUV output equals golden test data
- DUV output equals golden test data
** Failure: Error in test 'DUV output equals golden test data'
    actual: 1011011, expected: 1011010
```

Listing 20.3 shows the refactored version of the testbench. Note that the declarations were moved to the architecture declarative part because they need to be accessed from the two processes. A declaration was added for signal `test_vector_index`, which carries data and control information from the generator to the checker. In lines 29–30, the generator applies a new input to the DUV and then assigns the index of the current test vector to this new signal. This resumes the checker process, which then uses the index value to read the expected response from the test data table.

Listing 20.3 Decoupled and refactored version of the 7-segment decoder testbench

```vhdl
1    architecture testbench of bcd_to_7seg_tb is
2        signal bcd: std_logic_vector(3 downto 0);
3        signal seven_seg: std_logic_vector(6 downto 0);
4
5        type stimulus_response_type is record
6            bcd: std_logic_vector(3 downto 0);
7            seven_segment: std_logic_vector(6 downto 0);
8        end record;
9
10       type stimulus_response_array_type is array (natural range <>) of
11           stimulus_response_type;
12
13       constant TEST_DATA: stimulus_response_array_type := (
14           (x"0", "0111111"),
15           ...
16           (x"f", "0000000")
17       );
18
19       signal test_vector_index: natural range 0 to TEST_DATA'length-1 := 0;
20
21       constant DUV_DELAY: time := 12 ns;
22       constant TEST_INTERVAL: time := 20 ns;
23   begin
24       duv: entity work.bcd_to_7seg
25           port map(bcd => bcd, seven_seg => seven_seg);
26
27       stimuli_generator: process begin
28           for i in TEST_DATA'range loop
29               bcd <= TEST_DATA(i).bcd;
30               test_vector_index <= i;
31               wait for TEST_INTERVAL;
32           end loop;
33
34           report "End of testbench. All tests passed.";
35           std.env.finish;
36       end process;
37
38       response_checker: process begin
39           wait on test_vector_index'transaction;
40           wait for DUV_DELAY;
41           assert seven_seg = TEST_DATA(test_vector_index).seven_segment
```

```
42              report "Error: DUV.seven_seg /= TEST_DATA.seven_seg"
43              severity failure;
44      end process;
45   end;
```

Third Iteration—A Decoupled Testbench Separating stimulus generation from response checking was an improvement, but it is possible to take this idea one step further. We can decouple the two processes completely by making the checker process obtain all of its information from the signals that are connected to the DUV. In test-bench jargon, a component that observes the DUV inputs or outputs and translates them to higher level transactions is called a *monitor*. It is good practice to rely on moni-tors to provide all the inputs to the checker, rather than communicating ad hoc with the stimulus generator and other high-level processes. This ensures that the part of the testbench that performs the analysis will be reusable with different sources of stimulus. Because our testbenches are extremely simple, most of them will not use monitors as distinct processes, but this practice is still worth following.

However, decoupling the stimulus generator from the response checker causes another problem: if the checker can only see the raw DUV inputs and outputs, how can it know whether an output is correct? In the previous testbench, we used the array index passed from the stimulus generator to find the right stimulus-response pair in the table.

The answer is to use a *predictor* to obtain the expected response from a DUV input. There are several ways to implement a predictor. One is to use a golden model—a refer-ence implementation of the same algorithm performed by the DUV. Another alterna-tive is to implement a simpler model at a higher level of abstraction. In our case, all we have is a truth table, so our predictor can be a simple function that looks up the expected value corresponding to a given input value:

```
function predictor(bcd: std_logic_vector(3 downto 0))
    return std_logic_vector is
begin
    for i in TEST_DATA'range loop
        if TEST_DATA(i).bcd = bcd then
            return TEST_DATA(i).seven_segment;
        end if;
    end loop;
    report_error("Predictor error: input not found.");
end;
```

Here is how we would change the response checking process to use this predictor:

```
response_checker: process
    variable bcd_sample: std_logic_vector(3 downto 0);
begin
    -- Sample the DUV input when it changes
    wait on bcd;
    bcd_sample := bcd;

    -- Wait for the outputs to settle and check results
    wait for DUV_DELAY;
    assert_eq(seven_seg, predictor(bcd_sample),
        "DUV output equals golden test data");
end process;
```

It is not necessary to show the entire testbench code because the other parts are the same as in listing 20.3. Figure 20.8 shows a block diagram of the decoupled version of the testbench.

We can use this new testbench design to introduce another component called a *scoreboard*. A scoreboard is a verification component that correlates the input transactions on the DUV with the observed responses. There are two different interpretations for the role of a scoreboard in a testbench. The first is that the scoreboard is simply a data structure that holds the necessary information for the checker. The second is that the scoreboard *is* the checker, and it should include all the necessary mechanisms to compare an input transaction with an expected response, including a predictor (e.g., a golden reference model) and a comparator. The scoreboard shown in figure 20.8 adheres to this second definition.

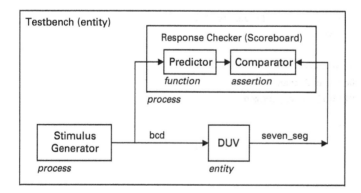

Figure 20.8
Testbench with decoupled stimulus generation and response checking.

Fourth Iteration—Adding Timing Checks The final improvement we will make to our testbench is to add timing checks. Remember that our specifications require the decoded value to be present at the output in at most 15 ns. We can add a check for this requirement as a separate process, using the attribute 'stable to ensure that the output value did not change after being sampled:

```
timing_checker: process begin
    wait on bcd;
    wait for TEST_PERIOD;
    assert_true(
        "DUV output is stable",
        seven_seg'stable(TEST_PERIOD - DUV_DELAY)
    );
end process;
```

If we run the modified testbench using the RTL model of the DUV, then the test will always pass because the functional model does not include any explicit delays. To see the new test at work, we should replace the RTL model with the synthesized gate-level netlist produced by the synthesis tool and set the DUV_DELAY constant to a value that is lower than the propagation delay of the synthesized model (say, 12 ns). When we run the simulation, we should see an output similar to the following:

```
# ** Failure: Error in test 'DUV output equals golden test data'
#      actual: 0111X11, expected: 0111111
#    Time: 12 ns  Iteration: 0  Instance: /bcd_to_7seg_tb
# ** Failure: Error in test 'DUV output is stable'
#      assertion expression is false
#    Time: 20 ns  Iteration: 0  Instance: /bcd_to_7seg_tb
```

20.3 Example #2: Testbench for a Clocked Circuit

Verifying a sequential circuit is usually a more complex operation because it requires more interaction between the DUV and testbench; it is not enough to provide and input and wait for an output. We will use a design of the receive half of a UART to demonstrate a testbench for a clocked circuit.

Design Specifications—UART Rx Module The uart_rx module (figure 20.9) receives a standard RS-232 signal at 115,200 bps and outputs the received data one byte at a time. The reception of a byte is initiated with a start bit (logic level 0) and concluded with a stop bit (logic level 1). When the reception is completed, the module pulses the strobe output high for one clock cycle. The module operates at 100 MHz.

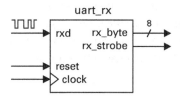

Figure 20.9
UART receive module (DUV).

This is a short description for this kind of circuit. In a real-life design, the specifications would be much more precise and elaborate. Nevertheless, it is enough for our purposes to demonstrate the design of a testbench. Here are some of the questions we could ask in our verification plan:

- Is the sequence of levels [low, high, high, low, low, high, low, high, low, high] with a bit period of 8.7 us interpreted as the byte "01010011"?
- Is a sequence rejected if it does not have a stop bit?
- Is every byte value from 0 to 255 received correctly?
- Are the bytes received correctly when the bit period is reduced by 5%?

The first question is an example of a *simplest success case* test.[12] This kind of test is useful to demonstrate that the basic functionality is correct, and it is a good way to start a testbench. Later, we can add corner cases and other tests aimed at stressing the design more thoroughly.

We could add many other questions about the functional correctness of the design and the functional coverage that we want to achieve in our verification process. The thoroughness of these questions depends on how much confidence we want to have that our design is working as specified. It is a good exercise to think of some other questions that could be added to the list.

First Iteration—A Linear Testbench   Our first iteration will be a linear testbench to test the simplest success case defined in our verification plan. In general, linear testbenches are adequate only for the simplest designs. Here we use it only as a stepping stone for a better structure that will arise in the next iterations.

In a linear testbench, it would be possible to treat the clock signal as any other input and generate the clock transitions directly in the stimulus process. However, this complicates the code and creates an undesired coupling between high-level tasks and low-level bit manipulation. In general, it is best to create a free-running clock that operates independently from the main test sequence.

Listing 20.4 shows the complete code for the linear testbench. It follows the same layout from the 7-segment decoder example; the main difference is the generation of the clock and reset signals (lines 14–15). Note that the clock signal needs to be initialized with a logic level (line 5), or it will not toggle: the negation of the default initial value 'U' is also 'U', which would cause the signal to keep this value indefinitely.

Listing 20.4 Linear testbench with a simplest success case for the UART Rx module

```
1    entity uart_rx_tb is
2    end;
3
4    architecture testbench of uart_rx_tb is
5        signal clock: std_logic := '0';
6        signal reset: std_logic := '0';
7        signal rxd: std_logic := '1';
8        signal rx_strobe: std_logic;
9        signal rx_data: std_logic_vector(7 downto 0);
10       constant BIT_LENGTH: time := 8.68 us;
11   begin
12       duv: entity work.uart_rx port map (clock => clock, reset => reset,
13           rx_data => rx_data, rx_strobe => rx_strobe, rxd => rxd);
14
15       clock <= not clock after 5 ns;
16       reset <= '1', '0' after 20 ns;
17
18       stimuli_and_checker: process
19           -- Start bit, data bits (LSB to MSB), and stop bit
20           constant TX_DATA: std_logic_vector := b"0_11001010_1";
21       begin
22           wait until not reset;
23
24           -- Drive Tx bits one at a time on DUV rxd line
25           for i in TX_DATA'range loop
26               rxd <= TX_DATA(i);
27               wait for BIT_LENGTH;
28           end loop;
29
30           -- Wait for DUV to acknowledge the reception
31           wait until rx_strobe for BIT_LENGTH;
32
33           assert_eq(rx_strobe, '1', "rx_strobe is 1");
34           assert_eq(rx_data, "01010011", "rx_data is 01010011");
35
36           print("End of testbench. All tests passed.");
37           std.env.finish;
38       end process;
39   end;
```

Because the DUV needs a reset pulse, the stimulus process cannot start applying data immediately. The *wait* statement in line 22 waits until the reset signal has been deasserted. The input stimulus is a single byte given by the constant TX_DATA in the main test process (line 20), which also includes the start and stop bits. The transmit data is driven on the DUV rxd line one bit at a time using a loop (lines 25–28). Next, the process waits until the DUV acknowledges receipt of the transmitted byte (line 31). Finally, the test performs two assertions to check the DUV outputs (lines 33–34).

A last point to note about the interaction with the reset signal is that in listing 20.4, the stimulus starts as soon as reset is deasserted. If the DUV uses a synchronous reset, then it is also possible (and often necessary) to make the *wait* statement synchronous to a clock edge by writing:

```
-- Wait until reset is low and a rising clock edge occurs:
wait until rising_edge(clock) and reset = '0';
```

Second Iteration—A Modular Testbench   One of the main problems with a linear testbench is that it is hard to reuse parts of the code for other tests. Another problem is that it is hard to see the overall sequence of operations because the task of defining the stimuli values is mixed with the task of performing the low-level pin manipulation to communicate with the DUV. One way to improve reusability and readability is to separate the tasks of providing high-level stimulus from interfacing with the DUV pins.

A component that translates higher level stimuli to low-level pin activity on the DUV interface is called a *driver*. The higher level stimuli are often called *transactions* or *commands*. Examples of transactions include writing a value to a memory address, performing an arithmetic operation on an ALU, or sending a network packet. In VHDL, a transaction is usually described with a record type because it typically involves several parameters. A stimulus generator encapsulates in a transaction all the knowledge needed to describe an operation and then sends the transaction to the driver. The driver will perform the low-level signaling required by the interface protocol, possibly using multiple clock cycles, and will notify the stimulus generator when the operation is completed.

The testbench for the UART Rx module is a good candidate for using a driver. We could use it to clean up the stimulus generator code by hiding from it any bit manipulation. In a larger design, a driver would be a proper entity instantiated in the testbench, making it reusable across tests and in different testbenches. In our simple testbench, the driver will be implemented as a process. Another possible implementation would be as a bus-functional model (BFM) with a procedural interface, as shown in the example of figure 12.5. In this way, the stimulus generator could request the transmission of a byte with a simple procedure call.

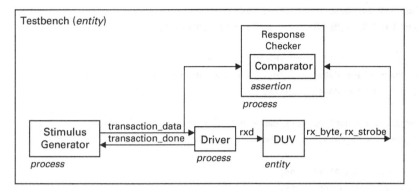

Figure 20.10
Modular testbench design for the UART Rx module.

Figure 20.10 shows a modular organization for the UART testbench. The stimulus generator has the single responsibility of producing higher level stimuli—in this case, a byte. Because the transaction is simple, there is no need for a record type. The driver encapsulates the knowledge about the serial protocol, converting the byte values to pin wiggles on the DUV rxd input. When the transmission is complete, the driver notifies the stimulus process by making an assignment to signal transaction_done. The rest of the testbench is similar to our previous examples. In this case, the response checker does not need a predictor because the same kinds of data are present on the stimulus transaction and DUV outputs.

Listing 20.5 shows the architecture of the modular UART testbench. Two signals were added for the communication between the stimulus generator and the driver (lines 7–8). We have also extended our testbench to send all bytes from 0 to 255, using a *for* loop (lines 20–25). When the stimulus process wants to send a byte, it assigns the value to be sent to signal transaction_data (line 22). This causes the driver process to resume (line 39). After sending the byte, the driver process makes an assignment to transaction_done, causing the stimulus process to resume. Note that the actual value of this assignment is irrelevant because the *wait* statements in the processes use the 'transaction attribute, which toggles on every assignment. In this example, we always assign the value true. In some occasions, it may be useful to make the handshake signals change or toggle on each transaction, making the events easily identifiable in a waveform viewer.

Listing 20.5 Architecture of the modular UART testbench

```
1 architecture testbench of uart_rx_tb is
2    ...
```

```
 3     constant BIT_INTERVAL: time := 8.7 us;
 4     constant BYTE_RX_TIMEOUT: time := 11 * BIT_INTERVAL;
 5
 6     -- Interface signals between the stimulus and driver processes
 7     signal transaction_data: std_logic_vector(7 downto 0);
 8     signal transaction_done: boolean;
 9 begin
10     duv: entity work.uart_rx port map(clock => clock, reset => reset,
11         rx_data => rx_data, rx_strobe => rx_strobe, rxd => rxd);
12
13     clock <= not clock after 5 ns;
14     reset <= '1', '0' after 20 ns;
15
16     stimuli_generator: process begin
17         wait until not reset;
18
19         -- Send to the driver all characters from 0-255, one at a time
20         for i in 0 to 255 loop
21             -- Writing to signal send_data resumes the driver process
22             transaction_data <= std_logic_vector(to_unsigned(i, 8));
23             -- Wait until driver has completed the transmit operation
24             wait on transaction_done'transaction;
25         end loop;
26
27         wait for BIT_INTERVAL;
28
29         print("End of testbench. All tests passed.");
30         std.env.finish;
31     end process;
32
33     -- Convert the byte value on send_data to transitions on the rxd pin.
34     -- Notifies completion with a transaction on send_done.
35     driver: process
36         variable tx_buffer: std_logic_vector(9 downto 0);
37     begin
38         -- Wait until stimulus generator requests a transmit operation
39         wait on transaction_data'transaction;
40
41         -- Add stop and start bits, and send value one bit at a time
42         tx_buffer := '1' & transaction_data & '0';
43         for i in tx_buffer'reverse_range loop
44             rxd <= tx_buffer(i);
45             wait for BIT_INTERVAL;
```

```
46        end loop;
47
48        transaction_done <= true;
49     end process;
50
51     response_checker: process
52        variable expected_byte: std_logic_vector(7 downto 0);
53     begin
54        -- Make a copy of the high-level stimulus sent to the driver
55        wait on transaction_data'transaction;
56        expected_byte := transaction_data;
57
58        -- Wait until the DUV acknowledges a reception or a timeout occurs
59        wait until rx_strobe for BYTE_RX_TIMEOUT;
60
61        -- Compare the actual and expected values of the DUV outputs
62        assert_eq(rx_strobe, '1', "rx_strobe is 1");
63        assert_eq(rx_data, expected_byte, "rx_data is " & to_string(rx_data));
64     end process;
65 end;
```

Note that the use of the attribute 'transaction causes a small terminology conflict: there are *testbench* transactions, which are high-level stimuli or responses, and *signal* transactions, which are values scheduled for a signal during a simulation. The same inconvenience happens with the term *driver*. Because the focus of this chapter is on testbenches, when these terms appear unqualified in the text, they will always have their testbench-related meanings.

20.4 Example #3: Testbench for an FSM

Finite state machines (FSMs) have certain characteristics that warrant special attention during verification. We will use a simple vending machine as an example of how to write a testbench for an FSM.

Design Specifications The vending machine interfaces with a coin acceptor, a display for the collected amount, and a dispenser. The machine accepts nickels (5¢ coins) and dimes (10¢ coins). When the amount collected is greater than or equal to 15¢, the FSM produces a single-clock pulse on the dispenser output. The external display accepts unsigned values from 0 to 15 and should show the current collected amount. The machine does not give change, and any excess amount is lost when an item is dispensed. When a nickel and a dime are inserted at the same time, the machine should ignore the coin of lesser value.

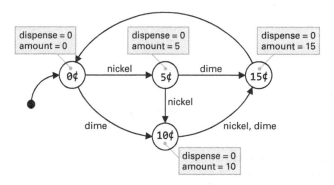

Figure 20.11
State diagram for the vending machine FSM.

How do we know whether the FSM works? We could try to verify an FSM like any other circuit, applying stimuli at its inputs and observing its outputs. However, this is not always the best approach. Besides the observable behavior at its outputs, an FSM has two coverage metrics that are worth tracking: *state coverage* and *transition coverage*. We will extend this example to use transition coverage in section 20.8. For now, we will only verify the functionality presented at the FSM outputs. In any case, we need to verify all the states, outputs, and transitions.

Here are some of the questions we could ask about the correctness of the FSM:

- Does the machine start with `amount = 0` and `dispense = '0'` upon reset?
- Do the `amount` and `dispense` outputs present the values [0, 5, 10, 15, 0] and [0, 0, 0, 1, 0], respectively, when three nickels are inserted at consecutive clock edges?
- Do the `amount` and `dispense` outputs present the values [0, 5, 15, 0] and [0, 0, 1, 0], respectively, when a nickel and a dime are inserted at consecutive clock edges?
- Does the FSM stay in the current state when no coins are inserted?
- Have all the states and transitions been tested?

The first question tests the machine state on reset. The next two questions are simple success cases for the FSM. To test the FSM thoroughly, we could manually construct a set of input sequences capable of exercising all the possible transitions. This would also require manually specifying the expected outputs and writing a directed test for each sequence. However, a better approach is to create a predictor that provides the expected response for any input conditions. We will write a linear testbench first to test the two simple test cases, and then we will create a predictor that will allow us to test the FSM exhaustively.

First Iteration—A Linear Testbench We will start with a linear testbench that tests the reset condition and the two simple success cases outlined in the previous section. This will also allow us to demonstrate how to perform several tests in the same testbench. In a linear testbench, this is done easily by encapsulating each test in a procedure and then calling the test procedures from the main test process.

Listing 20.6 shows the architecture for the linear testbench. The statement part of the `test_sequencer` process simply calls the three testcases in order (lines 64–77). The rest of the testbench is mainly support code for the testcases. The code also includes a record type (line 17–22) and the corresponding array type (line 24) to specify a sequence of inputs to be applied and the corresponding output values to be checked. The procedure `test_input_sequence` (lines 45–61) applies a sequence of coins and verifies the expected outputs.

In testbench code, it is good practice to invest in infrastructure code to make the code in the testcases simpler. With the data types and procedures created for this testbench, testing each sequence of coins is a simple and clearly readable operation (lines 66–77).

Another difference from the previous testbenches is that the reset input is under control of each test because we want the tests to start from a known condition. Finally, the clock input is free running, but we would like to sample the outputs always following a clock edge after all signals have settled. For this reason, we call the procedure `wait_clock_edge` (lines 26–29) every time we apply a new set of inputs and before doing the checks.

Listing 20.6 Linear testbench for the vending machine FSM

```
1   architecture testbench of vending_machine_fsm_tb is
2       signal clock: std_ulogic := '0';
3       signal reset: std_ulogic;
4       signal nickel_in: std_ulogic;
5       signal dime_in: std_ulogic;
6       signal dispense: std_ulogic;
7       signal amount: std_ulogic_vector(3 downto 0);
8
9       constant CLOCK_PERIOD: time := 20 ns;
10      constant COMBINATIONAL_DELAY: time := 6 ns;
11  begin
12      duv: entity work.vending_machine_fsm port map (...);
13
14      clock <= not clock after CLOCK_PERIOD / 2;
15
16      test_sequencer: process
17          type test_data_type is record
```

```
18              nickel: std_logic;
19              dime: std_logic;
20              dispense: std_logic;
21              amount: std_logic_vector(3 downto 0);
22          end record;
23
24          type test_data_array_type is array (natural range <>) of test_data_type;
25
26          procedure wait_clock_edge is begin
27              wait until rising_edge(clock);
28              wait for COMBINATIONAL_DELAY;
29          end;
30
31          procedure reset_with_default_inputs is begin
32              nickel_in <= '0';
33              dime_in <= '0';
34              reset <= '0';
35              wait_clock_edge;
36          end;
37
38          procedure test_reset is begin
39              print("test_reset");
40              reset_with_default_inputs;
41              assert_eq(dispense, '0', "dispense is 0 following reset");
42              assert_eq(amount, "0000", "amount is 0000 following reset");
43          end;
44
45          procedure test_input_sequence(name: string; seq: test_data_array_type) is
46          begin
47              print("test_input_sequence: " & name);
48              reset_with_default_inputs;
49
50              for i in seq'range loop
51                  nickel_in <= seq(i).nickel;
52                  dime_in <= seq(i).dime;
53                  wait_clock_edge;
54                  assert_eq(
55                      dispense, seq(i).dispense,
56                      "dispense is " & to_string(seq(i).dispense));
57                  assert_eq(
58                      amount, seq(i).amount,
59                      "amount is " & to_string(seq(i).amount));
60              end loop;
```

```
61              end;
62
63         begin
64            test_reset;
65
66            test_input_sequence( name => "nickel, nickel, nickel", seq => (
67                (nickel => '1', dime => '0', dispense => '0', amount => 4d"05"),
68                (nickel => '1', dime => '0', dispense => '0', amount => 4d"10"),
69                (nickel => '1', dime => '0', dispense => '1', amount => 4d"15"),
70                (nickel => '0', dime => '0', dispense => '0', amount => 4d"00")
71            ) );
72
73            test_input_sequence( name => "nickel, dime", seq => (
74                (nickel => '1', dime => '0', dispense => '0', amount => 4d"05"),
75                (nickel => '0', dime => '1', dispense => '1', amount => 4d"15"),
76                (nickel => '0', dime => '0', dispense => '0', amount => 4d"00")
77            ) );
78
79            print("End of testbench. All tests completed successfully.");
80            std.env.finish;
81        end process;
82     end;
```

Second Iteration—A Modular Testbench   The linear testbench from the previous section could benefit from a few improvements. First, it needs to be extended to verify the entire FSM, and not only the two simple success cases. Second, it would be tedious and error-prone to specify a large number of stimuli and expected outputs manually. Third, we would like to remove from the testcases the responsibility of manually controlling the low-level signals.

All those improvements can be made with a more modular testbench design (figure 20.12). The stimulus generator produces a transaction corresponding to a sequence of coins, and the driver applies the coins one at a time. The scoreboard monitors the DUV inputs and outputs. Inside the scoreboard, a predictor determines the expected values of the DUV outputs, which are checked against the actual outputs by the comparator. In the case of a mismatch, an assertion is violated.

The code for this modular testbench is shown in listing 20.7. The transaction type is defined as two 5-coin sequences (lines 4–5). Each sequence is an std_logic_vector, corresponding to the values of the coins applied at successive clock cycles. For example, if the nickels sequence is "10000" and the dimes sequence is "01000", then a nickel is inserted in the first clock cycle and a dime during the second. This is done in a loop in the driver process (lines 32–36). The job of the stimulus generator is greatly simplified

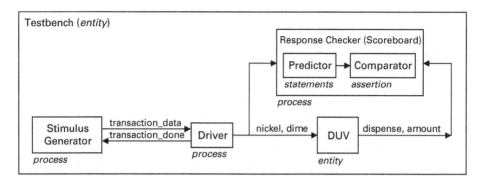

Figure 20.12
Modular testbench design for the vending machine FSM.

because it only needs to provide coin sequences as transactions (lines 18–23). In this case, we are testing all possible combinations of the two 1-bit inputs with a length of five clock cycles (a total of 2^{10} combinations).

Listing 20.7 Modular testbench code for the vending machine FSM

```
 1 architecture testbench of vending_machine_fsm_tb is
 2     ...
 3     type coins_transaction_type is record
 4         nickels: std_logic_vector(1 to 5);
 5         dimes: std_logic_vector(1 to 5);
 6     end record;
 7
 8     signal transaction_data: coins_transaction_type;
 9     signal transaction_done: boolean;
10 begin
11     ...
12     stimuli_generator: process
13         variable seq: std_logic_vector(1 to 10);
14     begin
15         wait until not reset;
16
17         -- Generate all possible 10-bit sequences
18         for i in 0 to 2**10-1 loop
19             seq := std_logic_vector(to_unsigned(i, 10));
20             -- Break down each 10-bit sequence into sequences of nickels and dimes
21             transaction_data <= (nickels => seq(1 to 5), dimes => seq(6 to 10));
22             wait on transaction_done'transaction;
23         end loop;
```

```
24
25        print("End of testbench. All tests completed successfully.");
26        std.env.finish;
27    end process;
28
29    driver: process begin
30        wait on transaction_data'transaction;
31
32        for i in 1 to 5 loop
33            nickel_in <= transaction_data.nickels(i);
34            dime_in <= transaction_data.dimes(i);
35            wait_clock_edge;
36        end loop;
37
38        transaction_done <= true;
39    end process;
40
41    response_checker: process
42        variable amount_predicted, amount_duv: natural range 0 to 15;
43        variable dispense_predicted: std_logic;
44    begin
45        wait_clock_edge;
46
47        -- Predictor for the 'amount' output
48        if reset = '1' or amount_predicted = 15 then
49            amount_predicted := 0;
50        elsif dime_in then
51            amount_predicted := minimum(amount_predicted + 10, 15);
52        elsif nickel_in then
53            amount_predicted := amount_predicted + 5;
54        end if;
55
56        -- Predictor for the 'dispense' output
57        amount_duv := to_integer(unsigned(testbench.amount));
58        dispense_predicted := '1' when amount_duv = 15 else '0';
59
60        -- Comparator - assert that DUV outputs and predictor outputs match
61        assert_eq(amount_predicted, amount_duv,
62            "amount is as expected: " & to_string(amount_duv));
63        assert_eq(dispense_predicted, dispense,
64            "dispense is as expected: " & to_string(dispense));
65    end process;
66 end;
```

The most interesting part of this testbench is the predictor, which concisely describes the behavior of the FSM outputs (lines 48–58). It is essentially a transfer function that mimics the DUV behavior at a higher level of abstraction. At every clock pulse, the DUV outputs are compared with the predictor, causing an assertion violation if they differ (lines 61–64).

This new testbench allows us to test all combinations of inputs for a length of five clock cycles. From visual inspection of the state diagram, this is enough to cover all the states and transitions. However, in a verification job, we want actual proof that the device has been verified under all the relevant conditions. We will see how this can be done for this FSM after we discuss functional coverage in section 20.6.

Peeking at State Values   Another important point about FSM verification is that in some cases it is more effective to inspect the machine state directly rather than looking only at its outputs. This can be done using VHDL external names (introduced in section 8.2.3). As an example, assuming that the DUV has a signal named current_state declared in its architecture, we could access this signal from our testbench with the construct:

```
<< signal duv.state: state_type >>
```

The previous construct can be used just like a normal signal, but it is good practice to create an alias if we are going to use it repeatedly in our testbench:

```
alias duv_state is << signal duv.state: state_type >>;
```

In this way, we can access the DUV state directly from the testbench and test it using assertions:

```
reset_with_default_inputs;          -- Apply reset and drive all inputs to 0
assert duv_state = zero_cent;       -- Ensure that initial state is 0¢

nickel_in <= '1';                   -- Hold nickel_in high for the next cycles
wait_clock_edge;
assert duv_state = five_cents;      -- Ensure state is now 5¢
wait_clock_edge;
assert duv_state = ten_cents;       -- Ensure state is now 10¢
wait_clock_edge;
assert duv_state = fifteen_cents;   -- Ensure state is now 15¢
```

Our access to the signal is not limited at reading its value. We can make assignments to it just like any other signal. However, this is probably not a good idea because the signal would have multiple drivers. Instead, we can use a *force assignment* (introduced in section 11.4) to override the signal value, followed by a *release assignment* to return control of the signal back to the model:

```
reset_with_default_inputs;        -- Apply reset and drive all inputs to 0
assert duv_state = zero_cent;     -- Ensure that initial state is 0¢

duv_state <= force fifteen_cents; -- Force FSM into state 15¢
wait_clock_edge;
assert duv_state = fifteen_cents; -- Ensure state is now 15¢
assert dispense = '1';            -- Ensure that dispense output is high

duv_state <= force ten_cents;     -- Force FSM into state 10¢
wait_clock_edge;
assert duv_state = ten_cents;     -- Ensure state is now 10¢
assert dispense = '0';            -- Ensure that dispense output is low

duv_state <= release;   -- Release control of the signal back to the model
```

We should be careful when doing assignments to external signals because we could force a model into an inconsistent state. Actually, we need to know a model in deep detail to force one of its signals without driving it into an invalid state.

This technique is available not only for FSMs but for any kind of model. This is a powerful tool, but it should not be used gratuitously. As explained in section 11.4, external names turn a testbench into a white-box test. If a test makes assumptions about the internal implementation of a model, then it is not verifying a design but rather verifying an implementation. This makes the test more fragile because changes in the implementation may require changing the testbenches as well.

20.5   Example #4: Testbenches with File I/O

Sometimes it is impractical or infeasible to compute the expected responses for a DUV using an alternative VHDL implementation. However, it may be easy to obtain a series of golden output vectors corresponding to the expected DUV outputs. For example, we may have a model available in another programming language, or we may have a published list of vectors with which the DUV must comply. In such cases, we can provide the expected inputs and outputs via files read during the simulation. We will see an example using a hash generator as the DUV.

Design Specifications   The circuit must calculate the hash function $h(x)$ for which the following reference vectors are given in the file *golden.txt*:

```
hartj09sq7shh60gxrupgwy2jus9mizc 1110000100100101100110001111000101
e6aifwgcyf42f8yucde5j4xdh1a3byu6 0111000001001101100000110110110101
qpreydipz0u9g8q6xgj6l5e8otuybxru 1001001010010010000100011111111010
```

otqnk1oj0csdpxlar49uaim0o48e3m5y 00110111001011101101100011110110
azi7ucov41g0nhtve2x50sx88i8eeymb 10010100101100111110000110111010
y0birz6zjhvib9rsd4uuqavpcd7p8mkw 01001101011000001101101010011111
3l80y48tfpsm5jq5jmnj8t27erd8onnv 10100111000111101111010101010101
fdu4daduvujr6ttrf1ecyavq02gmi5ey 01110011100111000110000100111011
oxaf7665xunyd36b2jb6fe0cck2kmy5c 01101001101101111001100100000011
bjwvsqt1x9aot0orrt4eckp5srg61zhh 11100001000111110000000000101111
...   ...

First Iteration—A Linear Testbench We can start by writing a linear testbench (listing 20.8). A single thread of execution (a single process) is responsible for reading the input file (lines 18–20), generating the stimulus (line 23), and checking the DUV response (lines 27–28).

Listing 20.8 Linear testbench for the hash generator

```
1  architecture testbench of linear_hash_calculator_tb is
2      constant DUV_DELAY: time := 10 ns;
3      constant TB_DATA_PATH: string := "...";
4
5      signal hash_duv: std_logic_vector(31 downto 0);
6      signal message_duv: string(1 to 32);
7  begin
8      duv: entity work.hash_calculator port map (...);
9
10     test_controller: process
11         file golden_data: text open read_mode is TB_DATA_PATH & "golden.txt";
12         variable current_line: line;
13         variable hash_golden: std_logic_vector(31 downto 0);
14         variable message_golden: string(1 to 32);
15     begin
16         while not endfile(golden_data) loop
17             -- Read input vector and expected respose
18             readline(golden_data, current_line);
19             read(current_line, message_golden);
20             read(current_line, hash_golden);
21
22             -- Apply stimuli
23             message_duv <= message_golden;
24             wait for DUV_DELAY;
25
26             -- Check DUV response
27             assert_eq(hash_duv, hash_golden,
```

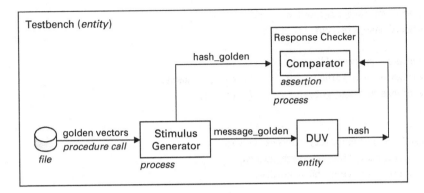

Figure 20.13
Modular testbench architecture for the hash generator.

```
28                     "hash for message " & message_golden & " matches golden output");
29          end loop;
30
31          print("End of testbench. All tests passed.");
32          std.env.finish;
33      end process;
34 end;
```

Second Iteration—A Modular Testbench We can make the testbench more modular by separating the generator and checker processes. Another improvement is to separate file operations from stimulus generation by moving the file operations to a procedure. In any case, because this is a simple testbench, the code will not benefit much. Figure 20.13 shows a block diagram of the new testbench architecture. The corresponding code is shown in listing 20.9.

Listing 20.9 Modular testbench code for the hash generator

```
1 architecture testbench of hash_calculator_tb is
2      constant DUV_DELAY: time := 10 ns;
3      constant GOLDEN_DATA_PATH: string := "...";
4
5      signal message: string(1 to 32);
6      signal hash: std_logic_vector(31 downto 0);
7      signal hash_golden: std_logic_vector(31 downto 0);
8
9      file golden_data: text open read_mode is GOLDEN_DATA_PATH & "golden.txt";
10
11     procedure read_golden_vectors(
```

```
12          signal message_sig: out string;
13          signal hash_sig: out std_logic_vector
14      ) is
15          variable current_line: line;
16          variable hash_golden: std_logic_vector(31 downto 0);
17          variable message_golden: string(1 to 32);
18      begin
19          -- Read input vector and expected respose
20          readline(golden_data, current_line);
21          read(current_line, message_golden);
22          read(current_line, hash_golden);
23
24          -- Return values
25          message_sig <= message_golden;
26          hash_sig <= hash_golden;
27      end;
28 begin
29      duv: entity work.hash_calculator port map (message => message, hash => hash);
30
31      stimulus_generator: process
32      begin
33          while not endfile(golden_data) loop
34              read_golden_vectors(message, hash_golden);
35              wait for DUV_DELAY;
36          end loop;
37
38          print("End of testbench. All tests passed.");
39          std.env.finish;
40      end process;
41
42      response_checker: process
43          variable hash_sample: std_logic_vector(31 downto 0);
44      begin
45          -- Sample DUV output
46          wait on message'transaction;
47          hash_sample := hash;
48
49          -- Check DUV response
50          wait for DUV_DELAY;
51          assert_eq(hash, hash_golden,
52              "hash for message " & message & " matches golden output");
53      end process;
54 end;
```

20.6 Random Stimulus and Functional Coverage

The increasing gap between verification complexity and design complexity has prompted a call for new verification approaches and methodologies. One of the biggest revolutions in functional verification was the introduction of coverage-driven constrained random verification. This section provides a brief introduction to functional coverage and random stimulus and shows how they can be used in VHDL testbenches.

20.6.1 Functional Coverage

Once we have a verification plan, the verification job becomes a matter of writing code to exercise the DUV according to the plan and check its behavior. The job is finished when we have written the code for all items in the plan.

The metric that tells us how much of our verification plan has been completed is called *functional coverage*. For each question or item in the verification plan, there will be a corresponding test in a testbench. As tests are performed, these items are ticked off from the list until coverage reaches 100%.

20.6.2 Directed Testing

There are two general approaches to achieving complete functional coverage. One is to write a large number of tests that check specific items from the verification plan. The test applies the necessary stimulus to put the DUV in the specified condition and then observes the responses. Each test is engineered for a targeted goal and uses a fixed set of stimuli. This approach is called *directed testing*.

20.6.3 Random-Based Verification

The other approach is to write a checking mechanism that can tell whether the DUV is operating correctly and then apply random stimuli until all the interesting conditions have been observed. Of course the stimuli are not completely random: we must specify a series of constraints to make them valid and relevant to our test cases. This approach is called *constrained random verification*. Although directed testing can also use functional coverage, random-based verification requires it. Otherwise there is no way to know what the test really did or when we are done testing. For this reason, this approach is also called *coverage-driven random-based verification*.

20.6.4 Intelligent Coverage

Random stimuli are usually created by generators that can be constrained to produce the cases described in the verification plan. When there is no communication between the coverage monitors and the random generators, many repeated sequences may be generated. *Intelligent coverage* is a technique that consists in feeding coverage information back into the random generators to focus on coverage holes and reduce the number of repeated tests.

20.6.5 Coverage Bins

When we run tests using random stimuli, we do not know a priori which values will be observed during a simulation. To know whether a specific point in the testbench has been tested with a value of interest, the testbench collects coverage information. Depending on our plan, the points of interest may be the DUV inputs, outputs, or internal values. Each point for which coverage information is collected is called a *coverage item* or *coverage point*.

For each point, we may specify the number of times that a value or range of values must be observed before the end of the simulation. This number is called a *coverage goal*, and each value or range is called a *coverage bin*. Each time an item is sampled, its value is taken, and the corresponding bin gets incremented. When all bins of all coverage points have met their coverage goals, the simulation is finished.

We can specify the range and coverage goal of each bin so that values that are functionally equivalent are grouped, and interesting values are guaranteed to appear with a certain frequency. Often the interesting values happen close to the limits of a parameter. Figure 20.14a shows a coverage model for a parameter named *Len*. The values between 8 and 63 are given different coverage goals. Values between 0 and 7 are illegal; if they appear in a simulation, then this will be considered an error.

It is also possible to collect information about two or more parameters simultaneously and to give their combined occurrence different coverage goals. This kind of coverage model is called *cross coverage*. Figure 20.14b shows an example of a cross-coverage model for two parameters named R_A and R_B. The corner cases were given a coverage goal of ten; the edge cases, five; and the remaining valid values have a goal of one. Cases where $R_A = R_B$ are invalid bins.

20.6.6 VHDL and Random-Based Verification

Unlike other hardware design languages that need to be extended with hardware verification languages (HVLs) to perform functional coverage and random-based

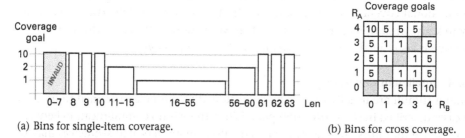

(a) Bins for single-item coverage.   (b) Bins for cross coverage.

Figure 20.14
Coverage bins for single-item coverage and for cross coverage.

verification, VHDL is powerful enough to provide all the required language features. With the appropriate set of packages, it is possible to add randomization and functional coverage to any testbench.

The Open Source VHDL Verification Methodology (OS-VVM) is a methodology for the creation of intelligent testbenches based on a set of open-source packages. The packages provide support for constrained random stimulus, functional coverage, and intelligent coverage. The packages work with VHDL-2002 and 2008 and can be downloaded from the project's website (www.osvvm.org).

Because the packages are written in pure VHDL, there is no need to learn a new verification language; moreover, the testbench code is readable by RTL designers as well as verification engineers, and the code works in any standard VHDL simulator.

20.7 Example #5: Testbench with Random Stimulus and Functional Coverage

We will illustrate the use of random-based stimulus and functional coverage in VHDL with a simple example. For the randomization and coverage features, we will use OS-VVM version 2.3.1. The basic concepts and usage details will be explained as needed.

Design Specifications Suppose we were given some obscure code that divides an unsigned integer by the constant value three using only shifts, additions, and multiplications by the constant values three and five (which can also be reduced to shifts and additions). We would like to test it with inputs from 1 to 1,000,000,000 to gain some confidence that it really works. (Actually, the code comes from Hank Warren's *Hacker's Delight*,[13] so it has been thoroughly tested, but that makes it no less mysterious.)

The code translated to VHDL is:

```
-- q is quotient, d is dividend
q := (d srl 2) + (d srl 4); -- q = d*0.0101 (approx)
q := q + (q srl 4);         -- q = d*0.01010101
q := q + (q srl 8);
q := q + (q srl 16);
r := resize(d - q * 3, 32); -- 0 <= r <= 15.
q := resize(q + (5 * (r + 1) srl 4), 32);
```

Because testing the algorithm exhaustively in the desired range could take an entire day, we will test it by dividing the input range in 20 bins, with a coverage goal of at least 100 values per bin.

Functional Coverage with OS-VVM The basic steps to collect functional coverage from a value in VHDL code are:

1. Create a *coverage item* object (an object of type `CovPType`).
2. Model the intended coverage (specify the bins).
3. Sample the item during the simulation to collect coverage information.
4. Use the coverage object to generate random data (if intelligent coverage is used) and decide when to stop the simulation.

The first choice in a coverage-driven testbench is what to monitor for coverage. In our example, we will monitor the input stimulus to the DUV. Using the stimulus as coverage items is a good idea because it allows the coverage information to influence the random number generation, thus achieving the coverage goals faster.

Because `CovPType` is a protected type, the coverage item object must be a variable. Because our testbenches are structured as a series of communicating processes, we need to declare the object as a shared variable in the architecture declarative part:

```
-- The point we want to monitor for coverage is declared as a shared variable so
-- that distinct processes can initialize it, use it, and check coverage results.
shared variable dividend_cover_item: CovPType;
```

We create bins using the function `GenBin`. To create an array of 20 equally spaced bins in the range from 0 to 1,000,000,000, we would write:

```
-- Return an array of 20 bins in the range 0 to 1,000,000,000
GenBin(Min => 0, Max => 1_000_000_000, NumBin => 20)
```

To associate the generated bins with our coverage item, we use the method `AddBins`:

```
-- Initialize cover item with 20 bins in the range 0 to 1,000,000,000
dividend_cover_item.AddBins(
    GenBin(Min => 0, Max => 1_000_000_000, NumBin => 0)
);
```

Now the coverage item is ready to start accumulating coverage results. We should choose an appropriate time to do the input sampling. In our case, we could accumulate whenever the DUV input `dividend` changes:

```
coverage_collector: process begin
    wait on dividend;
    -- Increment bin count corresponding to the current test value
    dividend_cover_item.ICover( to_integer(dividend) );
end process;
```

The last step is to use the coverage information to decide whether to stop the test-bench. We can test whether an item has been covered with the method `IsCovered`:

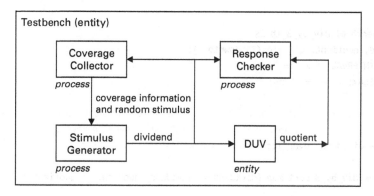

Figure 20.15
Testbench design with functional coverage and random stimulus for the divider model.

```
while not dividend_cover_item.IsCovered loop
    -- Apply some more stimulus...
end loop;
```

First Iteration—Basic Random Stimulus And Functional Coverage Now we can apply what we have learned to the design of our testbench (figure 20.15). The stimulus generator provides the dividend value, the single input to the DUV. The same value is also sampled by the response checker and coverage collector processes. The stimulus generator uses coverage information to decide whether to stop the simulation.

Listing 20.10 shows the complete testbench code. In this version, the stimulus generator uses coverage information only for deciding whether to stop the test (line 36) and not for stimulus generation. In other words, it does not use intelligent coverage. To generate the stimulus, the code uses a simple random variable (line 28), which is disconnected from the coverage information. This means that stimulus values can be repeated, and the simulation is not as efficient as it could be. The stimulus is generated using the RandInt method of type RandomPType (line 37).

Listing 20.10 Testbench code with functional coverage and random stimulus for the divider model

```
1 library ieee, osvvm;
2 use ieee.numeric_std.all;
3 use osvvm.CoveragePkg.all;
4 use osvvm.RandomPkg.all;
5 use work.testbench_utils_pkg.all;
6
7 entity div_by_3_tb is
8 end;
```

```
 9
10 architecture testbench of div_by_3_tb is
11    signal dividend, quotient: unsigned(31 downto 0);
12    constant TEST_INTERVAL: time := 20 ns;
13    constant DUV_DELAY: time := 10 ns;
14
15    -- Create cover item for the DUV input, declared as a shared variable so that
16    -- distinct processes can initialize it, use it, and check coverage results.
17    shared variable dividend_cover_item: CovPType;
18 begin
19    duv: entity work.div_by_3 port map (dividend => dividend, quotient => quotient);
20
21    coverage_collector: process begin
22        wait on dividend;
23        -- Increment bin count corresponding to the current DUV input
24        dividend_cover_item.ICover( to_integer(dividend) );
25    end process;
26
27    stimulus_generator: process
28        variable random_variable: RandomPType;
29    begin
30        -- Initialize cover item with 20 bins in the range 0 to 1,000,000,000
31        dividend_cover_item.AddBins(
32            GenBin(Min => 0, Max => 1_000_000_000, NumBin => 20)
33        );
34
35        -- Generate random stimuli until complete coverage is reached
36        while not dividend_cover_item.IsCovered loop
37            dividend <= to_unsigned(random_variable.RandInt(1_000_000_000), 32);
38            wait for TEST_INTERVAL;
39        end loop;
40
41        dividend_cover_item.SetMessage("Coverage for dividend input");
42        dividend_cover_item.WriteBin;
43        print("End of testbench. All tests passed.");
44        std.env.finish;
45    end process;
46
47    response_checker: process begin
48        wait on dividend;
49        wait for DUV_DELAY;
50        assert_eq(quotient, dividend / 3, to_string(to_integer(dividend)) &
51            " / 3 = " & to_string(to_integer(quotient)));
```

```
52    end process;
53 end;
```

When we run the testbench, we should see a report similar to the following:

```
# %% WriteBin: Coverage for dividend input
# %% Bin:(         0 to   49999999)   Count = 1  AtLeast = 1
# %% Bin:( 50000000 to   99999999)   Count = 3  AtLeast = 1
...
# %% Bin:(950000000 to 1000000000)   Count = 4  AtLeast = 1
# End of testbench. All tests passed.
```

The report tells us the number of samples that were collected for each bin (value Count in the example report). The second bin was exercised with three values and the last bin with four. Because our test plan requires only one instance for each bin, the simulation could still be optimized.

Second Iteration—Intelligent Coverage In the first version of the testbench, the random stimulus generator was independent from the coverage results. For this reason, some bins were covered more than once. We can optimize the testbench by using the coverage information to select only random values that have not been covered yet. This approach is called *intelligent coverage*.

A bin that is below its coverage goal is called a *coverage hole*. We can use the method RandCovPoint to return a random value belonging to a bin that has not met its coverage goal. All we need to change is the assignment expression inside the stimulus generator loop:

```
while not dividend_cover_item.IsCovered loop
    dividend <= to_unsigned(dividend_cover_item.RandCovPoint, 32);
    wait for TEST_INTERVAL;
end loop;
```

Now the coverage report shows only one value in each bin:

```
# %% Bin:(         0 to   49999999)   Count = 1  AtLeast = 1
# %% Bin:( 50000000 to   99999999)   Count = 1  AtLeast = 1
# %% Bin:(100000000 to  149999999)   Count = 1  AtLeast = 1
...
# %% Bin:(900000000 to  949999999)   Count = 1  AtLeast = 1
# %% Bin:(950000000 to 1000000000)   Count = 1  AtLeast = 1
```

20.8   Example #6: FSM Transition Coverage

In the testbench we created for the FSM of section 20.4, we relied on visual inspection of the transition diagram to conclude that it had been exercised completely. Now we will see how to use functional coverage to confirm that assumption.

Using single-item coverage, we could easily keep track of the visited states in an FSM. There would be only two differences from the way we collected coverage information in the previous example. First, the FSM state must be converted to an integer to be used with the GenBin and ICov functions. Second, the current state signal is not an output of the DUV, so we need VHDL external names to peek into the FSM, as described in "Peeking at State Values" in section 20.4.

However, just because all the states were visited, it does not mean we have exercised the entire FSM. For an FSM, a more interesting metric is *transition coverage*. Transition coverage is defined as tracking the occurrence of sequences of values in the same coverage item.

There is no direct way to sample a sequence of values over time, but we can use cross-coverage—sampling two different items at the same time—to the same effect. If we sample the FSM current state and next state using a cross-coverage item, we can measure its transition coverage. Alternatively, the sampling process could keep track of the previous value and then use the previous and current values as a cross-coverage item. The attribute 'last_value can be used for this purpose.

Cross-Coverage with OS-VVM   To create a coverage model for single-item coverage, we used the method AddBins, passing an array of bins as argument. This array is created by the function GenBin:

```
cover_item.AddBins( GenBin(Min => 0, Max => 15) );
```

To create a cross-coverage model, we use the method AddCross, passing multiple arrays of bins as arguments. The coverage model will be the cross product of the arrays:

```
operands_coverage.AddCross( GenBin(0, 15), GenBin(0, 15) );
```

The previous cross-coverage model could represent, for instance, an ALU that can use any combination of two registers from a register file with 16 locations. In the testbench code, we would sample the signals that select the two registers to accumulate the coverage information:

```
operands_coverage.ICover( (source_reg, dest_reg) );
```

If some of the combinations are invalid, then we can specify *illegal bins*. In single-item coverage, we can concatenate a call to GenBin with a call to IllegalBin to create coverage models that disallow specific bins. For example, if bins 0, 1, 3, and 4 were valid but 2 were invalid, we would write:

```
GenBin(0) & GenBin(1) & IllegalBin(2) & GenBin(3) & GenBin(4)
```

or:

```
GenBin(0, 1) & IllegalBin(2) & GenBin(3, 4)
```

Similarly, in a cross-coverage model, we could disallow combinations of bins by writing:

```
cross_cover.AddCross( GenBin(0), IllegalBin(0) &    GenBin(1) &    GenBin(2) );
cross_cover.AddCross( GenBin(1),    GenBin(0) & IllegalBin(1) &    GenBin(2) );
cross_cover.AddCross( GenBin(2),    GenBin(0) &    GenBin(1) & IllegalBin(2) );
```

This cross-coverage model disallows any combination in which the two items have the same value. We will use this kind of restriction to create the coverage model for the transitions of our FSM.

Using Cross-Coverage for FSM Transition Coverage The steps for modeling cross-coverage are the same as for single-item coverage: declare, model, accumulate, interact, and report.[14] Only the model and accumulate steps are somewhat different.

We would like to use illegal bins as an error-checking mechanism to detect invalid transitions. We can determine which bins should be illegal by looking at the FSM state diagram (figure 20.16a). For example, from state 0¢ the machine can move to 0¢, 5¢, and 10¢, which are shown as white cells in the matrix of figure 20.16b. A transition from 0¢ to 15¢ is not possible, so it is indicated with a shaded cell.

The following code sets up the cover item for cross-coverage collection. The code uses aliases to make the names shorter and easier to read as a matrix. With this layout, the correspondence between the code and the matrix of figure 20.16b is clear:

```
alias cov is fsm_transition_coverage;
alias legal is GenBin[integer return CovBinType];
alias illegal is IllegalBin[integer return CovBinType];
...
-- Setup the coverage model
cov.AddCross( GenBin(0),   legal(0) &   legal(1) &   legal(2) & illegal(3) );
cov.AddCross( GenBin(1), illegal(0) &   legal(1) &   legal(2) &   legal(3) );
cov.AddCross( GenBin(2), illegal(0) & illegal(1) &   legal(2) &   legal(3) );
cov.AddCross( GenBin(3),   legal(0) & illegal(1) & illegal(2) & illegal(3) );
```

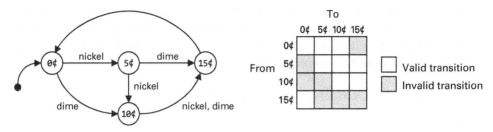

(a) State diagram.   (b) Matrix of valid and invalid transitions.

Figure 20.16
Determining valid and invalid transitions for the vending machine FSM.

The following code shows how the cross-coverage information is collected. The process declares aliases for the FSM current state and next state signals using external names. To collect coverage information, we use the ICover method, the same used in single-item coverage. Because ICover expects integer arguments, we use the 'pos attribute to convert the state values to integers.

```
coverage_collector: process
    alias cov is fsm_transition_coverage;
    alias current_state is << signal duv.current_state: state_type >>;
    alias next_state is << signal duv.next_state: state_type >>;
begin
    wait until rising_edge(clock);
    wait for COMBINATIONAL_DELAY;
    cov.ICover( (state_type'pos(current_state), state_type'pos(next_state)) );
end process;
```

Alternatively, we could sample a single signal (the current_state signal) at different moments in time to provide the cross-coverage information. This could be done with a process that keeps the previous signal value in a local variable:

```
sequential_coverage_collector: process
    alias cov is fsm_transition_coverage;
    alias current_state is << signal duv.current_state: state_type >>;
    variable previous_state: state_type := current_state;
begin
    wait until rising_edge(clock);
    cov.ICover( (state_type'pos(previous_state), state_type'pos(current_state)) );
    previous_state := current_state;
end process;
```

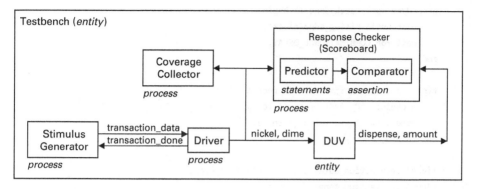

Figure 20.17
Block diagram of the testbench with transition coverage for the vending machine FSM.

Figure 20.17 shows the block diagram of the improved testbench for the vending machine FSM. The complete code is shown in listing 20.11.

Listing 20.11 Testbench code with transition coverage for the vending machine FSM

```
 1 library ieee, osvvm;
 2 use ieee.std_logic_1164.all;
 3 use ieee.numeric_std.all;
 4 use osvvm.CoveragePkg.all;
 5 use work.vending_machine_fsm_pkg.all;
 6 use work.testbench_utils_pkg.all;
 7
 8 entity modular_vending_machine_fsm_tb is
 9 end;
10
11 architecture testbench of modular_vending_machine_fsm_tb is
12     signal clock: std_ulogic := '0';
13     signal reset: std_ulogic := '0';
14     signal nickel_in: std_ulogic := '0';
15     signal dime_in: std_ulogic := '0';
16     signal dispense: std_ulogic;
17     signal amount: std_ulogic_vector(3 downto 0);
18
19     constant CLOCK_PERIOD: time := 20 ns;
20     constant COMBINATIONAL_DELAY: time := 6 ns;
21
```

```
22      procedure wait_clock_edge is begin
23          wait until rising_edge(clock);
24          wait for COMBINATIONAL_DELAY;
25      end;
26
27      type coins_transaction_type is record
28          nickels: std_logic_vector(1 to 5);
29          dimes: std_logic_vector(1 to 5);
30      end record;
31
32      signal transaction_data: coins_transaction_type;
33      signal transaction_done: boolean;
34
35      shared variable fsm_transition_coverage: CovPType;
36 begin
37      duv: entity work.vending_machine_fsm port map (
38          clock => clock, reset => reset, nickel_in => nickel_in,
39          dime_in => dime_in, dispense => dispense, amount => amount);
40
41      clock <= not clock after CLOCK_PERIOD / 2;
42      reset <= '1', '0' after 50 ns;
43
44      stimuli_generator: process
45          variable seq: std_logic_vector(1 to 10);
46          alias cov is fsm_transition_coverage;
47          alias legal is GenBin[integer return CovBinType];
48          alias illegal is IllegalBin[integer return CovBinType];
49      begin
50          -- Setup the coverage model
51          cov.AddCross( GenBin(0),   legal(0) &   legal(1) &   legal(2) & illegal(3) );
52          cov.AddCross( GenBin(1), illegal(0) &   legal(1) &   legal(2) &   legal(3) );
53          cov.AddCross( GenBin(2), illegal(0) & illegal(1) &   legal(2) &   legal(3) );
54          cov.AddCross( GenBin(3),   legal(0) & illegal(1) & illegal(2) & illegal(3) );
55
56          wait until not reset;
57
58          -- Generate all possible 10-bit sequences
59          for i in 0 to 2**10-1 loop
60              seq := std_logic_vector(to_unsigned(i, 10));
61              -- Break down each 10-bit sequence into sequences of nickels and dimes
62              transaction_data <= (nickels => seq(1 to 5), dimes => seq(6 to 10));
63              wait on transaction_done'transaction;
64          end loop;
65
```

```
66        fsm_transition_coverage.SetMessage("FSM transition coverage");
67        fsm_transition_coverage.WriteBin(WriteAnyIllegal => enabled);
68        print("End of testbench. All tests completed successfully.");
69        std.env.finish;
70    end process;
71
72    driver: process begin
73        wait on transaction_data'transaction;
74
75        for i in 1 to 5 loop
76            nickel_in <= transaction_data.nickels(i);
77            dime_in <= transaction_data.dimes(i);
78            wait_clock_edge;
79        end loop;
80
81        transaction_done <= true;
82    end process;
83
84    response_checker: process
85        variable amount_predicted, amount_duv: natural range 0 to 15;
86        variable dispense_predicted: std_logic;
87    begin
88        wait_clock_edge;
89
90        -- Predictor for the 'amount' output
91        if reset = '1' or amount_predicted = 15 then
92            amount_predicted := 0;
93        elsif nickel_in then
94            amount_predicted := amount_predicted + 5;
95        elsif dime_in then
96            amount_predicted := minimum(amount_predicted + 10, 15);
97        end if;
98
99        -- Predictor for the 'dispense' output
100       amount_duv := to_integer(unsigned(testbench.amount));
101       dispense_predicted := '1' when amount_duv = 15 else '0';
102
103       -- Comparator - assert that DUV outputs and predictor outputs match
104       assert_eq(amount_predicted, amount_duv,
105           "amount is as expected: " & to_string(amount_duv));
106       assert_eq(dispense_predicted, dispense,
107           "dispense is as expected: " & to_string(dispense));
108   end process;
109
```

```
110     coverage_collector: process
111         alias cov is fsm_transition_coverage;
112         alias current_state is << signal duv.current_state: state_type >>;
113         alias next_state is << signal duv.next_state: state_type >>;
114     begin
115         wait until rising_edge(clock);
116         wait for COMBINATIONAL_DELAY;
117         cov.ICover( (state_type'pos(current_state), state_type'pos(next_state)) );
118     end process;
119 end;
```

20.9 Conclusion

The examples worked in this chapter were intended to provide a gentle introduction on how to write testbenches in VHDL. Because it takes time and effort to write such tests, it is appropriate to close this chapter with an insight about their cost and benefit.

Writing these testbenches helped uncover several bugs in the models they verify. More important, all those bugs were caught before they could make it into a finished design or get anywhere near a hardware device, where they would be much harder to find and relate with the corresponding code.

Writing those testbenches also took time. The testbenches presented in this chapter have a 1.5 production code to test code ratio; in other words, there are 15 lines of testbench code for every 10 lines of DUV code. Note that this is still much less than the 4 to 1 ratio commonly found in real projects.

So how do we assess the benefit of writing tests? The first thing to bear in mind is that it is not writing the code that take times but rather thinking about and debugging the code. Writing tests helps us spend less time debugging, especially if the tests are written concurrently with the RTL code. If we write tests as we add new features, then we can keep the code free of bugs most of the time. We will also have only a small part of the code to look for bugs when they show up. Moreover, writing tests for individual modules reduces the time spent to integrate them into the system. The best way to reduce the integration effort is to ensure that all modules are bug-free before plugging them into the system.

In practice, most developers find that writing good tests takes less time than fixing code that was written without tests. One argument supporting this claim is that after getting used to testing, few developers go back to the practice of not writing tests. Most developers become, as is commonly said, "test-infected."[15]

Another important point (stressed throughout this book) is that having a good set of tests allows us to change the existing code safely when we find an opportunity for improvement. If we are afraid of touching the code, then those changes will not be made, and the code quality will deteriorate. Over time those small improvements keep the code quality high and prevent a decrease in developer productivity.

One last point is that writing tests for our code makes us see it from the client code perspective. If it is hard to write a test for module, then it probably has a bad interface and is also hard to use. As you write tests, you will find that many of the principles that make for a good design also make it easier to test. The reverse is also true: if a design is not easily testable, then there is usually an underlying design problem. Testing is easy when we have a good design.[16]

Notes

Chapter 1

1. IEEE, *1076-2008 Standard VHDL Language Reference Manual* (New York: IEEE, 2009), p. iv.
2. Steve McConnell, *Code Complete: A Practical Handbook of Software Construction*, 2nd ed. (Redmond, WA: Microsoft Press, 2004), p. 399.

Chapter 2

1. Thomas J. McCabe, "A Complexity Measure," *IEEE Transactions on Software Engineering* SE-2, no. 4 (1976): pp. 308–320.
2. Michael Feathers, "The Deep Synergy between Testability and Good Design" (lecture, Norwegian Developers Conference, NDC, 2010).
3. Michael Keating and Pierre Bricaud. *Reuse Methodology Manual for System-On-a-Chip Designs* (Norwell, MA: Kluwer, 2002), p. 7.
4. Robert L. Glass, "Frequently Forgotten Fundamental Facts about Software Engineering," *IEEE Software* 18 (2001): pp. 110–112.
5. Philip B. Crosby, *Quality Is Free: The Art of Making Quality Certain* (New York: McGraw-Hill, 1979), p. 172.
6. McConnell. *Code Complete*, p. 30.
7. Ruth Malan and Dana Bredemeyer, "Less Is More with Minimalist Architecture," *IEEE IT Professional Magazine* 4, no. 5 (2002): pp. 46–48.
8. ISO/IEC/IEEE, *42010:2011-ISO/IEC/IEEE Systems and Software Engineering–Architecture Description* (New York: IEEE, 2011).

Chapter 3

1. Edsger W. Dijkstra, "The Humble Programmer," *Communications of the ACM* 15, no. 10 (1972): 859–866. This is Dijkstra's acceptance speech for the 1972 ACM Turing Award. Dijkstra was one of the first to point out that no one's skull is large enough to hold an entire program: "The competent programmer is fully aware of the strictly limited size of his own skull; therefore he approaches the programming task in full humility, and (...) avoids clever tricks like the plague."

2. Robert C. Martin, *Agile Principles, Patterns, and Practices in C#* (Upper Saddle River, NJ: Prentice Hall, 2007), p. 105.

3. Sandy Metz, *Practical Object-Oriented Design in Ruby* (Reading, MA: Addison-Wesley, 2012), p. 16.

4. Martin Fowler, *Refactoring—Improving the Design of Existing Code* (Reading, MA: Addison-Wesley, 1999), p. xvi.

5. Chris Sterling, *Managing Software Debt* (Reading, MA: Addison-Wesley, 2010), p. 16.

6. The seminal work in this area is George A. Miller's paper, "The Magical Number Seven, Plus or Minus Two: Some Limits on Our Capacity for Processing Information," *Psychological Review* 63 (1956): 19–35. More recently, Nelson Cowan published "The magical number 4 in short-term memory: A reconsideration of mental storage capacity," *Behavioral and Brain Sciences* 24 (2001): 87–114, where he updates the numeric limits with various experiments that were not available when Miller published his paper.

7. Jeff Johnson, *Designing with the Mind in Mind, Second Edition* (Burlington, VT: Morgan Kaufmann, 2014), p. 121.

8. Donald Norman, *The Design of Everyday Things* (New York: Basic Books, 2002), p. 79.

Chapter 4

1. "The design entity is the primary hardware abstraction in VHDL. It represents a portion of a hardware design that has well-defined inputs and outputs and performs a well-defined function." IEEE, *1076-2008 VHDL Language Reference Manual*, p. 7.

2. Ed Yourdon and Larry Constantine. *Structured Design* (Englewood Cliffs, NJ: Prentice-Hall, 1979), p. 73. The book offers a detailed explanation of how good modularity helps reduce a system's costs.

3. McConnell, *Code Complete*, p. 105.

4. The original reference for the kinds of cohesion is the paper, "Structured Design" by Stevens, Myers, and Constantine, *IBM Systems Journal* 13, no. 2 (1974): 115–139. For a detailed explanation on the types of cohesion, see Ed Yourdon and Larry Constantine's book, *Structured Design*, pp. 105–140.

5. Microsoft Patterns & Practices Team, *Microsoft Application Architecture Guide* (Redmond, WA: Microsoft Press, 2009). There are many other definitions for the Single Responsibility Principle, but this one is relevant because it explicitly says that a module can still have a single responsibility if it aggregates closely related functionality.

6. Andrew Hunt and David Thomas, *The Pragmatic Programmer–From Journeyman to Master* (Reading, MA: Addison-Wesley, 1999), p. 37.

7. Hunt, *The Pragmatic Programmer*, p.26.

8. Kent Beck and Cynthia Andres, *Extreme Programming Explained: Embrace Change*, 2nd ed. (Reading, MA: Addison-Wesley, 2004), p. 108.

Chapter 5

1. Janick Bergeron, *Writing Testbenches—Functional Verification of HDL Models*, 2nd ed. (New York: Springer, 2003), p. 3. Bergeron states that 70% of the design effort is consumed with verification; when a design is completed, up to 80% of the total code volume is composed of testbenches.

Chapter 7

1. "An expression is a formula that defines the computation of a value." IEEE, *1076-2008 VHDL Language Reference Manual*, p. 117.

Chapter 8

1. IEEE, *1076.6-2004 Standard for VHDL Register Transfer Level (RTL) Synthesis*, p. 29.
2. Dustin Boswell and Trevor Foucher, *The Art of Readable Code* (Sebastopol, CA: O'Reilly, 2011), p. 70.
3. McConnell, *Code Complete*, p. 440.

Chapter 9

1. "Within a given simulation cycle, an implementation may execute concurrent statements in parallel or in some order. The language does not define the order, if any, in which such statements will be executed. A description that depends upon a particular order of execution of concurrent statements is erroneous." IEEE, *1076-2008 VHDL Language Reference Manual*, p. 169.

Chapter 10

1. McConnell, *Code Complete*, p. 445.
2. Graham Bolton and Stuart Johnston, *IfSQ Level-2 A Foundation-Level Standard for Computer Program Source Code*, 2nd ed. (Cambridge: IfSQ, Institute for Software Quality, 2009), p. 26. The Institute for Software Quality (IfSQ) compiles a list of defect indicators in software. Indicator SP-2, *Nesting too deep*, refers to code in which "statements involving a condition have been nested to a depth of more than 4."
3. Philippe Garrault and Brian Philofsky, *Xilinx White Paper #231: HDL Coding Practices to Accelerate Design Performance* (San Jose, CA: Xilinx, 2006), p. 20.
4. Boswell and Foucher, *The Art of Readable Code*, p. 72.
5. McConnell, *Code Complete*, p. 355.
6. Software Productivity Consortium. *Ada 95 Quality and Style: Guidelines for Professional Programmers* (Herndon, VA: Software Productivity Consortium, 1995), p. 88.
7. McConnell, *Code Complete*, p. 373.
8. McConnell, *Code Complete*, p. 385.
9. European Space Agency (ESA), *VHDL Modelling Guidelines* (Noordwijk: European Space Agency, 1994), p. 8.

Chapter 13

1. "The declarations of all predefined types are contained in package STANDARD." IEEE, *1076-2008 VHDL Language Reference Manual*, p. 35.

2. IEEE, *1076-2008 VHDL Language Reference Manual*, p. 447. The archive file is described in annex A and is available at http://standards.ieee.org/downloads/1076/1076-2008/.

3. "A synthesis tool shall interpret the following values as representing a logic value 0: The BIT value '0'; The BOOLEAN value FALSE; The STD_ULOGIC values '0' and 'L'. It shall interpret the following values as representing a logic value 1: The BIT value '1'; The BOOLEAN value TRUE; The STD_ULOGIC value '1' and 'H'." IEEE, *1076-2008 VHDL Language Reference Manual*, p. 278.

4. IEEE, *754-2008 Standard for Floating-Point Arithmetic* (New York: IEEE, 2008).

Chapter 14

1. Mentor Graphics, *ModelSim® User's Manual Software Version 10.1e* (Wilsonville, OR: Mentor Graphics, 2013), p. 289.

Chapter 15

1. An inspection of all VHDL projects marked with "design done" status at the popular Open-Cores.org website, on July 1, 2013, revealed that only 50 projects used functions or procedures (or 45% of the total).

2. Peter Ashenden and Jim Lewis, *The Designer's Guide to VHDL*, 3rd ed. (Burlington, VT: Morgan Kaufmann, 2008), p. 207.

3. Martin, *Clean Code*, p. 39.

4. Martin, *Clean Code*, p. 35.

5. The default values were observed in SciTools' Understand 3.1 and Verisoft's Testwell CMT++ 5.0. In *Clean Code*, Robert Martin defends that "Functions should not be 100 lines long. Functions should hardly ever be 20 lines long." Martin, *Clean Code*, p. 34.

6. "Routines longer than 150 lines (excluding comments and blank lines) have been shown to be less stable, more subject to change, and more expensive to fix than shorter routines." Bolton and Johnston, *IfSQ Level-2*, p. 24.

7. Richard W. Selby and Victor R. Basili, "Analyzing Error-Prone System Structure," *IEEE Transactions on Software Engineering* SE-17, no. 2 (1991), pp. 141–152.

8. McConnell, *Code Complete*, p. 172.

9. IEEE, *1076-2008 VHDL Language Reference Manual*, p. 20.

10. IEEE, *1076.6-2004 Standard for VHDL Register Transfer Level (RTL) Synthesis*, p. 20.

11. McConnell, *Code Complete*, p. 178.

12. McConnell, *Code Complete*, p. 174.

13. McConnell, *Code Complete*, p. 349.

Chapter 16

1. "Don't be afraid to make a name long (...) A long descriptive name is better than a short enigmatic name. A long descriptive name is better than a long descriptive comment." Martin, *Clean Code*, p. 39.
2. McConnell, *Code Complete*, p. 283; Martin, *Clean Code*, p. 21.
3. Kernighan and Plauger, *The Elements of Programming Style* (New York: McGraw-Hill, 1978), p. 21.
4. McConnell, *Code Complete*, p. 22.
5. McConnell, *Code Complete*, p. 22.
6. Martin, *Clean Code*, p. 25.
7. McConnell, *Code Complete*, p. 260.
8. "Name Things What You Call Them," revision 29. Ward's Wiki at Cunningham & Cunningham, Inc., last modified January 26, 2005, http://c2.com/cgi/wiki?NameThingsWhatYouCallThem.
9. McConnell, *Code Complete*, p. 263.
10. McConnell, *Code Complete*, p. 286.
11. George Orwell, *A Collection of Essays* (Orlando: Harcourt, 1970), p. 156.
12. Pete Goodlife, *Code Craft: The Practice of Writing Excellent Code* (San Francisco: No Starch Press, 2006), p. 45.
13. Krzysztof Cwalina and Brad Abrams, *Framework Design Guidelines: Conventions, Idioms, and Patterns for Reusable .NET Libraries*, 2nd ed. (Reading, MA: Addison-Wesley, 2008), p. 66.
14. Martin, *Clean Code*, p. 11. This passage paraphrases Ward Cunningham on clean code: *"You know you are working on clean code when each routine you read turns out to be pretty much what you expected."*

Chapter 17

1. Kernighan and Plauger, *The Elements of Programming Style*, p. 144.
2. McConnell, *Code Complete*, p. 793.
3. Goodlife, *Code Craft*, p. 82.
4. William Zinsser, *On Writing Well, 30th Anniversary Edition: The Classic Guide to Writing Nonfiction* (New York: Harper Perennial, 2006), p. 12.
5. This algorithm was introduced by Wilco Dijkstra in 1996 for use in 32-bit microcontrollers. The FPGA version vas adapted by Matti T. Tommiska and is explained in the paper, "Area-efficient implementation of a fast square root algorithm." doi: 10.1109/iccdcs.2000.869869.

Chapter 18

1. Richard J. Miara et al. 1983. "Indentation and Comprehensibility." *Communications of the ACM* 26, 11 (November): 861–867.
2. Ted Tenny. 1988. "Program Readability: Procedures versus Comments." *IEEE Transactions on Software Engineering* SE-14, no. 9 (September): 1271–1279.

3. Dawn Lawrie, Henry Feild, and David Binkley. "An Empirical Study of Rules for Well-Formed Identifiers." *Journal of Software Maintenance and Evolution* 19, no. 4 (July): 205–229.

4. According to McConnell, "The Fundamental Theorem of Formatting is that good visual layout shows the logical structure of a program." McConnell, *Code Complete*, p. 732.

5. Richard J. Miara et al. 1983. "Indentation and Comprehensibility." *Communications of the ACM* 26, 11 (November): 861–867.

6. Boswell and Foucher, *The Art of Readable Code*, p. 39.

7. ESA, *VHDL Modelling Guidelines*, p. 4.

8. Software Productivity Consortium, *Ada 95 Quality and Style*, p. 10.

9. Microsoft Corporation. "Capitalization Conventions." Microsoft Design Guidelines for Developing Class Libraries (Microsoft .NET Framework), accessed November 30, 2014, http://msdn.microsoft.com/en-us/library/vstudio/ms229043.

Chapter 19

1. Andrew Rushton, *VHDL for Logic Synthesis*, 3rd ed. (Chichester, West Sussex: Wiley, 2011), p. 7.

2. Frank Vahid, *Digital Design* (Chichester, West Sussex: Wiley, 2006), p. 225.

3. Hubert Kaeslin, *Digital Integrated Circuit Design: From VLSI Architectures to CMOS Fabrication* (Cambridge: Cambridge University Press, 2008), p. 291.

4. Michael Keating and Pierre Bricaud, *Reuse Methodology Manual for System-on-a-Chip Designs*, 3rd ed. (Norwell, MA: Kluwer, 2002).

5. "A driver can be visualized as a logical hardware device that always sources a value onto a signal." Ben Cohen, *VHDL Coding Styles and Methodology*, 2nd ed. (Norwell, MA: Kluwer, 1999), p. 117.

6. Volnei Pedroni, *Circuit Design and Simulation with VHDL*, 2nd ed. (Cambridge, MA: MIT Press, 2010), p. 169.

7. Mohit Arora, *The Art of Hardware Architecture: Design Methods and Techniques for Digital Circuits* (New York: Springer, 2011), p. 40.

Chapter 20

1. Bruce Wile, John Goss, and Wolfgang Roesner. *Comprehensive Functional Verification–The Complete Industry Cycle* (Burlington, MA: Morgan Kaufmann, 2005), p. 21.

2. Bergeron, *Writing Testbenches*, p. xv.

3. Bergeron, *Writing Testbenches*, p. 10.

4. Andrew Piziali, *Functional Verification Coverage Measurement and Analysis* (New York: Springer, 2004), p. 12.

5. Mark Glasser, *Open Verification Methodology Cookbook* (New York: Springer, 2009), p. 4.

6. Steve Freeman and Nat Pryce, *Growing Object-Oriented Software Guided by Tests* (Reading, MA: Addison-Wesley, 2009), p. 5.

7. Kent Beck, *Test-Driven Development: By Example* (Reading, MA: Addison-Wesley, 2002), p. 1.

8. Beck, *Test-Driven Development*, p. 24.

9. Beck, *Test-Driven Development*, p. ix.

10. Lars Asplund, "VUnit – Automated Unit Testing for VHDL." Accessed April 1, 2015. https:// github.com/LarsAsplund/vunit.

11. Rushton, *VHDL for Logic Synthesis*, p. 302.

12. Freeman and Pryce, *Growing Object-Oriented Software Guided by Tests*, p. 41.

13. Henry S. Warren. *Hacker's Delight,* 2nd ed. (Reading, MA: Addison-Wesley, 2012), p. 254.

14. Jim Lewis. "Functional Coverage Using CoveragePkg. User Guide for Release 2014.07." Open Source VHDL Verification Methodology. Accessed November 30, 2014. http://osvvm.org/wp -content/uploads/2014/10/CoveragePkg_user_guide.pdf.

15. "... most people who learn [test-driven development] find that their programming practice changed for good. *Test Infected* is the phrase Erich Gamma coined to describe this shift." Beck, *Test-Driven Development*, p. xii.

16. "[Testing] tends to synergize well with design. We can learn a lot by doing our testing, and those things themselves tend to make our design better. Testing isn't hard; testing is easy in the presence of good design." Michael Feathers, "The Deep Synergy between Testability and Good Design" (lecture, Norwegian Developers Conference, NDC, 2010).

Bibliography

On VHDL

Ashenden, Peter. *The Designer's Guide to VHDL*, 3rd ed. Burlington, MA: Morgan Kaufmann, 2008.

Rushton, Andrew. *VHDL for Logic Synthesis*, 3rd ed. Chichester, West Sussex: Wiley, 2011.

Pedroni, Volnei A. *Circuit Design and Simulation with VHDL*, 2nd ed. Cambridge, MA: MIT Press, 2010.

Ashenden, Peter, and Jim Lewis. *VHDL 2008–Just the New Stuff*. Burlington, MA: Morgan Kaufmann, 2007.

Cohen, Ben. *VHDL Coding Styles and Methodologies*, 2nd ed. Norwell, MA: Kluwer, 1999.

On Code

McConnell, Steve. *Code Complete: A Practical Handbook of Software Construction*, 2nd ed. Redmond, WA: Microsoft Press, 2004.

Martin, Robert. *Clean Code: A Handbook of Agile Software Craftsmanship*. Upper Saddle River, NJ: Prentice Hall, 2008.

Goodlife, Pete. *Code Craft: The Practice of Writing Excellent Code*. San Francisco, CA: No Starch Press, 2006.

Boswell, Dustin, and Trevor Foucher. *The Art of Readable Code*. Sebastopol, CA: O'Reilly, 2011.

Hunt, Andrew, and David Thomas. *The Pragmatic Programmer: From Journeyman to Master*. Reading, MA: Addison-Wesley, 1999.

Kanat-Alexander, Max. *Code Simplicity: The Fundamentals of Software*. Sebastopol, CA: O'Reilly, 2012.

Feathers, Michael. *Working Effectively with Legacy Code*. Upper Saddle River, NJ: Prentice Hall, 2004.

On Circuit and Chip Design

Keating, Michael, and Bricaud, Pierre. *Reuse Methodology Manual for System-On-a-Chip Designs*. Norwell, MA: Kluwer, 2002.

Bergeron, Janick. *Writing Testbenches: Functional Verification of HDL Models*. New York, NY: Springer, 2003.

Bergeron, Janick. *Writing Testbenches using SystemVerilog*. New York: Springer, 2006.

Pedroni, Volnei A. *Finite State Machines in Hardware: Theory and Design (with VHDL and SystemVerilog)*. Cambridge, MA: MIT Press, 2013.

Standards

IEEE. *1076-2008 Standard VHDL Language Reference Manual*. New York: IEEE, 2008.

IEEE. *1076.6-2004 Standard for VHDL Register Transfer Level (RTL) Synthesis*. New York: IEEE, 2004.

Index